Phillips Memorial
Library
Providence College

Chemical Atomism in the Nineteenth Century

From Dalton to Cannizzaro

Alan J. Rocke

Ohio State University Press: Columbus

Copyright © 1984 by the Ohio State University Press
All Rights Reserved

Library of Congress Cataloging in Publication Data
Rocke, Alan J., 1948–
Chemical atomism in the nineteenth century.
Bibliography: p.
Includes index.
1. Chemistry—History. 2. Atomic theory—History.
I. Title
QD15.R63   1984   541.2′4   83-25082
ISBN 0-8142-0360-4

*Dedicated to my parents,
Sol J. and Verva C. Rocke*

Contents

PREFACE		xi
ACKNOWLEDGMENTS		xvii
1.	PROLOGUE	1
	Eighteenth-Century Background	1
	The Study of Elemental Combining Proportions	6
	Stoichiometric Laws and Atomic Theories	10
2.	THE RISE OF DALTON'S CHEMICAL ATOMIC THEORY	21
	From Mixed Gases to Chemical Atoms, 1790–1803	22
	Historiographic Digression	27
	The Discovery of Multiple Proportions	29
	The Pursuit of Chemical Atoms, 1803–1810	33
	The Formula Problem	35
	Gay-Lussac's Law of Combining Volumes	40
3.	THE EARLY RECEPTION IN GREAT BRITAIN: 1804–1815	49
	Thomson	49
	Prout	52
	Davy	55
	Wollaston	61
	Berzelius	66
	Retrospect	79

4.	**THE FRENCH RESPONSE, 1808–1836**	99
	Berthollet	99
	The Theory of Avogadro and Ampere	101
	The Law of Dulong and Petit	107
	Gay-Lussac	109
	Dumas	115
5.	**ROMANTIC INTERLUDE: ATOMISM ENCOUNTERS GERMAN IDEALISM**	125
	Romanticism and *Naturphilosophie*	125
	Gilbert	128
	Schweigger	130
	Kastner	133
	Döbereiner	134
	Meinecke	137
	Retrospect	141
6.	**THE CONSOLIDATION OF BERZELIAN ATOMISM**	153
	Mitscherlich and Isomorphism	154
	Berzelius' Elaboration of Chemical Atomism	156
	The Emergence of Isomerism	167
	The Older Radical Theory and the Berzelian Hegemony	174
	Gmelin's Insurgency	177
7.	**ORGANIC CHEMISTRY IN FRANCE**	191
	Hydracids, Substitution Theory, and Types	191
	Laurent and Gerhardt	200
	The Reform of Equivalents	205
8.	**NEWER TYPES AND RADICALS**	215
	Williamson's Water Type	215
	Gerhardt's Generalization of Types	223
	Frankland, Kolbe, and the Isolation of Radicals	229
	Conflict and Confluence	236

9.	THE EMERGENCE OF VALENCE AND STRUCTURE	251
	Radicals Engulf Types	251
	Kekulé on the Nature of Carbon	261
	Priority Controversies	273
10.	THE KARLSRUHE NEXUS	287
	Kinetic Theories in Physics	288
	The Karlsruhe Congress of 1860	292
	Conceptions of Valence and Equivalence after Karlsruhe	299
	Retrospect	304
11.	EPILOGUE	313
	The English Debates	313
	The French Debates	321
	The German Debates	326
	Dénouement	331
Bibliography		341
Index		375

Preface

It has been said, with some justice, that every philosopher is at heart either an Aristotelian or a Platonist. Other dichotomies are possible. There is, for instance, the eternal dialogue between idealists who invoke continuities in nature and mechanist-materialists who prefer more realistic and concrete conceptions. As early as the fifth century B.C., the ancient atomists opposed the perceived obscurity of Parmenides' monism with a thoroughgoing reductionistic materialism, and the intellectual heirs of both factions have enlivened the history of philosophy down to the present day.

Reverberations of the debate were strong in the nineteenth century. The discovery at the beginning of the century of stoichiometry—a group of regularities governing the weights of substances undergoing chemical reactions—led John Dalton and others to the development of a theory of chemical atoms. One could choose to assume, as Dalton did, that those chemical atoms were also ultimate physical atoms, the true indivisible primary particles of matter, there being as many types of atoms as elements. It was also possible to accept and utilize Dalton's chemical atomism, but advocate a less materialistic and more unitary microphysical theory. Subsequent experience has of course discredited Daltonian physical atomism, which has been replaced by theories involving varieties of "elementary" particles, a sort of atomism on a smaller level. The last word—if there is one—has surely not been spoken on the subject of physical atomism.

Nineteenth-century controversies over physical atoms have been intensively examined by historians of science, beginning with a 1959 article by Gerd Buchdahl that first drew scholarly attention to a persistent current of antiatomism during the entire century; the best recent studies in this area have been carried out by William Brock, David Knight, and Mary Jo Nye.[1] Most of these efforts concentrate on the years after 1860, when atomic skepticism became more open and vocal. As interesting as this period is, it is not in my central focus. Indeed, it is known that by the 1860s chemists had achieved a remarkably uniform consensus regarding concepts and systems of *chemical* atomism. For this reason I do not treat two significant developments of the late 1860s and 1870s: the periodic law and stereochemistry. I view these developments as more a result than a cause of that consensus, though we will see in chapter 11 that the success of stereochemical theories was used as one of the atomists' more effective replies to their opponents.

I am not aware that any earlier historian has sought to explore the consequences of a proper distinction between physical and chemical atomic theories—a distinction I regard as a necessary starting point before good sense can be made of the history of chemical atomism. Here I would add that simply using the words "chemical atomism" is not the same as carefully defining the set of ideas and meaningfully distinguishing it from physical considerations.[2] The second paragraph of this preface adumbrates definitions of chemical and physical atoms, which are elaborated in detail in chapter 1 and passim. The failure to examine the distinction between the two has resulted in a general failure to perceive that the chemical theory was as successful and as uncontroversial as the physical theory was reviled and often rejected.

Indeed, little recent work has been done on post–Daltonian chemical atomism,[3] which is surprising considering the circumstance that atomism has formed the conceptual basis of chemical theory since Dalton's day. One of my intentions is in fact to show that the nineteenth-century chemical-atomist discussions were nearly exclusively devoted to debating the merits of differing conventional systems, all of which were ontologically equivalent theoretical constructs. The historical puzzle here is not the sources of skepticism, but rather the sources of confidence in atomism. In terms of practice—and ignoring rhetoric—virtually all chemists after Dalton were chemical atomists. (I hasten to add that I intend the word *practice* to include theoretical as well as empirical procedures.)

Analyzing what a scientist actually does as well as his explicit claims regarding his own work—and really trusting the former body of evidence over the latter in cases of conflict—can lead to potential gains in understanding that are worth some risks. Albert Einstein exhorted historians of science: "If you want to find out anything from the theoretical physicists about the methods they use, I advise you to stick closely to one principle: don't listen to their words, fix your attention on their deeds."[4] Among recent historians, I. B. Cohen and Gerald Holton have been two of the most successful in following this principle. Holton properly interprets "deeds" widely enough to include the development of ideas in the scientist's own mind, and he distinguishes between this study and the more conventional examination of how those ideas are received in the scientific community. He designates these two aspects of the development of science as S_1 and S_2 respectively and believes that both are proper subjects for the historian. "A science of S_1," he avers, "must be possible."[5]

Some have argued, against Holton and others, that historical method ought to be purely descriptive, with no admixture of analytical, cognitive, or philosophical judgments.[6] Maurice Mandelbaum disagrees: "To trace the essential phenomenological and inferential steps by means of which . . . our modern conception of the world has come to be what it is, remains one of the most comprehensive and challenging tasks for historians of science. . . . [S]uch a task . . . [is] not only of historical but of epistemological significance."[7] Following D. C. Williams, Mandelbaum designates one of the most interesting of those inferential techniques as *transdiction*: deducing properties of unobservable from those of observable entities. Mandelbaum correctly notes that all scientific inference involves at least a weak sort of transdictive procedure; the philosophical problems surrounding induction illustrate the impossibility of a meaningful science created solely from direct experience. A stronger form of transdiction is used to investigate a realm that is unobservable even in principle. Descartes' and Newton's postulation of rectilinear inertia, for instance, asserts the existence of a sort of motion that not only is never seen, but can never even theoretically be seen except in a thought experiment. The utility and heuristic value of inertia, however, fully justify its postulation.[8]

One of the strongest forms of transdiction has been used for theories of matter. It is well known that Newton applied transdictive procedures to microphysics, but with no great success. Throughout the eighteenth century this endeavor remained attractive but elusive. It was only in the nineteenth century that transdiction was widely and successfully applied to the microcosm, and in no field earlier or more dramatically than in chemistry. This book thus describes the maturation of a fundamental technique that today is used confidently and even unconsciously whenever a chemist calculates a formula, observes a reaction, or interprets a spectrogram.

1. G. Buchdahl, "Sources of Scepticism in Atomic Theory," *Brit. J. Phil. Sci.* 10 (1959): 120-34; W. H. Brock, ed., *The Atomic Debates: Brodie and the Rejection of the Atomic Theory* (Leicester: Leicester University Press, 1967); D. M. Knight, *Atoms and Elements: A Study of Theories of Matter in England in the Nineteenth Century* (London: Hutchinson, 1967); D. M. Knight, *The Transcendental Part of Chemistry* (Folkestone: Dawson, 1978); M. J. Nye, *Molecular Reality: A Perspective on the Scientific Work of Jean Perrin* (New York: Science History, 1972); M. J. Nye, "The Nineteenth-Century Atomic Debates and the Dilemma of an 'Indifferent Hypothesis'," *Stud. Hist. Phil. Sci.* 7 (1976): 245-68.

2. For instance, Knight in *Atoms and Elements* uses the term, but his discussion—as with most of the recent writers on nineteenth-century atomism—focuses nearly exclusively on physicists or on chemists dealing essentially with physical issues.

3. The most extensive treatment of nineteenth-century chemical atomism is found in Hermann Kopp's *Entwickelung der Chemie in der neueren Zeit* (Munich, 1873). It is understating the case to suggest that this work is badly dated. One example of the modern neglect of chemical atomism is the critical atomic weight reform of Gerhardt and Laurent. S. C. Kapoor's biography of Laurent in the *Dictionary of Scientific Biography*, for instance, has only a short paragraph on this question. The biography of Gerhardt in the same reference work, by M. Crosland and J. Brooke, devotes a column to the matter, but is by no means definitive.

4. A. Einstein, *Ideas and Opinions* (New York: Bonanza, 1954), p. 270.

5. I. B. Cohen, *The Newtonian Revolution* (Cambridge: Cambridge University Press, 1980), pp. 9-16 and passim; G. Holton, *Thematic Origins of Scientific Thought* (Cambridge, Mass: Harvard University Press, 1973), pp. 17-29 and passim. There is a growing body of literature on this subject; see for example Karin D. Knorr et al., eds., *The Social Process of Scientific Investigation* (Dordrecht: Reidel, 1981), and the various writings of Hans Reichenbach.

6. E.g., David B. Wilson, in a commentary on a paper by E. N. Hiebert, in D. H. D. Roller, ed., *Perspectives in the History of Science and Technology* (Norman: University of Oklahoma Press, 1971), pp. 93-97.

7. M. Mandelbaum, *Philosophy, Science and Sense Perception* (Baltimore: Johns Hopkins University Press, 1964), p. 245.

8. Ibid., pp. 61-117; Alexandre Koyré, "Galilée et la loi d'inertie," *Études galiléennes* (Paris, 1939).

Acknowledgments

It is a pleasure to acknowledge the indispensable help I have received during the ten-year incubation period of this manuscript. My doctoral advisor, Aaron Ihde, introduced me to the history of chemistry and has always represented for me a professional and personal exemplar. It is difficult fully to identify those other teachers and colleagues who have most influenced my work, but at least Erwin Hiebert, Victor Hilts, David Lindberg, Robert Schofield, and Daniel Siegel deserve mention and thanks. Hiebert, Schofield, Siegel, Theron Cole, June Fullmer, Larry Laudan, Seymour Mauskopf, and Mary Jo Nye were kind enough to read individual chapters and give me the benefit of their considerable expertise (such gracious colleagues must not be blamed for the obstinacy of the author, who on occasion foolishly failed to follow their generous counsel). I also owe a great deal to editors and collegial correspondents who more often than I like to admit have directed me away from the intellectual shoals and towards safe harbor, especially William Brock, Nicholas Fisher, Russell McCormmach, Robert Multhauf, and Arnold Thackray. Invaluable bibliographic and linguistic assistance was rendered by Dennis Hill, John Neu, Kärsten Olsson, and C. C. Rom.

On an institutional level I wish to thank the staffs of Memorial Library, University of Wisconsin-Madison; Freiberger and Sears Libraries, Case Western Reserve University; the Library of Congress; the British Library; the Royal Society; the Royal Institution; the Chemical Society of London; D. M. S. Watson

Library, University College, London; the Bayerische Staatsbibliothek and the Bibliothek des Deutschen Museums in Munich; and the Royal Swedish Academy of Sciences in Stockholm. The Deutscher Akademischer Austauschdienst enabled me to spend a year at the Institut für die Geschichte der Naturwissenschaften, Munich, where I received much kind help from H. Gericke, O. Krätz, I. Schneider, and M. Nida-Rümelin. Professor K. Hafner and Dr. A. Schmidt-Glenewinckel greatly assisted me at the August Kekulé Sammlung, Institut für Organische Chemie, Technische Hochschule, Darmstadt. The American Council of Learned Societies sponsored summer research in England, where Maurice Crosland and especially William Brock generously gave assistance and permitted me access to their own collections of resource materials. Permission to reproduce passages from unpublished manuscripts was kindly granted by William Brock, the Kekulé Sammlung, the Royal Institution, the Royal Society, the Chemical Society, and University College. Parts of the first three chapters of this book first appeared in an earlier form in volume 9 of *Historical Studies in the Physical Sciences*, edited by Russell McCormmach and Lewis Pyenson and published in 1978 by the Johns Hopkins University Press, and early versions of parts of chapters 4 and 5 appeared as articles in *Isis* in 1978 and 1979. For their expertise and professionalism I am also grateful to the staff of the Ohio State University Press.

Finally, and most importantly, I wish to thank my wife Cristine for her loving encouragement during the writing of this book.

1

Prologue

THE EIGHTEENTH-CENTURY BACKGROUND

It is a commonplace that eighteenth-century science was permeated by the exceedingly rich and often ambiguous legacy of Isaac Newton. The shifting and manifold senses of what it meant to be a Newtonian during the century following the publication of *Philosophiae naturalis principia mathematica* and *Opticks*[1] have provided historians with a convoluted but fruitful and important field for their endeavors.[2] Probably the most central theme in this plexus is the theory of matter as conceived of by Newton and as interpreted and elaborated by Enlightenment natural philosophers.

Newton's concept of the intimate structure of matter has been described appropriately as "second-order mechanism" or "dynamic corpuscularity."[3] Newton adhered to the dominant seventeenth-century vision that all phenomena could be explained by reference to the size, shape, number, and motion of the particles that compose all bodies. But to this first-order mechanical philosophy, Newton added another crucial ingredient: the powers or forces exerted by these corpuscles. It is well known that Newton repeatedly publicly disclaimed speculations on the physical cause of such powers and asserted that it would be absurd to consider them as "innate, inherent and essential to Matter." Nonetheless, his eminently successful mathematical use of universal gravitation to explain with quantitative precision the "system of the world" almost inevitably led to the unquestioning assimilation of action at a distance in successive generations of Newtonian natural philosophers.[4]

Newton also sought, with no great success and largely in unpublished writings, to apply this conception that had proved so powerful in the macroscopic realm of celestial mechanics to the microscopic realm of chemistry as well. In the preface to the first edition of the *Principia*, Newton wrote:

> I wish we could derive the rest of the phenomena of Nature by the same kind of reasoning from mechanical principles, for I am induced by many reasons to suspect that they may all depend upon certain forces by which the particles of bodies, by some causes hitherto unknown, are either mutually impelled towards one another, and cohere in regular figures, or are repelled and recede from one another. These forces being unknown, philosophers have hitherto attempted the search of Nature in vain; but I hope the principles here laid down will afford some light either to this or some truer method of philosophy.[5]

Many eighteenth-century chemists were dominated by this Newtonian search for those certain short-range forces analogous to gravitation possessed by the undifferentiable and inertially homogeneous primary particles of matter. Those scientists who sought to apply this abstract and mathematical program have been described as reductionists or physicalists.[6]

In chemistry, physicalism found its most successful development in the theory and measurement of "chemical affinities"; in physics, the tradition culminated in what Robert Fox calls the Laplacian program, which was enormously influential in the first fifteen years of the nineteenth century. The Laplacians attempted to apply transdictive procedures, occasionally with success; consistent with their Newtonian heritage, they focused on the microscopic forces rather than the presumed ultimate material particles themselves.[7]

Other approaches to microphysics besides physicalism could, by appropriate citation and exegesis, be derived from the rich Newtonian legacy. For instance, Newton wrote in Query 31 of the *Opticks*: "It seems probable to me, that God in the Beginning form'd matter in solid, massy, hard, impenetrable, moveable Particles, of such Sizes and in such Proportion to Space, as most conduced to the End for which he form'd them ... it may also be allow'd that God is able to create Particles of Matter of several Sizes and Figures, and in several Proportions to Space, and perhaps of different Densities and Forces. . . . At least, I see nothing of Contradiction in all this."[8] Such a view of relatively simple-minded atomism could also be discerned in Proposition 23 of Book 2 of the *Principia*, where Newton showed that a gas composed of particles repelling each other by a force inversely proportional to distance would obey Boyle's law.[9] From such origins arose, particularly in Great Britain, the textbook and lecture tradition of popular Newtonianism, with its predilection for straightforward mechanical models. Though the adherents of this tradition professed similar ideals as the mathematical physicists, their protestations generally "lacked compulsion," as Thackray put it.[10] They had little patience for mathematical subtleties and sought concrete rather than abstract representations for microscopic phenomena.

A third Newtonian tradition, often linked with both of the preceding ones but in many ways distinct from them, has been termed nonreductionist, empiricist, or materialist.[11] Like those in the textbook tradition, the adherents of this viewpoint refrained from denigrating the physicalists' search for the primary particles and forces of matter. However, they believed that such a search was probably still premature, hence would continue to be rather sterile for the foreseeable future. They generally took Newton's professed positivism earnestly, regarded speculations on the fine structure of matter as metaphysical, and sought instead more proximate, cautious, and qualitative explanations for chemical phenomena. Of course, this antimetaphysical attitude led, as it always does, directly to a *different* metaphysics, one that was more closely related to scholastic forms than to seventeenth-century mechanical philosophy. Many of these materialists explained the phenomena of nature by reference to the presence or absence of substances bearing (and conferring on neutral matter to a degree proportional to quantity) the property in question. Temperature was referred to a caloric fluid, charge to an electric fluid, combustibility to a phlogistic fluid, and so on. Schofield has shown that these ideas derive in some part from Newton's published speculations on his very versatile ethereal medium, and that this

Newtonian-materialist viewpoint dominated British natural philosophy from about 1740 to the turn of the century.[12]

The diverse eighteenth-century conceptions regarding what constituted truly primary or elemental substances were closely related to these Newtonian Enlightenment traditions.[13] Newton's own viewpoint was derived partly from that of his older friend Robert Boyle; both referred to the primary "particles of the first composition," and very much doubted that any true elements were yet isolated, or perhaps were even isolable. It was probable, thought Boyle and Newton, that every known substance contained the true elements combined in a complex hierarchical structure, that the sensible components of real substances were composed of particles of the nth composition, as it were.[14] This mechanist viewpoint influenced many eighteenth-century chemists and survived well into the next century; examples include Freind, Keill, Buffon, Guyton de Morveau, Cavendish, Priestley, and Davy.

However, an empiricist option derived from the writings of G. E. Stahl, the German chemist, was also in evidence from early in the eighteenth century. Stahl did not so much controvert the mechanists' conception of simple substances as assert its sterility for chemistry. Instead, he sought to define more empirical and chemically useful "principles" that could be regarded as composing, and importantly, conferring properties on substances. This interpretation gradually became more influential as the century wore on; nonetheless, throughout the century it had to battle against what Thackray refers to as 'an entrenched, prestigious, and sterile physicalism."[15]

There were in Stahl's views two separable criteria for simple substances: the more modern operational criterion that elements were to be viewed as simply the last point of analysis and the more Aristotelian (or Paracelsian) notion of property-bearing principles. Several scholars have recently shown how the two ideas were only slowly and incompletely distinguished during the course of the eighteenth century and how in fact the second was still in evidence, though muted, in the work of such "moderns" as Lavoisier, Davy, and Berzelius.[16] But the operational criterion of elementarity gradually insinuated itself into the consciousness of chemists, so that by the time Lavoisier first clearly and unambiguously stated it in his classic *Traité élémentaire de chimie*, it could provoke but little controversy.[17]

Here to the standard notice of Lavoisier's empiricism we must add two caveats. First, as firmly and properly as he is to be classified as an empiricist, he by no means rejected the mechanist vision. Indeed, he thought it probable that the chemists' elements were in fact composed of simpler forms of matter. However, he was concerned to define his terms more consistent with instrumentality than with an ultimate but nonexperiential reality. There are two important consequences of this standpoint. First, Lavoisier's list of elements was explicitly provisional and subject to alteration whenever chemistry improved so that a hitherto undecomposed "element" could be analyzed. This led to a good deal of uncertainty and controversy (as well as discomfort in some minds) concerning which substances were truly elemental.[18] Second, Lavoisier denied his elements any ontological status as building blocks of nature. They were not elements in any Aristotelian or Daltonian sense, but were in fact undergirded by an implicit reductionism.[19] This did not, however, reduce their ontological appeal for many of the followers of Lavoisier.

The second caveat has to do with the operational criterion itself: "the last point which analysis is capable of reaching," as Lavoisier put it. It has been too little noted that there is a certain circularity here if we start from the obvious definition of analysis as "resolution of a complex into simpler substances." How do we know which substances are complex and which simpler? This is not a trivial question: the Stahlians obviously regarded strong heating (in air!) as the best method of analysis, and it would be ahistorical for us to denigrate their consequent belief in the elemental nature of metallic calxes. In brief, Lavoisier's putative operational definition of "element" only becomes meaningful—indeed, only becomes operational—when one applies some operational criterion for "analysis." For uncertain reasons the second half of the eighteenth century saw increasing acceptance both of the concept of weight as a unique measure of quantity of matter and (implicitly) of the axiom of conservation of mass; as a consequence the balance began to be used routinely for quantitative gravimetric analysis.[20] It was the *linkage* of this new gravimetric criterion for analysis with Lavoisier's empirical definition of elements that first conferred true operationalism on that definition and permitted a provisional consensus on a list of "simple" substances.[21]

THE STUDY OF ELEMENTAL COMBINING PROPORTIONS

This new late eighteenth-century concern with precise gravimetric analysis not only provided criteria for deciding on a list of elements, but also led surprisingly quickly and smoothly to the recognition and establishment of the regularities in the relative proportions by weight in which those elements enter into combination with each other. The most basic of these stoichiometric relations is the law of definite proportions, which asserts that elements can combine only in a small number of fixed proportions by weight.

Most historians who have dealt with the development of stoichiometry have asserted that this law was a commonplace assumption well before it was first explicitly stated by J. L. Proust and unsuccessfully opposed by C. L. Berthollet.[22] Mauskopf has, on the other hand, recently argued that "the assumptions behind gravimetric analysis are not necessarily the same as the fundamental tenet of the law of definite proportions," and that the precise relationship "remains obscure."[23] He has also sought to clarify the relations between this and two other late eighteenth-century laws: the "principle of constant saturation proportions" defended, for instance, by Guyton de Morveau and Thomas Thomson and the "doctrine of fixed mineral species" of such mineralogists as Romé de l'Isle, Haüy, and Dolomieu.[24]

At this time the position of leading chemists toward what we would call definite proportions was indeed not simple. Lavoisier, who is rightly regarded as the most adept and consistent gravimetric analyst of his day, seems normally to have assumed that compounds have fixed compositions, but suggested indefinite gradations of oxygenation for many substances.[25] It was precisely this newly discovered phenomenon of multiple oxides of a single element that seemed to pose the greatest difficulty for the assumption (or law) of constant composition, for if any two elements always combined only in a *single* proportion it would perhaps never have come into question at all. From 1794 — the year that Lavoisier died — Proust sought to demonstrate that each metal forms at most only two oxides, and that each oxide (indeed, every well-defined substance) has a fixed composition by weight. Proust argued that the known instances of apparently continuous series of oxides were only mixtures in varying proportions of two definite oxides. Solutions, glasses, alloys, and amalgams were also characterized as mere physical mixtures rather than true chemical compounds.[26]

Claude-Louis Berthollet was the first prominent convert to Lavoisier's antiphlogistic chemistry. Nonetheless, his tendency toward physicalism was somewhat opposed to his older colleague's dominant empiricism, and this orientation strongly influenced Berthollet's scientific predilections. For one thing, Berthollet participated in that important Newtonian theme of Enlightenment chemistry, the theory of affinities. Historians have followed this theme from Newton through the work of Geoffroy, Buffon, Cullen, Black, Guyton de Morveau, Bergman, W. Higgins, Kirwan, Wenzel, and Richter.[27] Berthollet significantly modified the theory by denying the "elective" (determinative) characteristic of chemical affinity, replacing it with a more complex and relative version that anticipated modern ideas on chemical equilibrium and mass action. Affinity was not an absolute force, according to Berthollet, but could be modified by such factors as relative concentration of reactants, solubilities, "cohesion," and "elasticity" (precipitation and volatilization of products).[28]

Berthollet's physicalist ideas on chemical equilibrium led almost inexorably to opposition to the work of the empiricist Proust. The gentlemanly polemic between these two Frenchmen in the opening years of the nineteenth century has often been described.[29] Berthollet viewed Proust's ideas as a recrudescence of the elective affinities that he had quite properly and successfully demolished, and he exposed a vicious circularity in Proust's crucial distinction between chemical compounds and physical mixtures (whatever failed to exhibit definite proportions Proust *defined* as a mixture). He was able to explain undoubted examples of definite proportions as merely special cases where such phenomena as precipitation or gas formation ended the process of equilibration between reactants and products. Proust, for his part, was able to score important points and to point out inconsistencies in Berthollet's theory.[30] But there was no immediate decisive victory for definite proportions, as has often been asserted. Rather, the dispute simply faded away around 1807. It was in fact the rise of the theory of chemical atomism about this time that effectively conferred orthodoxy on Proust's ideas.[31]

Two years before Proust first stated the law of definite proportions, a Prussian chemist by the name of J. B. Richter coined the term *stoichiometry* and used it in the title of his book.[32] Richter assumed definite proportions in the compounds he analyzed; moreover, he was convinced that chemistry must properly be viewed as a branch of applied mathematics, and, as Partington put it, he "busied himself in finding regularities among the combining proportions where Nature has not provided any."[33] But Richter also drew attention to a remarkable and quite seminal observation: the quantities of any two acids that precisely saturate a given base—or of two bases saturating a given acid—form a ratio that is independent of the neutralizing substance. Consequently, if an appropriate standard substance is assigned an arbitrary dimensionless standard number, all other acids and bases can be assigned fixed numbers representing relative quantities that are equivalent in neutralizing power.

Richter himself used the cumbersome approach of writing separate tables for each acid and base. But a Berlin *Gymnasium* professor, E. G. Fischer, collated Richter's data into a single table of twenty-one acids and bases, all related to the standard of sulfuric acid equals 1000. He published the table as a note in his translation of Berthollet's *Recherches sur les lois de l'affinité*, and Berthollet incorporated Fischer's note into his *Essai de statique chimique*.[34] Since Richter's own publications were little known, even in Germany, this was the first opportunity for Richter's concept of chemical equivalency to become widely disseminated. Richter afterwards published an expanded table of forty-eight acids and bases, following Fischer's example.[35]

It might seem surprising that this phenomenon, which every

modern chemist immediately views as an obvious and direct expression of chemical atomism, was introduced with absolutely no reference to corpuscular theories of matter. But modern chemists are born and bred in the atomist metaphysics. Berthollet and Richter, on the other hand, were both committed to versions of the Newtonian physicalist tradition that sought to discern and quantify the short-range effects of chemical affinity rather than to weigh atoms, a tradition, in other words, that focused on forces rather than on ponderable matter itself.[36] Richter believed that matter was infinitely divisible, thus metaphysically excluding any sort of atomism.[37] Similarly, Berthollet interpreted Richter's neutralization series simply as measures of relative affinities. He thought that there was a simple proportionality between affinity (reactivity) and saturation capacity (defined as inversely proportional to equivalent weight).[38] However logical this appeared, his contemporaries pointed out that empirically this did not hold, that if anything affinities are found to be *inversely* proportional to saturation capacities.[39] In 1839 Gay-Lussac wrote: "A few years later Berthollet certainly would never have proposed as a measure of affinity a method which gives nothing but the atomic weights or equivalents."[40] However, Berthollet was able to justify his apparently contrafactual relation in the same way as he justified opposition to definite proportions: he suggested that the true affinity series was obscured by the interference of particular reaction conditions and the physical states of the products.[41]

There is a second way in which physicalism barred the conceptual route that could lead to chemical atomism. For any true chemical atomic theory, an assumed list of ontological elements is at least heuristically convenient. This almost necessitates a nonreductionist commitment, not just the sort of positivistic nonreductionism characteristic of Lavoisier behind which could lurk a thoroughgoing physicalism; rather, it asks for a more or less ontological nonreductionism. Berthollet tended to view any list of elemental substances as a mere expedient. It should also be noted that Richter's equivalents were almost exclusively acid-base equivalents. To be sure, in certain places he did speak of equivalents of what we now know as elements, for instance of the metals.[42] However, Richter's "elements" were not ours. As a single example, he retained phlogiston in his theoretical views and thus never regarded metals as truly elemental.[43] Chemical atomism was well camouflaged by such presuppositions and metaphysical commitments.

STOICHIOMETRIC LAWS AND ATOMIC THEORIES

It is time now to clarify the precise relationships between physical and chemical atomic theories and between chemical atomism and the laws of stoichiometry.[44] The chemical atomic theory forms the conceptual basis for assigning relative weights to elements and assigning molecular formulas to compounds. I will argue in this study that this theory was universally (if implicitly and often unknowingly) accepted throughout the course of the nineteenth century. The physical atomic theory, on the other hand, was controversial and far from universally accepted, since it made what seemed to many to be rather doubtful statements about the intimate mechanical nature of substances: for instance, that matter consists of hard, massy, unsplittable, impenetrable, spherical atoms surrounded by a caloric fluid, exerting forces on their neighbors and cohering into definite arrangements.[45] To be sure, the two types of atomistic theories were often closely related; Dalton, for one, expounded and defended both. Throughout the century, evidence of their compatibility gradually accumulated, leading to their ultimate unification in the early years of the present century. Unfortunately, few scientists distinguished between the two theories, and as a consequence every attack on physical atomism impugned, by association, the scientific worth of chemical atomism.[46]

The theory of chemical atoms was frequently in the nineteenth century (and even more frequently is still today) wrongly identified with the laws of stoichiometry, and a sharp distinction here is crucial to my interpretation of the history of nineteenth-century chemical theory. The keystone of stoichiometry is the law of equivalent proportions, which, in one of its forms, states that *all chemical reactions take place in proportions by weight represented by elemental "equivalent weights."*[47] Of course, this proposition requires a lemma characterizing the central concept: *To each element may be assigned one or more dimensionless "equivalent weights," which for a given element form a series of small integral submultiples of a characteristic number (often, though not always, itself an equivalent weight).* For example, since one part by weight of hydrogen combines with eight or with sixteen parts by weight of oxygen—to form water or hydrogen peroxide—eight and sixteen are equivalent weights of oxygen relative to hydrogen as unity.

The pivotal importance of the case of water and the ubiquity of oxides has given rise to a convenient operational definition of *equivalent weight* as *that quantity of an element that combines with, or replaces, one part of hydrogen or eight parts of oxygen*. To determine the equivalent of some element A in one of its oxides AO_x, the chemist simply analyzes or synthesizes the oxide and expresses the composition in terms of parts of A per eight parts of oxygen. Clearly, the concept is only meaningful for the analysis of compounds composed of only two elements, usually the test element and the standard. Defined in this way, elemental chemical equivalents have three important characteristics: first, since they are calculated directly from the analytical data, they are empirical entities; second, the equivalent weight depends on the compound analyzed, and most elements exhibit several different values; and third, equivalent weights for a given element are usually identical to, or exact submultiples of, the atomic weight.[48]

Another common modern definition of elemental equivalent weight, *atomic weight divided by valence*, is not operational. This definition, of course, had to await the theory of valence; it is first found in the writings of Williamson, Kekulé, Erlenmeyer, and Hofmann in the 1860s. It is not often noted that these two definitions are not identical; that is, the equivalent weights for a given element specified by the two rules do not always coincide. The operational rule is better in that it is more general, more empirical, and specifies a wider variety of equivalents for a given element. For example, the nonoperational definition suggests that the equivalent weights of nitrogen are 4.7 and 2.8 (14/3 and 14/5), whereas the operational rule specifies equivalent weights of 14, 7.0, 4.7, 3.5, and 2.8 (resulting from analyses of N_2O, NO, N_2O_3, NO_2, and N_2O_5, respectively). Unless otherwise noted, we will always use the operational definition.

The term *equivalent* was popularized by Wollaston in an important paper published in 1814.[49] Several commentators have correctly perceived that Wollaston used the concept here in a fundamentally different sense than the modern one,[50] but the consequences of this confusion have never been fully explored. Wollaston chose a single equivalent for each element, established it as invariant, and then used that value to calculate molecular formulas for compounds that the element was known to form. The choice of equivalent was dictated by an assumed formula for the lowest oxide of the element in question. By using the assumed formula to fix an otherwise variable parameter, Wollaston transferred the whole operation from the realm of experiment to that of theory, and I will argue (especially in chapter 3) that Wollaston's quantities are operationally identical with chemical atomic weights.

It is necessary to state precisely what I mean by chemical atoms. The laws of stoichiometry describe certain regularities in the weight proportions of the elements that combine to form chemical compounds. The chemical atomic theory suggests an explanation for these regularities: *there exists for each element a unique "atomic weight", a chemically indivisible unit, that enters into combination with similar units of other elements in small integral multiples*. In spite of the similarity of this definition to that of elemental equivalents, certain characteristics of atomic weights distinguish them from equivalents. First, atomic weights are presumed to be invariant. Second, there is no single chemical operational definition as for equivalents. Third, the primary chemical method of establishing atomic weights (at least in the early nineteenth century) is to use assumed formulas for the simple oxides of the elements. Finally, once chosen, atomic weights are used to deduce formulas and molecular weights for all known compounds. Wollaston's "equivalents" exhibit all of these characteristics of "chemical atomic weights."

But an important distinction can still be made between Wollaston's and Dalton's elemental weights. Dalton was well aware that his weights were based upon assumed formulas and thus had an uncertain foundation, but he regarded his assumptions as very probable.[51] Wollaston, on the other hand, regarded his choice of formulas as a convention, that is, as an arbitrary selection based on practical criteria such as simplicity and convenience; he was apparently unconcerned (at least initially) with questions of molecular reality. In general, nineteenth-century chemists associated with either the "realists" or the "conventionalists," and throughout the century there was considerable tension between them. Whether they realized it or not, however, both schools were advocating chemical atomism as I have defined it. Both groups posited invariant elemental weights to explain the inelegant variability of empirical combining proportions; both used these chemical building blocks to construct schematically all known compounds. This is chemical atomism.

The argument, of course, rests on the validity of my definition of chemical atomism, a definition that avoids all reference to the ultimate structure of matter. This is consistent with my thesis that it is useful and meaningful, both in a cognitive and in an historical sense, to distinguish clearly between chemical and physical atomism. But it is then necessary to show that the atomic theory shorn of all physical aspects is more than a mere statement of scientific laws. In other words I want to argue that, contrary to the belief of many nineteenth-century chemists and twentieth-century historians, a purely chemical atomism has greater content than stoichiometry. I have already urged that chemical atomism properly ought to be regarded as a theoretical and not a legal construct. But some readers may feel uncomfortable labeling as theoretical a scheme that appears explicitly to eschew any visualizable mechanism or model as an explanatory vehicle.

However, a model is not a necessary component of a scientific theory. To be more exact, models are arguably essential only if defined in a wide sense, and I contend that chemical atomism satisfies this definition. Rom Harré has offered a useful distinction between what he calls "reticular" and "explanatory" theories, which explain corresponding laws in a minimal and in a maximal sense respectively. Examples of reticular theories are geometric optics and Newtonian dynamics. Such a theory consists of "a set of relationships between refined observa-

Sci. 4 (1974): 89-136. Elizabeth Garber, "Molecular Science in Late-Nineteenth-Century Britain," *Hist. Stud. Phys. Sci.* 9 (1978): 265-97 (265 n) emphasizes the last point. For Newtonian transdiction (or "transduction"), see J. E. McGuire, "Atoms and the 'Analogy of Nature': Newton's Third Rule of Philosophizing," *Stud. Hist. Phil. Sci.* 1 (1970): 3-58.

8. Newton, *Opticks*, pp. 400-404.

9. Newton, *Mathematical Principles*, p. 300.

10. Thackray, *Atoms and Powers*, pp. 234-78 (238).

11. The last is used preferentially by Schofield in *Mechanism and Materialism*. It must be noted that these three traditions overlap in more than one respect; exclusive pigeonholes cannot be applied to categorize eighteenth-century Newtonians.

12. Ibid., passim.

13. In addition to the works previously cited, see M. B. Hall, "The History of the Concept of Element," in D. S. L. Cardwell, ed., *John Dalton and the Progress of Science* (Manchester, England: Manchester University Press, 1968), pp. 21-39; and R. Siegfried and B. J. Dobbs, "Composition, A Neglected Aspect of the Chemical Revolution," *Ann. Sci.* 24 (1968): 275-93.

14. Cohen, *Papers and Letters*, p. 258 (1710); Newton, *Opticks*, p. 394 (1706). On Boyle's theory of matter, see M. B. Hall, *Robert Boyle on Natural Philosophy* (Bloomington: Indiana University Press, 1965), pp. 68-74, 84-87, 210-30.

15. Thackray, *Atoms and Powers*, pp. 161-98 (197).

16. Ibid.; Hall, "History"; M. Daumas, *Lavoisier: Théoricien et expérimentateur* (Paris, 1955); Siegfried and Dobbs, "Composition"; and M. P. Crosland, "Lavoisier's Theory of Acidity," *Isis* 64 (1973): 306-25.

17. A. L. Lavoisier, *Traité élémentaire de chimie* (Paris, 1789): *Elements of Chemistry*, trans. R. Kerr (Edinburgh, 1790; rpt. New York: Dover, 1965), p. xxiv.

18. Siegfried and Dobbs, "Composition"; R. Siegfried, "The Phlogistic Conjectures of Humphry Davy," *Chymia* 9 (1964): 117-24; Siegfried, "Lavoisier's Table of Simple Substances," *Ambix* 29 (1982):29-48.

19. Thackray, *Atoms and Powers*, pp. 270-75.

20. Ida Freund, *The Study of Chemical Composition: An Account of its Method and Historical Development* (Cambridge: Cambridge University Press, 1904; rpt. New York: Dover, 1968), pp. 58-75, 101-6; J. R. Partington, *A History of Chemistry*, 4 vols. (London: Macmillan & Co., 1961-70), 3:376-78.

21. Siegfried and Dobbs, "Composition," pp. 288, 292; Schofield, *Mechanism and Materialism*, p. 215. D. M. Knight asserts — I believe wrongly — that there can be no operational definition of analysis, hence of element: *Atoms and Elements: A Study of Theories of Matter in England in the Nineteenth Century* (London: Hutchinson, 1967), pp. 21-24, 39. Davy was probably the first to mention the gravimetric criterion for analysis: J. Davy, ed., *Collected Works of Sir Humphry Davy*, 9 vols. (London, 1839-40), 9:385; in this he was followed much later by W. Ostwald, *Grundriss der allgemeinen Chemie*, 3d ed. (Leipzig, 1899), p. 12.

Prologue 17

22. H. Kopp, *Geschichte der Chemie*, 4 vols. (Brunswick, 1843-47), 2:367; H. Kopp, *Die Entwickelung der Chemie in der neueren Zeit* (Munich, 1873), pp. 219-23, 268; E. von Meyer, *A History of Chemistry From Earliest Times to the Present Day*, trans. G. M'Gowan (London: Macmillan & Co., 1891), p. 175; A. N. Meldrum, "The Development of the Atomic Theory: (1) Berthollet's Doctrine of Variable Proportions," *Mem. Lit. Phil. Soc. Manchester* 54:7 (1910): 1-16; H. Guerlac, "Quantification in Chemistry," *Isis* 52 (1961): 194-214; Partington, *History*, 3:644; and S. C. Kapoor, "Berthollet, Proust, and Proportions," *Chymia* 10 (1965): 53-110.

23. S. Mauskopf, "J. L. Proust," in *Dictionary of Scientific Biography*, 16 vols. (New York: Scribner's, 1970-1980), 11 (1975): 166-72.

24. S. Mauskopf, "Thomson Before Dalton: Thomas Thomson's Considerations of the Issue of Combining Weight Proportions Prior to his Acceptance of Dalton's Chemical Atomic Theory," *Ann. Sci.* 25 (1969): 229-42; S. Mauskopf, "Minerals, Molecules and Species," *Arch. int. hist. sci.* 23 (1970): 185-206.

25. Lavoisier, *Elements*, pp. 185-89; Crosland, "Lavoisier's Theory," pp. 317-21; Partington, *History*, 3:644; Kapoor, "Berthollet," p. 87.

26. See Mauskopf, "J. L. Proust," and citations therein.

27. E.g., H. Guerlac, "The Background to Dalton's Atomic Theory," in Cardwell, *John Dalton*, pp. 57-91, esp. 73-84.

28. C. L. Berthollet, *Recherches sur les lois de l'affinité* (Paris, 1801), and more esp. *Essai de statique chimique*, 2 vols. (Paris, 1803). Good secondary studies are F. L. Holmes, "From Elective Affinities to Chemical Equilibria: Berthollet's Law of Mass Action," *Chymia* 8 (1962): 105-45; and Michelle Sadoun-Goupil, *Le chimiste Claude-Louis Berthollet (1748-1822): Sa vie, son oeuvre* (Paris: Vrin, 1977).

29. E.g. in Kapoor, "Berthollet," pp. 84-108; or Partington, *History*, 3:644-53.

30. Kapoor, "Berthollet," pp. 96-108.

31. Meldrum, "Berthollet's Doctrine"; Mauskopf, "Proust"; Mauskopf, "Thomson Before Dalton."

32. J. B. Richter, *Anfangsgründe der Stöchyometrie*, 3 vols. in 4 (Breslau and Hirschberg, 1792-94). Richter also published his views on combining proportions in his *Ueber die neuern Gegenstände der Chymie*, 11 vols. (Breslau and Hirschberg, 1791-1802); and in D. L. Bourguet, ed., *Chemisches Handwörterbuch* 3-6 (Berlin, 1803-5), to which he contributed important articles.

33. Partington, *History*, 3:674-86. Partington plausibly suggests that this conviction derived from Richter's education at Königsberg, where Kant was teaching. See also Partington, "Jeremias Benjamin Richter and the Law of Reciprocal Proportions," *Ann. Sci.* 7 (1951): 173-98, and 9 (1953): 289-314.

34. Berthollet, trans. E. G. Fischer, *Ueber die Gesetze der Verwandtschaft* (Berlin, 1802), pp. 229-30n; *Essai de statique chimique* 1: 116-20, 134-38.

35. Richter, in Bourguet, ed., *Handwörterbuch* 3 (1803): 164.

36. Thackray, *Atoms and Powers*, passim. S. Mauskopf, in "Haüy's Model of Chemical Equivalence: Daltonian Doubts Exhumed," *Ambix* 17 (1970): 182-91 (191), writes: "Haüy's interpretation of [Richter's] law of neutrality

demonstrates that the concept of atomic weights was not, in itself, necessarily deducible from Richter's data. . . . As long as it was affinity forces upon which chemical theory focused, even Richter's work would be interpreted dynamically, as Haüy's model demonstrated."

37. Richter, *Anfangsgründe* vol. 1, pt. 1 (1792); xx–xxx, cited by Partington, "Richter," pp. 189–90.

38. Berthollet, *Essai de statique chimique* 1: 125.

39. As Bergman, Kirwan, and Richter had argued before 1803; see T. Thomson, *History of Chemistry*, 2 vols. (London, 1830–31), 2:163–64; A. Ladenburg, *Lectures on the History of the Development of Chemistry Since the Time of Lavoisier*, 2d ed., trans. L. Dobbin (Edinburgh: Alembic Club, 1900), p. 37; Holmes, "Chemical Equilibria," pp. 124–25; Partington, "Richter," pp. 297–99; Partington, *History* 4: 578–80.

40. J. L. Gay-Lussac, "Considérations sur les forces chimiques," *Ann. chim. phys.* [2] 70 (1839): 407–34 (417).

41. Berthollet, *Essai de statique chimique* 1:124–28.

42. Richter, *Gegenstände* 9: 116f., 137–41; 10: 45f., 86f., 134f., 153f., 168f.; cited in Partington, "Richter," pp. 184–85, 305–12, and *History* 3: 677–78, 684–85.

43. Richter, *Gegenstände* 3: 179–97; Partington and D. McKie, "Historical Studies on the Phlogiston Theory. – IV. Last Phases of the Theory," *Ann. Sci.* 4 (1939): 113–49, esp. 130–35, 146–48. Richter called any homogeneous substance an "element": Partington, *History*, 3: 680.

44. This is the goal of my "Atoms and Equivalents: The Early Development of the Chemical Atomic Theory," *Hist. Stud. Phys. Sci.* 9 (1978): 225–63.

45. David Knight (*Atoms and Elements*, pp. 32–34) and others have made this distinction, but it has not been much studied.

46. For nineteenth-century atomic skepticism, see M. J. Nye, "The Nineteenth-Century Atomic Debates and the Dilemma of an 'Indifferent Hypothesis'," *Stud. Hist. Phil. Sci.* 7 (1976): 245–68; M. J. Nye, *Molecular Reality* (New York, Science History, 1972); and W. H. Brock, ed., *The Atomic Debates* (Leicester: Leicester University Press, 1967).

47. The equivalent weight concept has several applications besides its use for elements in the law of equivalent proportions; there are acid-base equivalents (à la Richter), molecular equivalents, radical equivalents, electrochemical equivalents, and oxidation-reduction equivalents. As these have only a peripheral relationship to chemical atomism, they will not be dealt with here.

48. Before the advent of direct physical methods around 1914, the most precise determinations of atomic weights were accomplished by determining the equivalent weights through careful "wet" chemical analysis and then multiplying by the appropriate integer. See Freund, *Chemical Composition*, pp. 200–225.

49. William Wollaston, "A Synoptic Scale of Chemical Equivalents," *Phil. Trans. Roy. Soc.* 104 (1814): 1–22. The word had been used by Henry Cavendish, *Phil. Trans. Roy. Soc.*, 57 (1767): 102, and by Wollaston himself, *Phil. Trans. Roy. Soc.* 98 (1808): 100–101, as applied to acids and bases.

50. For example, A. Laurent, *Chemical Method*, trans. W. Odling (Lon-

don, 1855), pp. 1-16; A. Kekulé, "Aequivalent und Aequivalenz," in H. von Fehling, ed., *Neues Handwörterbuch der Chemie* (Brunswick, 1871), pp. 77-89; A. Wurtz, "Discours Préliminaire: Histoire des doctrines chimiques depuis Lavoisier," in *Dictionnaire de chimie pure et appliquée* 1 (Paris, 1868), pp. xiii-xiv, xxi-xxii, lix-lx; G. Salet, "Équivalents," *Dictionnaire de chimie pure et appliquée*, 1 (1868), pp. 1,252-53; A. Ladenburg, *Vorträge über die Entwicklungsgeschichte der Chemie in den letzten hundert Jahren*, 1st ed. (Brunswick, 1869), pp. 67-68; F. J. Moore, *History of Chemistry* (New York: McGraw-Hill, 1918), p. 81; and D. C. Goodman, "Wollaston and the Atomic Theory of Dalton," *Hist. Stud. Phys. Sci.* 1 (1969): 37-59, esp. 44-50. See also n. 44.

51. J. Dalton, *A New System of Chemical Philosophy*, pt. 2 (Manchester, 1810), p. 276.

52. R. Harré, *Matter and Method* (London: Macmillan & Co., 1964), pp. 9-18, 57-58.

53. W. Whewell, *Philosophy of the Inductive Sciences*, 2 vols., 2d ed. (London, 1847), 2:65; *History of the Inductive Sciences*, 3 vols., 3d ed. (London, 1857), 3:131.

54. P. Duhem, *Aim and Structure of Physical Theory*, trans. P. Weiner from 2d French ed. of 1914 (Princeton: Princeton University Press, 1954), pp. 69-104.

55. M. B. Hesse, *Models and Analogies in Science* (London: Sheed and Ward, 1963), p. 21.

56. N. Campbell, *Physics: The Elements* (Cambridge: Cambridge University Press, 1920), pp. 129-58; D. H. Mellor, "Models and Analogies in Science: Duhem *versus* Campbell," *Isis* 59 (1968): 282-90.

57. Laurent argued in 1854 that true equivalents may not be added to each other to derive molecular weights: *Chemical Method*, pp. 1-16; M. Donovan noted that chemical atomism was but poorly camouflaged by the term "equivalent": *Treatise on Chemistry* (London, 1832), pp. 399-400; S. Cannizzaro argued in much the same fashion: "Considerations on Some Points of the Theoretic Teaching of Chemistry," *Faraday Lectures*, ed. C. S. Gibson and A. J. Greenaway (London, 1928), pp. 17-43, esp. pp. 19-22; see also E. Divers, "The Atomic Theory without Hypotheses," *Rep. Brit. Assoc. Adv. Sci.* 72 (1902): 557-75.

2

The Rise of John Dalton's Chemical Atomic Theory

In a classic historical paper, Gerald Holton explored the role of the Michelson-Morley experiment in the formulation of Einstein's special theory of relativity. Holton's intent was to refute the widely accepted view that there was a strong genetic link between the experiment and the theory. In the course of the discussion, he introduced a proper distinction between "private" and "public" science; that is, between how a scientist wrestles with a problem in his own mind and how ideas are ultimately accepted or rejected in the scientific community. He was concerned to demonstrate, at least for this case, "the experimenticist fallacy of imposing a logical sequence" between experimental discovery and theoretical creativity, and instead adopted Einstein's own view that "the struggle with [the scientists'] problems, their trying everything to find a solution which comes at last often by very indirect means, is the correct picture."[1]

There are close parallels here to the relationship between the discovery of the law of multiple proportions and the origin of Dalton's chemical atomic theory. As with the origins of special relativity: (1) a logically impeccable experimenticist version was widely accepted in the nineteenth century, and is still found today; (2) Dalton's own path to chemical atomism was in fact far more indirect, and the relationship to the "crucial experiment" far more problematical, than the inductivists would like to admit; and (3) a strict distinction between private and public science is a necessary starting point. The account that follows attempts to synthesize and elaborate the state of the art of this fascinating historical problem.[2]

FROM MIXED GASES TO CHEMICAL ATOMS, 1790-1803.

Dalton was by inclination and training a meteorologist interested in the physics of gases and, in particular, in such widely debated questions as the state of water vapor in the air and how to account for the fact that the atmosphere does not separate out into layers of gases according to their densities, but rather is fully homogeneous. The prevailing opinions, from the 1780s well into the nineteenth century, were that water vapor is dissolved in air like a salt in water and that homogeneity is produced by a weak chemical combination between the various gases.[3] As early as 1793, Dalton argued that an entirely mechanical effect was operating in both cases; he thought that air is "an intimate mixture of various elastic fluids or *gasses*," water vapor being a "fluid *sui generis*, diffused among the rest"; in short, each gas exists and acts independently and purely physically rather than chemically.[4]

In 1810 Dalton presented to the Royal Institution a first-person account of the reasoning that led him to this belief.[5] He related how he began by attempting "to reconcile or rather to adapt [the prevailing] chemical theory of the atmosphere to the Newtonian doctrine of repulsive atoms or particles." He constructed schematically ("upon paper") an appropriately proportioned mixture of three gases, NOAq, NO, and N, but noted that such substances would tend to separate out according to specific gravity.[6] To remedy this defect, he tried extending the caloric envelope of the heavier particles in order to make them larger and thus less dense; but this stratagem resulted in the atmosphere being precisely the density of nitrogen gas, a circumstance that could "not for a moment be tolerated." He decided that the chemical theory was irreconcilable with the phenomena. It was thus necessary to suppose that Aq, N, and O formed peculiar particulate gases, affecting each other by mechanical repulsion but not by chemical attraction. This was the model underlying his *Meteorological Observations and Essays* of 1793. But this too was unsatisfactory, he related, as the fact of homogeneity was still unaccounted for.

However, "In 1801 I hit upon an hypothesis which completely obviated these difficulties. According to this, we were to suppose that the atoms of one kind did *not* repel the atoms of another kind, but only those of their own kind. This hypothesis most effectually provided for the diffusion of any one gas through another, whatever might be their specific gravities, and perfectly reconciled any mixture of gases to the Newtonian theory."[7] Dalton immediately published a brief sketch of this theory in the *Journal of Natural Philosophy, Chemistry and the Arts* and a detailed account in the *Memoirs of the Manchester Literary and Philosophical Society*.[8] In the latter he wrote: "When two elastic fluids, denoted by A and B, are mixed together, there is no mutual repulsion amongst their particles; that is, the particles of A do not repel those of B, as they do one another. Consequently, the pressure or whole weight upon any one particle arises solely from those of its own kind."[9] This is essentially the law of partial pressures, prefigured by Dalton in 1793 but explicitly stated here for the first time as a direct consequence of his "first theory of mixed gases."

As only a few alert historians have noted, these ideas are undergirded by certain important elements of Dalton's mature theory of chemical atoms. For instance, the homogeneity of the atmosphere is anomalous for physical theories of mixed gases *only* if the atoms of different elements are presumed to have characteristic and different weights and if equal volumes of any two gases under the same conditions of temperature and pressure contain equal (or approximately equal) numbers of particles. If the distinguishing characteristic of atoms of different elements were, say, volume or shape rather than weight, the atmosphere would remain homogeneous even in the absence of chemical attractions. Indeed, it is becoming clear that this first assumption was by no means unique to Dalton.[10]

That Dalton believed in the second supposition (and hence necessarily in the first as well) is strongly implied by a diagram at the end of his "Experimental Essays" of October 1801. His purpose there was to explicate his theory of mixed gases by illustrating the atmosphere as consisting of particles of oxygen, nitrogen, water, and gaseous carbonic acid equidistant from similar particles but not equidistant from nonsimilar ones. Significantly, the numbers of particles of each gas are shown proportional to their partial pressures, hence implying the equal volumes-equal numbers (EVEN) hypothesis.[11] We also have Dalton's direct testimony regarding his belief in EVEN at this time.[12]

There is still another element of Dalton's mature atomic theory discernible in these deliberations: the belief that compound particles are composed only of small numbers of constituent atoms. Dalton could easily envisage such simple molecules as NO or NOAq. But he gave up the chemical theory of the atmosphere rather than proceeding to the logical conclusion that air consisted of a single homogeneous chemical compound, perhaps because that would involve such complicated molecules as $N_{52}O_{14}Aq$.[13]

Dalton's physical theory of mixed gases was widely publicized; it was also widely criticized by such luminaries as Thomas Thomson, John Gough, John Murray, William Henry, and Claude-Louis Berthollet. Dalton was quick and quite effective with his replies.[14] It was in defending and elaborating his ideas on mixed gases that Dalton was led to his chemical atomic theory. This happened by two different pathways: in investigating his analytical method for determining the proportion of oxygen in any mixture of gases and in extending his theory to include the solution of gases in water.

The latter approach began with Dalton's quantitative study of carbonic acid, both in the atmosphere and dissolved in water. In 1801 or 1802 he discovered that the quantity in solution is always proportional to the pressure of carbonic acid above the water. William Henry was provoked by his opposition to Dalton's theory of gases to embark on a similar research program, and he reached a similar but more general conclusion ("Henry's law"): the weight of any gas dissolved in water is proportional to superincumbent pressure, or, equivalently, the volume in solution is independent of pressure. As Henry soon realized, this law suggests that solution of a gas is a purely physical rather than a chemical effect, thus providing indirect support for Dalton's physical theory of mixed gases. Much more significant was a surprising aspect of this phenomenon: when dealing with mixtures of gases, the solubility of a given component of the mixture was determined by the partial pressure of that component alone, not by the total pressure. For instance, oxygen gas, no matter what the pressure, could not keep any carbonic acid gas in solution, or vice versa. This seemed to validate the fundamental postulate of Dalton's theory, viz., that the particles of nonsimilar gases did not repel each other.[15]

However, the fact that all gases are not equally soluble was a problem since it seemed to indicate some interaction with water that was not purely mechanical. Dalton wrote:

> This question I have duly considered, and though I am not yet able to satisfy myself completely, I am nearly persuaded that the circumstance depends upon the weight and number of the ultimate particles of the several gases: those whose particles are lightest and single being least absorbable and the others more according as they increase in weight and complexity. An enquiry into the relative weights of the ultimate particles of bodies is a subject, as far as I know, entirely new: I have lately been prosecuting this enquiry with remarkable success. The principle cannot be entered upon in this paper; but I shall just subjoin the results, as far as they appear to be ascertained by my experiments.[16]

There followed immediately the first published "Table of the relative weights of the ultimate particles of gaseous and other bodies:"

Hydrogen	1	
Azot	4.2	
Carbone	4.3	
Ammonia	5.2	
Oxygen	5.5	
Water	6.5	
Phosphorus	7.2	
Phosphuretted hydrogen	8.2	
Nitrous gas	9.3	[9.7]
Ether	9.6	
Gaseous oxide of carbone	9.8	
Nitrous oxide	13.7	[13.9]
Sulphur	14.4	
Nitric acid	15.2	
Sulphuretted hydrogen	15.4	
Carbonic acid	15.3	
Alcohol	15.1	
Sulphureous acid	19.9	

Sulphuric acid	25.4
Carburetted hydrogen from stag. water	6.3
Olefiant gas	5.3

It should be noted that the values for nitrous gas and nitrous oxide are inconsistent with "azot = 4.2", and that the last two entries were added between the reading of the paper (21 October 1803) and its publication (November 1805).[17] It is likely, however, that the rest of the table accurately reflects Dalton's data as originally read.

This conclusion is reinforced by passages from Dalton's laboratory notebook published by Roscoe and Harden (the notebooks themselves were destroyed in an air raid in 1940). Under the date 6 September 1803, Dalton made some "observations on the ultimate particles of Bodies, & their combinations."[18] This included the following table of atomic and molecular weights:

Ult. at.	Hydrogen	1
	Oxygen	5.66
	Azot	4.—
	Carbon (Charcoal)	4.5
	Water	6.66
	Ammonia	5.—
	Nitrous Gas	9.66
	Nitrous Oxide	13.66
	Nitric Acid	15.32
	Sulphur	17
	Sulphureous acid	22.66
	Sulphuric acid	28.32
	Carbonic Acid	15.8
	Oxide of Carbone	10.2

It is evident from the notebook how Dalton arrived at these numbers. For instance, he cited Lavoisier's data for the composition of water (85% oxygen and 15% hydrogen); the unstated but obvious assumption of a "binary" water molecule (HO) places the weight of oxygen as 5.66 relative to hydrogen as unity, since $^{85}/_{15}$ = 5.66. Similarly, by applying William Austin's analysis of ammonia (80% nitrogen) to a binary formula, an atom of nitrogen must weigh about 4. Assuming "ternary" formulas for the higher oxides of carbon and sulfur and relying on analyses published by Lavoisier and Richard

Chenevix, Dalton calculated atomic weights for carbon and sulfur of 4.5 and 17.[19] And so on. Such reasoning lies at the heart of every chemical atomic theory.

Dalton's procedure makes implicit use of a convention that he codified in 1808 — the so-called "rule of greatest simplicity." According to this rule, if only one compound is known of two given elements A and B, the molecules are presumed to be binary (AB); if two such compounds are known, then one is binary and the other ternary (AB and A_2B); if three are known, one is binary and two are ternary (AB, A_2B, and AB_2); if four are known, one and only one is quaternary; and so on. In brief, one always assumes the simplest formulas, i.e., those with the least possible numbers of atoms.[20]

AN HISTORIOGRAPHIC DIGRESSION

So much is certain; we know when and how Dalton calculated the first atomic weights. What is still controverted, however, is why Dalton carried out his calculations; this important private aspect of Dalton's science has continued to fascinate historians and generate discussion from the early nineteenth century to the present day.

The prevailing nineteenth-century explanation, given no less than three times by Thomson, was that Dalton was led to the atomic theory by his discovery of multiple proportions in the two known hydrocarbons, methane and ethylene.[21] This inductivist version, so concordant with the Victorian model of heroic science, was discredited when Roscoe and Harden showed that Dalton's first analyses of these two gases were not carried out until August 1804. Roscoe and Harden thought that a "second theory of mixed gases" provided the motive for the September 1803 calculation, an idea they derived from Dalton himself that will be discussed below. The only problem was that Dalton said the "second theory" occurred to him in 1805, so Roscoe and Harden were compelled to assume that Dalton made a clerical error and meant to write 1803 instead.[22]

Roscoe and Harden's version was in turn opposed by Meldrum, who argued from a previously uncited notebook entry that the "second theory" dated from 1804 rather than 1803, hence could not possibly be the crucial event. For

Meldrum the turning point was Dalton's discovery of multiple proportions in the oxides of nitrogen in August 1803, which led him to apply a physical atomic theory to chemistry.[23] The importance of the nitrogen oxides to Dalton had been stressed by several earlier authors — including Dalton himself — and will be further discussed below. Yet another version, doubtfully noted by Dalton's first biographer and ably defended a number of years ago by Henry Guerlac, suggests that Dalton may have devised his theory after seeing Richter's table of equivalents in Berthollet's *Essai de statique chimique* of 1803. Recent work, has, however, cast severe doubt on this possibility.[24]

Another historiographic model was provided in a classic 1956 article by Leonard Nash.[25] Nash proposed a two-step mechanism whereby the first 1803 weight calculations were regarded merely as a preliminary, ad hoc, and undeveloped excursion into chemical atomism, to serve solely as an adjunct to Dalton's physical theories. A second step was then required to demonstrate to Dalton — and to provide Dalton the wherewithal to demonstrate to his colleagues — the fecundity of a purely chemical atomic theory. With one exception all versions proposed since 1956 have suggested such a two-step model; it is in fact difficult to avoid, considering the inevitable conflict of dates provided by Dalton himself.

Specifically, Nash argued that Dalton's gas solubility studies led him to investigate the "weight and complexity" of ultimate particles of matter with the hope of correlating this parameter with solubility. We have seen that Dalton publicly suggested such a train of events in his 1803 paper. Nash believed that the impetus for the further development of Dalton's chemical atomic theory was a combination of his discovery of multiple proportions in the hydrocarbons, a stimulating visit by Thomson the same month (August 1804), and the second theory of mixed gases (dated, securely it was thought, as September 1804).

Ten years later Thackray argued convincingly that the second theory was, as Dalton had said, devised only in late 1805. Thackray accepted both Nash's 1803 gas-solubility pathway and Roscoe and Harden's 1805 mixed-gas pathway, stressing and skillfully supporting the important thesis that Dalton's chemical atomism evolved only slowly, by a "gradual internal process," and remained almost completely undeveloped until 1805. Cole, the most recent participant in the saga, has followed Meldrum and Debus in placing the greatest significance on Dalton's study of the nitrogen oxides with a mind appropriately conditioned by certain ideas about physical atoms. This scenario combines elements of the Thomson inductivist and the Roscoe-Harden deductivist versions. Cole has ably argued that Dalton's partial (and ultimately unsuccessful) correlation between gas solublity and particle weights was most probably discovered adventitiously, only after he had a set of weights to manipulate. He has thus urged a return to the point of view prevalent before Nash's work, namely, that the gas-solubility correlation was simply the first application of a chemical atomic theory derived from other considerations.

THE DISCOVERY OF MULTIPLE PROPORTIONS

So we are left at the end of this digression with the same question we had at the beginning: what induced Dalton to turn rather abruptly to chemistry to calculate weights for his physical atoms? I believe the version of the story most recently advocated is also the most satisfactory, and the following account draws heavily from the work of Debus, Meldrum, and Cole.

Like all of the competing versions, this one derives from the protagonist himself. Dalton wrote in 1811: "I remember the strong impression which at a very early period of these inquiries was made by observing the proportion of oxigen to azote as 1, 2, and 3 [sic for 4], in nitrous oxide, nitrous gas, and nitric acid, according to the experiments of Davy."[26] Davy's analyses were published in 1800, and Dalton was familiar with the work at least by December 1803.[27] Those analyses indicate approximately precise multiple proportions for oxygen and nitrogen in nitrous oxide, nitric oxide ("nitrous gas"), and gaseous "nitric acid" (modern formulas N_2O, NO, and NO_2).[28]

But since early in 1803, Dalton had been carrying out his own studies of nitrogen oxides in conjunction with his use of Joseph Priestley's celebrated "nitrous air test" for the percentage of oxygen in atmospheric air. The basis for this test is the reaction of nitric oxide with oxygen, which Dalton discovered can take place in two different proportions forming an exact two-to-one ratio:

$$4NO + O_2 = 2N_2O_3$$
$$\text{and} \quad 2NO + O_2 = 2NO_2 \tag{2.1}$$

or, in more Daltonian terms:

$$2NO + O = N_2O_3$$
$$\text{and} \quad NO + O = NO_2 \tag{2.2}$$

or even, since twice the nitrous gas is equivalent to half the oxygen:

$$NO + O = NO_2$$
$$\text{and} \quad NO + 2O = NO_3 \tag{2.3}$$

It was natural for Dalton to want to study this reaction in connection with his work on mixed gases. His first results indicated nitrous gas–oxygen ratios for the two reactions of 2.7 and 1.7 instead of 4 and 2. The absence of an integral ratio between the two was not anomalous even given Dalton's atomistic proclivities, for at this time the product of the first reaction, nitrous acid, was not known to be a distinct compound, but rather was regarded as a mixture of nitric acid and nitric oxide.[29]

But by 4 August 1803, it seems Dalton had discovered the simple relationship outlined above:

> It appears, too, that a very rapid mixture of equal parts com. air and nitrous gas, gives 112 or 120 residuum. Consequently that oxygen joins to nit. gas sometimes 1.7 to 1, and at others 3.4 to 1 [sic for 1 to 1.7, 1 to 3.4].

Dalton probably obtained the 1.7 figure, representing complete oxidation of nitrous air to nitric acid, from Lavoisier. It is the 3.4 ratio that emerges from the data cited; but, significantly, *only from one of the two experimental findings* (112 parts by volume remaining from 200). That Dalton cited only the one result of calculation and did not use an average of the two is strong presumptive evidence that he was looking for an even multiple.[30] The case becomes even stronger when it is considered how difficult it is to obtain these results with the specified experimental procedure.[31]

Dalton referred to this pair of reactions as a clear instance of multiple proportions in his paper on the composition of the atmosphere — it was in fact the first such instance ever discovered and noticed. The paper was read in November 1802 and published three years later; so Dalton must have silently revised it before publication. The experiments on the nitrogen oxides cited there were not the 4 August 1803 results, but a new series performed between 10 October and 13 November.[32] He also explicitly mentioned this instance of multiple proportions in his gas solubility paper in such a way as to imply that the passages were there as originally read, but since the date of reading was 21 October 1803, he could have been referring to either series of experiments.[33] Dalton surely interpreted his 4 August 1803 experiments as multiple proportions after the second series was completed, but there is no *decisive* evidence that he interpreted them as such at the time, or, more to the point, that they led to his September atomic weight calculations.

Nash attempted to show that it is very unlikely that they did. His most telling point was that there is no indication, either in his notebooks or in his publications, that up until at least 6 September 1803 he saw anything in the 3.4 to 1.7 ratio other than another datum among many. But Nash himself pointed out that Dalton "was not at this time particularly interested in chemical phenomena" and that this work with the nitrogen oxides was in fact Dalton's first purely chemical investigation.[34] It was noted above that Dalton's pre-1803 atomistic speculations presupposed some important elements of his mature chemical atomic theory, especially that atoms of different elements differ in weight and that they combine one-to-one wherever possible. Perhaps his discovery of multiple proportions on 4 August 1803 was such an obvious consequence of these axioms that it simply failed to impress him as anything worth reporting or pursuing. The situation here is analogous to Einstein's reaction to the Michelson-Morley experiment: not an astonished "Good gracious!" but rather a complacent "Of course!" A chemist at that time might have used the former exclamation; Dalton was not yet really a chemist.

Dalton's notebooks provide more direct evidence of the importance of the nitrogen oxides to his 6 September 1803 atomic weight calculations.[35] Dalton first sought to verify the series N_2O, NO, NO_2, and NO_3 (note that this series is the most consistent possible with the simplicity axiom). Roscoe and Harden claim that this trial calculation is inconsistent with the formulas for nitrous and nitric acid implied by the 4 August results, if those results were interpreted as evidence for multiple proportions. This was their most important reason for opposing the view that Dalton did interpret them in that manner. But this oxide series *is* consistent with reactions (2.3) above, whereas Roscoe and Harden obviously had (2.2) in mind (as have all other historians to date). That Dalton may indeed have been thinking of oxygen rather than nitrous gas as the multiple is perhaps indicated by the careless slip in the 4 August notation, cited above.

In any case, Dalton quickly concluded from his analyses that the highest oxide, nitric acid, had to be NO_2 rather than NO_3. This must have been somewhat disappointing to Dalton, as it meant that the next highest oxide, nitrous acid, had to be formulated as N_2O_3. This accorded well with the analytical data available to Dalton, but he hesitated to write it in his first manuscript list of atomic weights as it contradicted his simplicity assumptions. It first appeared in an annotation dated 12 October 1803.[36]

Dalton then proceeded to apply his chemical theory to other simple compounds, using his simplicity assumptions to determine formulas. Ammonia, for instance, was assumed to be binary (NH). It might seem surprising that Dalton failed to uncover a contradiction here between the atomic weight of nitrogen calculated from ammonia and from nitrous gas, since a modern chemist using Dalton's formulas would calculate N = 4.7 and N = 7.0 in the two cases.[37] But his reliance on analyses that ultimately proved to be inaccurate and his willingness to round off in the most favorable direction successfully masked the contradiction; ultimately (from 1806) he settled on "about 5" as the best number.[38]

If he had taken nitrous oxide as NO rather than N_2O, he could have solved the problem of an excessively complicated formula for nitrous acid — that is, his oxide series would have been NO, NO_2, NO_3, and NO_4 rather than N_2O, NO, N_2O_3, and NO_2. But this would have put the atomic weight of nitrogen at around 11 given his value for oxygen (from 1806 it was 7), clearly far out of line from his nitrogen figure calculated from ammonia (less than 5). Dalton indicated twice in later years that he settled very early and very definitely on nitrous oxide as ternary and nitrous gas as binary.[39] His stated reason for the choice was the fact that nitrous gas has the lowest vapor density in the series.[40]

THE PURSUIT OF CHEMICAL ATOMS, 1803-1810

This last observation brings up yet another controversial area of Daltonian scholarship: his relationship to the equal volumes-equal numbers (EVEN) hypothesis. Clearly he held to this assumption in 1801, in fact probably well before that year. As early as 6 September 1803, however, he had some doubts. He noted that oxygen gas is denser than water vapor, even though the latter gas must consist of heavier particles (HO versus O); the conflict with EVEN seemed to him inescapable.[41] Thus the simplest and most logical means of measuring atomic and molecular weights, vapor densities, could not be relied on as a general method, and this forced Dalton to turn to chemical analyses and simplicity assumptions as a *pis aller*.

Nagging doubts about the EVEN assumption were transformed into conviction of its falsity (or so it would seem) when Dalton devised a new theory of mixed gases to replace his 1801 theory. Dalton wrote:

> Upon reconsidering this subject, it occurred to me that I had never contemplated the effect of a *difference of size* in the particles of elastic fluids. . . . And if the *sizes* be different, then on the supposition that the repulsive power is heat, no equilibrium can be established by particles of unequal size pressing against each other.

[This difference in size, he suggested, would initiate and maintain an "intestine motion" that would ensure a homogeneous mixture of the two gases.]

This idea occurred to me in 1805. I soon found that the *sizes* of the particles of elastic fluids *must* be different. [Here follows an argument similar to that cited from his notebook for 6 September 1803.]

The different *sizes* of the particles of elastic fluids under like circumstances of temperature and pressure once established, it became an object to determine the relative *sizes* and *weights*, together with the relative *number* of atoms in a given volume. This led the way to the combinations of gases and to the *number* of atoms entering into such combinations. . . . Thus a train of investigation was laid for determining the *number* and *weight* of all chemical elementary principles which enter into any sort of combination one with another.[42]

That this first-person account of the origin of his chemical atomic theory cannot be taken at face value is obvious given the manifest contradiction in dates. Yet it probably embodies a large part of the truth. Nothing prevents us from assuming that Dalton's 1805 "second theory of mixed gases" provided the impetus, if not for the creation of chemical atomism, at least for a realization of its importance and a greatly expanded application of it at the hands of its creator. When recounting the origin of the theory years later, it would be natural to remember this precipitating event rather than the first tentative calculations of 1803.[43]

Dalton by then had also discovered a second instance of multiple proportions in his analyses of ethylene and methane ("olefiant gas" and marsh gas or "carburetted hydrogen"). He was able to emend his first published table of particle weights to include molecular weights for these two compounds corresponding to the formulas CH and CH_2.[44] Just three weeks after these analyses, in August 1804, Dalton for the first time met his sometime adversary, Thomas Thomson, who happened to be visiting his former pupil William Henry in Manchester. Dalton told Thomson about his recent discovery and about the chemical atomic theory in general. Thomson was enthusiastic, jotted down a short memorandum, and became the first man to publicize the theory when he gave an account of it in the third edition of his textbook, published three years later.[45]

After 1805 Dalton clearly recognized the radical implications for chemistry of an atomic theory. In March 1807 he delivered a well-received series of lectures on his theory in Edinburgh. While there he visited with Thomson again and learned from him of Richter's work, probably for the first time. Later that year Thomson's summary of Dalton's theory appeared. Dalton himself had been preparing a full account of his ideas at least since 1805; the first volume of his *New System of Chemical Philosophy* was published in two parts in 1808 and 1810. In January 1808 Thomson and W. H. Wollaston independently announced additional examples of multiple proportions, pointing out that they were simple consequences of Dalton's chemical atomism. The theory was finally well launched.[46]

THE FORMULA PROBLEM

In the absence of physical evidence, a formula is essential for determining any atomic weight (we have examined one case study of this, the nitrogen oxides, in some detail). Consequently, Dalton's most significant innovation, the foundation upon which his entire system of chemistry rested, was his procedure for deciding on the most probable formulas. This was his rule of greatest simplicity, briefly mentioned at the end of Part 1 of the *New System*, and applied in detail to specific cases throughout Part 2. We will see that in 1807 Thomson regarded "Mr. Dalton's hypothesis" as nothing more nor less than the simplicity rule itself. This was understandable considering the similarity between Dalton's and Thomson's pre-1803 assumptions about physical atoms, which meant that what has traditionally been regarded as the essence of Dalton's theory, the combinations of atoms differing in weight, simply would not appear all that novel to Thomson.[47]

But critics observed that the simplicity rule had no empirical basis. John Bostock wrote: "When bodies unite only in one proportion, whence do we learn that the combination must be binary? Why is it not probable, that water is formed of two atoms of oxigen and one of hidrogen, of two atoms of hidrogen and one of oxigen, or in short of any assignable number of atoms of hidrogen and oxigen? I do not perceive that Mr. Dalton has given any reason in support of this binary combination, in preference to all the rest, and I am unable to conjecture what reason can be urged in its favor."[48] In answer to Bostock's objections, Dalton gave a cogent rationale, which had not appeared in his *New System*, as a justification of his rule. He argued that the fewer mutually repellent atoms there are in a molecule, the greater the mechanical stability of the system. The application of elementary geometric principles shows that a maximum of twelve equal spheres can be close-packed around a single sphere, so Dalton suggested that in a compound of elements A and B, the formula AB_{12} represents the highest possible ratio of B to A. However, the most stable entity would clearly be AB, the binary formula, in which there would be no repulsive forces at all. When only one compound of elements A and B was known, it was therefore eminently probable that the atoms combined according to this most stable formula.[49] The assertion that Dalton's rule was arbitrary or was merely connected in some vague way with a naive faith in the simplicity of nature misses the point;[50] in Dalton's eyes the rule was well grounded in accepted principles of the physics of gases.[51]

It is seen that the rule was not as arbitrary as critics have often portrayed it. However, a point often missed is that it is quite ambiguous: it did not stipulate operations.[52] The difficulty became apparent for cases of multiple proportions where it failed to specify which compound had the simpler formula. We have seen that Dalton had no decisive reason to choose the nitrogen oxide series he did. Again, the two oxides of carbon could be assigned the formulas C_2O and CO just as easily as CO and CO_2. In other words, the rule was silent on the critical question of which compound was binary and which was ternary.

Dalton was thus forced to turn to chemical and physical evidence for the answer to this question. For the oxides of nitrogen and carbon, he considered vapor densities to be the "best criterion," so he assigned the binary formulas to the lower oxide of carbon, CO ("carbonic oxide"), and to the second oxide of nitrogen, nitrous gas.[53] Dalton's procedure here involved a hidden EVEN assumption, even though he had expressly and repeatedly rejected that assumption since 1803.[54] In his discussion of the formula for "carburetted hydrogen," the assumption was no longer hidden: he stated explicitly that "there are the same number of atoms of hydrogen and of carburetted hydrogen in the same volume."[55] The use of volumes to indicate relative numbers of particles was central to his experimental investigation of the formulas for both hydrocarbons.

The best way to reconcile these contradictions is to assume, with Nash, that although Dalton did not regard EVEN as exact, he inferentially accepted it and used it as a serviceable approximation.[56] This view is supported by the fact that both in refuting and in accepting EVEN, Dalton usually added qualifications. In 1803 he did not deny "some relation" between vapor densities and atomic weights; in 1812 he admitted that there was "something wonderful" in Gay-Lussac's approximations to integral volumes; in 1827 he urged skepticism "till some reason can be discovered." In his 1810 table of the diameters of sixteen atoms and molecules, no fewer than ten are equal in size within a 5 percent error, and all of the others are within about 25 percent of this figure.[57] Conversely, when using EVEN to determine formulas, he noted that a smaller vapor density "is rather an indication than a proof" of a simpler formula, and he usually made efforts to adduce other chemical and physical arguments to support his formula assignments.[58]

Obviously one consideration holding Dalton back from accepting EVEN as exact was his second theory of mixed gases, whose fundamental postulate was an inequality of particle diameters. But this bar was removed when he returned decisively to his first theory of mixed gases, by 1826 at the latest. He wrote, "I have long been inclined to adopt [this theory] as most consistent with the phenomena," although he admitted that the theory still had "difficulties."[59] In view of this change of opinion, it is perhaps not surprising that Dalton returned unreservedly also to the EVEN axiom in 1841 — or so at least it would appear. He wrote at the time:

> It is my opinion that the simple atoms are *alike, globular*, and all of the same *magnitude* or *bulk*, whether of *hydrogen* 1, or *lead*, 90.
> My friend Mr. [Peter] Ewart, at my suggestion, made me a number of equal balls, about an inch in diameter, about 30 years ago; they have been in use ever since, I occasionally showing them to my pupils. . . .

I had no idea at that time (30 years ago) that the atoms were all of a *bulk*; but for the sake of illustration I had them made alike.[60]

In short, Dalton was by no means dogmatic on this question, vacillating between acceptance and rejection of this assumption from 1801 to 1841. This becomes particularly evident when we take not only his explicit statements but also his methodology as evidence; he relied heavily—though not exclusively—on vapor densities when his simplicity rule could give no guidance in the establishment of formulas. In his mature theory, Dalton used both chemical and physical evidence to cast light on purely chemical questions. It was this aspect of Dalton's work that led to the "empirical" modifications of Davy and Wollaston.

It is significant that Dalton was studying the *physics* of gases both when he first dabbled with chemical atomism in 1803 and when he was led to the recognition of its importance after 1805. It is rather unlikely that a *chemist* could then have taken the same path, for at least three reasons. First, as was discussed in the previous chapter, chemists were still uncertain which substances were true elements and which simply had not yet been decomposed. Dalton, outside the mainstream of chemical thought, accepted Lavoisier's list uncritically. Accordingly, he suggested atomic weights for substances that were later shown to be compound (such as the alkalies and the alkaline earths) and devised a formula for a "compound" (oxymuriatic acid, later called chlorine) that was soon shown to be elementary.[61]

Daltonian atoms have often been portrayed as similar to simple unsplittable billiard balls. This is somewhat overdrawn. Dalton accepted Lavoisier's operational definition of elements, recognized the provisionality of any list of elements, and never ruled out the possibility of subatomic structure.[62] However, he also expressed his belief that: "We might as well attempt to introduce a new planet into the solar system, or to annihilate one already in existence, as to create or destroy a particle of hydrogen. All the changes we can produce, consist in separating particles that are in a state of cohesion or combination, and joining those that were previously at a distance."[63] Furthermore, provisionality of elemental status did not mean for Dalton that his list of elements was anything less than highly probable.

Dalton transformed Lavoisier's provisional elements into ontological elements, though both were equally derived from the operational rule based on gravimetric criteria. This transformation—really the *psychological* key to chemical atomism—was facilitated by Dalton's adherence to the British school of popular Newtonianism and to a relatively straightforward materialist metaphysics.[64]

The second factor favoring Dalton's physical approach was the initial apparent absence of any scientific utility for a chemical atomic theory. It is important to realize that Dalton's own discovery of multiple proportions in the nitrogen oxides was probably the first stoichiometric relation known to him.[65] Without stoichiometry chemical atoms have little heuristic value, and one interesting experimental result could do no more than pique Dalton's interest in pursuing the idea for the light it might shed on his physical studies. In short, uncertainty concerning the usefulness of the theory for chemistry was no obstacle for the physicist Dalton. After 1807, with all three laws of stoichiometry now known and largely accepted—and seen as direct consequences of a chemical atomic theory—the heuristic power of the theory was appreciated. Again it must be emphasized that chemical atomism is not merely a restatement of stoichiometry. The theory is predictive and can be tested for consistency.[66] We have seen, for example, how Dalton regarded his atomic weight for nitrogen calculated from the oxides to be verified by the weight calculated from the hydride, ammonia—rather too optimistically, we would say today.

The third reason why a physical approach was fruitful was because chemists then had no approach to the determination of formulas. It was, in fact, Dalton's rule of greatest simplicity that first gave an entrée into chemical atomism, and it was essentially that rule that was being tested in the example just cited. Consequently, Thomson's identification of Dalton's theory with the simplicity rule was perceptive, and this recognition of Dalton's essential breakthrough has recently been stressed.[67] To reject Dalton's rule—as the chemist Bostock did—was to reject the theory; there was then no other alternative.

However tentatively and provisionally, Dalton had begun the process of extending the power of science to the microscopic material realm; he had begun to realize the Newtonian dream. As Guerlac put it: "Dalton's was surely the most important step ever taken in the quantification of the chemical theory. And it is the first successful example of the kind of transdictive procedure which Newton was hoping for."[68]

GAY-LUSSAC'S LAW OF COMBINING VOLUMES

As Dalton's relationship to the EVEN hypothesis could not help but be affected by the discovery of the law of combining volumes of gases, a brief account of this event will be provided at this time.

On 31 December 1808 Joseph Louis Gay-Lussac read what was to be perhaps his most important chemical memoir to the Société Philomatique of Paris.[69] On the basis of both synthetic and analytic data, Gay-Lussac revealed that gases combine only in very simple volume proportions, such that one part by volume of one reactant will combine with exactly one, two, or three parts by volume of another gas, to form one or two parts by volume of product (if, of course, it happens to be gaseous). His data included such reactions as (using modern notation):

$$100 \ BF_3 + 100 \ NH_3 = \text{solid product}$$
$$100 \ HCl + 100 \ NH_3 = \text{solid product}$$
$$100 \ CO_2 + 200 \ NH_3 = \text{solid product}$$
$$100 \ CO + 50 \ O_2 = 100 \ CO_2$$
$$100 \ N_2 + 50 \ O_2 = 100 \ N_2O$$
$$100 \ N_2 + 100 \ O_2 = 200 \ NO$$
$$100 \ N_2 + 300 \ H_2 = 200 \ NH_3$$

Only a few of the data were determined by Gay-Lussac himself; he relied heavily on previously published work by such scientists as C. L. Berthollet, A. Berthollet, Davy, Humboldt, Biot, Arago, and Bérard, and this increased the credibility of the new relation here disclosed. But Gay-Lussac often resorted to rounding off in order to maintain integral ratios.

By this time Dalton's theory was well known to Berthollet's Arcueil circle, of which Gay-Lussac was a member. Berthollet had written a long critical introduction to the French translation of the third edition of Thomson's *System of Chemistry*, much of which concerned the new atomic theory first revealed there.[70] By November 1808 Berthollet also had a personal copy of the first part of Dalton's *New System* and had read Thomson's and Wollaston's papers in the *Philosophical Transactions* adducing further examples of multiple proportions. Berthollet even confirmed Wollaston's results, though he opposed the atomic theory itself on several grounds.[71] Crosland has shown that Gay-Lussac's law was an outgrowth not so much of Gay-Lussac's and Humboldt's excellent 1805 investigation of the volumetric composition of water, nor even less of Daltonian atomism, but rather of his interest in vapor phase neutralization reactions, especially involving a new substance just discovered by Gay-Lussac and Thenard, "fluoboric gas." Wollaston's use of eudiometry to determine multiple proportions in the carbonates was apparently also a factor.[72]

Nonetheless, the discovery put Gay-Lussac in a somewhat ticklish position with his antiatomist mentor, patron, and friend Berthollet, particularly since the research was published in Berthollet's house organ. Although he noted that Dalton's work had "no relation" with his own and that it was "not entirely exact," he could not help but remark that Dalton's idea was "ingenious," that Thomson's and Wollaston's experiments "appear to confirm" it, that it has "a very large number of facts in its favor," and that his own results "are also very favorable to the theory." But in the end he remained faithful to Berthollet, suggesting that the two theories, apparently so different, are in fact reconcilable, and that "chemical action is exerted more powerfully when the elements are in simple ratios or in multiple proportions among themselves."[73]

Whether this tactful maneuver corresponded to Gay-Lussac's true feelings is impossible to say. Thomson provided an indication of how it appeared to contemporaries when he wrote to Dalton that the paper "is highly favorable to [your atomic theory], and it is easy to see that Gay-Lussac admits it, though respect for Berthollet induces him to speak cautiously." But Thomson clearly went too far when he asserted twenty years later that: "The object of Gay-Lussac's paper was to confirm and establish the new atomic theory, by exhibiting it in a new point of view.... We have only to consider a volume of gas to represent an atom, and then we see that in gases one atom of one gas combines either with one, two, or three atoms of another gas, and never with more."[74]

Dalton, of course, adamantly refused to accept Gay-Lussac's law, largely because it appeared to contradict his second theory of mixed gases. He attempted to refute Gay-Lussac's data with some care, suggesting for instance that the hydrogen to oxygen volume ratio in water was more like 1.97 than exactly 2.[75] As Nash has written:

> It is a little amusing to observe Dalton, whose experiments were generally very crude, quoting his own values as a refutation of those of Gay-Lussac, who was an acknowledged experimental virtuoso. Had this been an instance where Dalton would have expected to find a simple 1:2 ratio he would certainly have accepted 1:1.97 as a very satisfactory check. Here, however, armed with the staff of his theory, he enjoyed full confidence that Gay-Lussac *could not* be right, and the deficiency of 1.5 percent assumed a very large importance in his eyes.[76]

Nevertheless, Dalton and Gay-Lussac were not as far apart as it might seem. Dalton acknowledged "something wonderful in the frequence of the approximation"; he understood that "Gay-Lussac and the French chemists . . . [do not] deny the existence of atoms and their combinations, [but] they go no further than the expressions of facts," and we have seen that Dalton ultimately returned to belief in the EVEN assumption.[77] For his part, Gay-Lussac was careful to stress the "very large number of facts" in favor of Daltonian atomism, and we will see that at least by 1823 Gay-Lussac had become a chemical atomist in the full sense of the word—though not a *Daltonian* atomist.

---

1. G. Holton, "Einstein, Michelson, and the 'Crucial Experiment'," *Isis* 60 (1969): 133–97 (196–97); the latter quote is from R. S. Shankland, "Conversations with Albert Einstein," *Amer. J. Phys.* 31 (1963): 47–57. For other examples of similar historical methodology, see K. D. Knorr et al., eds., *The Social Process of Scientific Investigation* (Dordrecht: Reidel, 1981).

2. A few of the most valuable books and papers on this subject are: H. E. Roscoe and A. Harden, *A New View of the Origin of Dalton's Atomic Theory* (London: Macmillan & Co., 1896); H. Debus, "Die Genesis von Daltons Atomtheorie," *Z. physik. Chem.* 20 (1896): 359–76; 24 (1897): 325–52; and 29

(1899): 266-94; A. N. Meldrum, "The Development of the Atomic Theory," *Memoirs and Proceedings of the Manchester Literary and Philosophical Society* (hereafter cited as *Manch. Mem.*), 54:7 (1910), and 55:3, 4, 5, 6, 19, and 22 (1911), total 104 pp., paginated separately; L. K. Nash, "The Origin of Dalton's Chemical Atomic Theory," *Isis* 47 (1956): 101-16; J. R. Partington, *A History of Chemistry*, 4 vols. (London: Macmillan & Co., 1961-70), 3: 755-822; A. Thackray, "The Origin of Dalton's Chemical Atomic Theory: Daltonian Doubts Resolved," *Isis* 57 (1966): 35-55; A. Thackray, *John Dalton: Critical Assessments of His Life and Science* (Cambridge, Mass.: Harvard University Press, 1972); and T. Cole, Jr., "Dalton, Mixed Gases, and the Origin of the Chemical Atomic Theory," *Ambix* 25 (1978): 119-30.

3. Partington, *History*, 3: 763-64, 774-78.

4. John Dalton, *Meteorological Observations and Essays* (London, 1793), pp. 132-35.

5. Roscoe and Harden, *New View*, pp. 13-18 (MS lecture of 27 January 1810, since destroyed).

6. Ibid., pp. 14-15. Here I use modern symbols and conventions to express Dalton's verbal descriptions. NOAq represents a "compound particle" consisting of one atom each of nitrogen, oxygen and water; NO represents a "binary" compound of nitrogen and oxygen.

7. Ibid., p. 15.

8. John Dalton, "A New Theory of the Constitution of Mixed Aeriform Fluids and particularly of the Atmosphere," *J. Nat. Phil. Chem. Arts* (hereafter cited as *Nicholson's J.*) 5 (1801): 241-44, dated 14 September 1801; "On the Constitution of Mixed Gases," the first of four "Experimental Essays," *Manch. Mem.* 5 (1802): 535-50, read to the Society on 2 October 1802. The theory was probably devised no earlier than August 1801: Meldrum, "Development," 55:4, p. 12.

9. Dalton, "Mixed Gases", p. 536.

10. Mauskopf and Thackray have argued that W. Higgins and (more securely) T. Thomson held this view, hence they have concluded that it was probably a rather general one of the day. See S. H. Mauskopf, "Thomson Before Dalton: Thomas Thomson's Considerations of the Issue of Combining Weight Proportions Prior to his Acceptance of Dalton's Chemical Atomic Theory," *Ann. Sci.* 25 (1969): 229-42; Thackray, "Origin," p. 37.

11. Dalton, "Mixed Gases," table on p. 602. See Debus, "Genesis," II, 328-30 and III, 275; Partington, *History*, 3: 768; Thackray, "Origin," p. 39; and Cole, "Dalton," p. 123.

12. Dalton, *A New System of Chemical Philosophy*, pt. 1 (Manchester, 1808); pt. 2 (1810), vol. 2 (1827); pt. 1, p. 188.

13. This highly plausible suggestion has been made by Debus, "Genesis," II, 328-30, and has recently been well defended by Cole, "Dalton," pp. 123-24.

14. For a thorough review and defense, see Dalton, *New System*, pt. 1, pp. 150-87. Also see Partington, *History*, 3: 773-78; and Thackray, "Origin," p. 40.

15. Dalton, *New System*, pp. 182-84; W. Henry, "Experiments on the Quantity of Gases absorbed by Water, at different temperatures, and under

different Pressures," *Phil. Trans. Roy. Soc.* 93 (1803): 29-42; Dalton, "On the Absorption of Gases by Water and other Liquids," *Manch. Mem.* [2] 1 (1805; read 21 October 1803): 271-87, rpt. in *Foundations of the Atomic Theory* (Edinburgh: Alembic Club Rpt. no. 2, 1923), pp. 15-26.

16. Dalton, "On the Absorption of Gases," pp. 25-26. After the sentence ending with the word "complexity," Dalton added the following footnote, apparently sometime after the paper was read: "Subsequent experience renders this conjecture less probable."

17. Meldrum, "Development," 55:6, p. 3; Roscoe and Harden, *New View*, pp. 29, 62-63. The experiments from which the data in the last two entries were derived were not performed until August 1804.

18. Roscoe and Harden, *New View*, pp. 26-28.

19. Ibid., pp. 84-89. Here and throughout I use the words "binary," "ternary," in Dalton's sense, i.e., composed of two (or three) atoms.

20. Dalton, *New System*, pt. 1, pp. 213-15.

21. T. Thomson, *An Attempt to Establish the First Principles of Chemistry by Experiment*, 2 vols. (London, 1825), 1:11-12; T. Thomson, *History of Chemistry*, 2 vols. (London, 1830-31), 2:291; T. Thomson, "Biographical Account of the late John Dalton," *Proc. Phil. Soc. Glasgow* 2 (1845): 79-88 (85-86).

22. Roscoe and Harden, *New View*, pp. vii-ix, 16-17, 25-26.

23. Meldrum, "Development," 55:5, pp. 15-18; 55:6, pp. 9-18.

24. W. C. Henry, *Memoirs of the Life and Scientific Researches of John Dalton* (London, 1854), pp. 63, 84-86; H. Guerlac, "Some Daltonian Doubts," *Isis* 52 (1961): 544-54; Thackray, "Origin," p. 40; W. A. Smeaton, "Berthollet's *Essai de statique chimique* and its Translations: A Bibliographic Note and a Daltonian Doubt," *Ambix* 24 (1977): 149-58; Smeaton, "Berthollet's *Essai de statique chimique*: A Supplementary Note," *Ambix* 25 (1978):211-12.

25. For this and the following references, see n. 2 above.

26. Dalton, "Observations on Dr. Bostock's Review of the Atomic Principles of Chemistry," *Nicholson's J.* 29 (1811): 143-51 (143-45). Thomson also suggested in 1850 that Dalton devised his theory as a response to his discovery of multiple proportions in the nitrogen oxides ("Biographical Account of Dr. Wollaston," *Proc. Roy. Phil. Soc. Glasgow* 3 [1850]: 135-44 [140]).

27. Partington (*History*, 3:794) plausibly argues that Dalton only learned of these analyses in December 1803, as the first relevant entry in his notebook is from that month (Roscoe and Harden, *New View*, pp. 43-44). H. Davy, *Researches, Chemical and Philosophical, chiefly concerning Nitrous Oxide* (London, 1800), in J. Davy, ed., *Collected Works of Sir Humphry Davy*, 9 vols. (London, 1839-40), 3; Dalton, "Experimental Enquiry into the Proportion of the Several Gases or Elastic Fluids, Constituting the Atmosphere," *Manch. Mem.* [2] 1 (1805; read 12 November 1802): 244-58; rpt. in *Foundations of the Atomic Theory*, pp. 5-15.

28. Davy (*Works*, 3: 22, 84, 192; summarized 3:335) gave the following compositions:

The Rise of John Dalton's Chemical Atomic Theory 45

	% nitrogen	% oxygen
nitrous oxide	63.3	36.7
nitrous gas	44.05	55.95
nitric acid	29.5	70.5

Setting $^{36.7}/_{63.3} = x$, we find $^{55.95}/_{44.05} = 2.19\ x$ and $^{70.5}/_{29.5} = 4.12\ x$; or alternatively, setting $^{44.05}/_{55.95} = x$, we have $^{63.3}/_{36.7} = 2.19\ x$ and $^{29.5}/_{70.5} = x/_{1.88}$. The second calculation corresponds to modern formulas, the first to the series NO, NO_2, and NO_4, all within 10% experimental error. Davy later said that he strongly suspected that these figures gave Dalton his first notions of the atomic theory (W. C. Henry, *Memoirs*, p. 217).

29. Roscoe and Harden, *New View*, p. 34; Davy, *Works*, 3: 22-24.

30. Roscoe and Harden, *New View*, p. 38; Debus, "Genesis," 3:268-70; Cole, "Dalton," p. 125.

31. Nash, "Origin," p. 105; Partington, *History*, 3: 791.

32. Dalton, "Experimental Enquiry"; *Foundations*, pp. 8-9; Roscoe and Harden, *New View*, pp. 34-35.

33. Dalton, "Absorption of Gases"; *Foundations*, pp. 16n., 18n.

34. Nash, "Origin," pp. 104-8.

35. Roscoe and Harden, *New View*, pp. 26-46.

36. Ibid., p. 46. It also did not appear in the first *published* weight table.

37. The reader will recall that Dalton's water molecule was HO, making O = 8 (assuming accurate analysis), half the modern value. Hence the value for nitrogen calculated from Dalton's (correct) formula for nitric oxide will also be half the modern one, N = 7. Assuming only a third the modern number of hydrogen atoms in ammonia would make the atomic weight of nitrogen one third the modern weight, N = 4.7.

38. Dalton, *New System*, pt. 2, pp. 316-31, 436; Nash, "Origin," pp. 66-67; Partington, *History*, 3: 809.

39. In 1830: "So far back as the year 1803 I had resolved in my mind the various combinations then known of azote and oxygen, & had determined almost without doubts, that nitrous gas is a binary compound and nitrous oxide a ternary." In 1835: "*Nitrous oxyde* is composed of two particles of azote and one of oxygen. This was one of my earliest atoms. I determined it in 1803, after long and patient consideration and reasoning. Chemistry began then to take on a new appearance" (Thackray, *Critical Assessments*, pp. 91-92, 100; Dalton, quoted in *Manchester Times*, 24 October 1835, cited by E. C. Patterson, *John Dalton and the Atomic Theory* [Garden City, N. J.: Doubleday, 1970], p. 103).

40. Dalton, *New System*, pt. 2, p. 317.

41. Roscoe and Harden, *New View*, p. 27.

42. Ibid., pp. 16-17.

43. Thackray has shown that the theory really should be dated 1805 and that a passage in Dalton's notebook parallel to this 1810 account cited by Meldrum and dated 14 September 1804 (55:5, pp. 15-17) was almost certainly misdated ("Origin," pp. 44-45). Nash, Cole, and especially Thackray have emphasized the gradualness of Dalton's realization of the utility of chemical

3

The Early Reception in Great Britain: 1804–1815

THOMAS THOMSON

In the year 1804, on the 26th of August, I spent a day or two at Manchester, and was much with Mr. Dalton. At that time he explained to me his notions respecting the composition of bodies. I wrote down at the time the opinions which he offered, and the following account is taken literally from my journal of that date:

The ultimate particles of all simple bodies are *atoms* incapable of further division. These atoms (at least viewed along with their atmospheres of heat) are all spheres, and are each of them possessed of particular weights, which may be denoted by numbers. For the greater clearness he represented the atoms of the simple bodies by symbols. . . .

It was this happy idea of representing the atoms and constitution of bodies by symbols that gave Mr. Dalton's opinions so much clearness. I was delighted with the new light which immediately struck my mind, and saw at a glance the immense importance of such a theory, when fully developed.

The Scottish chemist Thomas Thomson thus described his first meeting with Dalton; published extracts from his diary certify its general accuracy.[1] Despite the fact that Thomson had respectfully criticized Dalton's first theory of mixed gases in the first two editions of his *System of Chemistry* and had advocated a corpuscular interpretation of Berthollet's conception of mass action, Thomson had much in common with Dalton. Both adhered to the British tradition of popular Newtonianism; both reasoned on the basis of corpuscular ideas; both assumed as axiomatic that the atoms of different elements differ in weight.[2]

It is understandable that Thomson immediately perceived the importance of the chemical atomic theory.

Much later he reproduced some of the atomic weights that he said Dalton had communicated to him at the time (H = 1, O = 6.5, C = N = 5). These seem however to fit better with Dalton's atomic weight values around early to mid-1806 rather than August 1804. It is likely that Thomson failed to write down any specific atomic weights at their first meeting, and that the weights given here derived from subsequent correspondence with Dalton in connection with the fourthcoming publication of the third edition of his *System* (1807).[3]

That edition was noteworthy for containing the first published description of Dalton's theory. Thomson wrote: "We have no direct means of ascertaining the density of the atoms of bodies; but Mr. Dalton, to whose uncommon ingenuity and sagacity the philosophic world is no stranger, has lately contrived an hypothesis which, if it prove correct, will furnish us with a very simple method of ascertaining that density with great precision."[4] The hypothesis referred to was not that of atoms, but merely Dalton's rule of greatest simplicity. Thomson described it in detail, concluding that the rule was "consonant to experiment" and quite possibly correct. He repeatedly returned to "Mr. Dalton's hypothesis" throughout the remainder of the volume, adducing in its favor many examples of multiple proportions. He also cited Richter's law of neutrality as "a decisive proof that the theory of Dalton, which has been repeatedly illustrated in this book, is something more than a hypothesis."[5] He provided atomic weights for twenty-three elements and molecular weights and formulas for thirty-eight compounds; his numbers resemble, but do not coincide with Dalton's.[6]

Nevertheless, throughout the volume Thomson appeared to assume the existence of atoms and to treat Dalton's theory merely as a method for the calculation of atomic "densities." This is significant for two reasons. First, it shows that at least one contemporary chemist regarded the essence of Dalton's theory to be not the combination of atoms differing in weight, but rather a method for determining the formulas of compounds — the sine qua non of chemical atomism. Second, Thomson's use of the word "density" as synonymous with weight strongly implies that he regarded all atoms and molecules to be (at least approximately) equal in size. Thomson may well have gotten this notion from Dalton himself (we have seen that Dalton was by no means dogmatic on the point); or it may represent Thomson's own belief. In any case Dalton repudiated the implication in 1814.[7]

Early in 1808 Thomson and Wollaston communicated to the Royal Society the independent discoveries of additional examples of multiple proportions.[8] Their papers were important in persuading their colleagues of the value of Dalton's ideas.[9] Thomson now no longer begged the question of the existence of physical atoms as he had the previous year. He enunciated as a law what I have characterized as the chemical atomic theory, asserting that it represented the results of "numerous and decisive experiments," and added: "From the knowledge of this curious law, it is difficult to avoid concluding that each of [the] elements consist [sic] of atoms of determinate weight, which combine according to certain fixed proportions, and that the numbers above given, represent the relative weights of these atoms respectively."[10]

Thomson considerably expanded on his view of Daltonian atomism in a major paper published serially in his own journal between July 1813 and August 1814.[11] Thomson considered "Mr. Dalton's theory of definite proportions" to be "the greatest step which chemistry has yet made as a science." He claimed to be expressing only his own views, but acknowledged that those views originated directly or indirectly entirely with Dalton. He thought it to be the "general opinion" that: "The ultimate elements of bodies . . . consist of *atoms*, or minute *solids*, incapable of farther division. . . . They must, I think, be physical points, as minute as you will, but still possessed of length, breadth, and thickness. This opinion, I say, is generally received by philosophers; and I cannot, for my part, conceive any other. It is taken for granted as the foundation of the Daltonian theory; and, I presume, will be readily admitted by everyone without hesitation."[12]

Thomson accepted Dalton's simplicity rule and derived it, as Dalton did, from the model of self-repulsive particles of gases.

He did not reject out of hand the possibility that the ultimate particles of oxygen consist of two or more inseparably linked atoms, but thought it very unlikely considering the accepted model. Furthermore, he pointed out that it is operationally indifferent whether one calculates a weight for an atom assumed to be simple, or for a "compound atom" which has structure but can never be dissected in chemical operations.[13]

So far this sounds like orthodox Daltonian atomism. But Thomson added a non-Daltonian factor by fully accepting Gay-Lussac's law of combining volumes. However, this did not entail subscribing to the notion that equal volumes of gases contain equal numbers of particles; in fact, Thomson explicitly rejected this axiom. He maintained Dalton's assumption that the water molecule is binary, which, combined with Gay-Lussac's law, resulted in the statement that there are precisely twice as many atoms of oxygen as hydrogen in a given volume.[14]

If acceptance of Gay-Lussac's law did not lead Thomson to equate relative atomic weights with relative vapor densities, it did enable him to follow Gay-Lussac in using vapor densities as an indirect means of investigating composition. For instance, the densities of oxygen and hydrogen gas were known with rather good precision. One only had to halve the density of oxygen—since there are twice as many atoms per volume—and then divide the result by the density of hydrogen to calculate the composition of water and the relative weights of hydrogen and oxygen atoms. This indirect method doubled the attainable precision compared to the measurement of actual combining weights.[15] Analogous reasoning could be applied to other compounds whose components are gaseous.

Thomson's long table of weights of "integrant particles" included 45 elements and 251 compounds, indeed, virtually every substance then known and well characterized.[16] Thomson followed Wollaston and Berzelius in choosing the standard O = 1.000. His table was the first to list precise nonintegral weight values and to cite all sources for his analytical data.

WILLIAM PROUT

At this time Prout was a brilliant young physician strongly inclined toward chemistry. Recent work has shown that he was committed to reductionism, unitary theories of matter, and the complexity of the elements as early as 1810, when he was still a medical student at Edinburgh. Apparently he was much impressed by Dalton's atomic theory, by Gay-Lussac's law of combining volumes, and by the unitary ideas contained in the last chapter of Humphry Davy's *Elements of Chemical Philosophy* (1812).[17]

In a paper published anonymously in Thomson's *Annals of Philosophy* in 1815, Prout sought to show that all atomic weights are integral multiples of that of hydrogen, i.e., that the weights are all integers in the system that sets H = 1. Proceeding from the assumption that Gay-Lussac's law was mathematically precise, he used previously published volumetric and gravimetric data to revise existing atomic weight values. Some of his methods were quite resourceful, if not completely original. For instance, recognizing the inaccuracies inherent in weighing hydrogen gas itself, he calculated its vapor density indirectly, from that of ammonia and nitrogen, assuming Gay-Lussac's integral combining ratios. Appropriate choices of data made the hypothesis appear verified and exact. A few of his more important atomic weights are H = 1, O = 8, C = 6, N = 14, S = 16, and Cl = 36, all precise integral integral values.[18] His later claim that he held to the equal volumes-equal numbers hypothesis is not supported by these figures.[19]

The integral-multiples hypothesis was for Prout but a stepping-stone to a thoroughgoing reductionist conception. Prout speculated that perhaps all the atoms of elements were ultimately composed only of hydrogen and oxygen, or perhaps of hydrogen alone. If true, this would constitute the realization of the ancient quest for the "protyle," or ultimate single building-block of all matter. Prout's dual hypotheses of integral multiples and the protyle proved to be two of the most controversial questions in chemical theory until the discovery of isotopes a century later.[20]

But the reaction only became animated after Thomson began to champion the integral-multiples hypothesis. In May 1816 he published a paper supporting Prout's data and atomic weights and revealing the the hitherto anonymous author's name. He suggested that if one adopted the convention of setting both the

ments he employed I do not know; but they must have been convincing ones, for Davy ever after became a strenuous supporter of the atomic theory."[25]

Like many of Thomson's anecdotes, this account must be treated with some caution. Certainly Davy never became a strenuous supporter of *Dalton's* version of the atomic theory. Davy's predilections were much more oriented to the romantic traditions of Kantian idealism and the dynamical view of nature, a tendency that he may have imbibed from his mentor, Thomas Beddoes, and his close friend, the poet S. T. Coleridge, around the turn of the century.[26] This put him at some distance from the British Enlightenment tradition of popular Newtonian corpuscularity in which Dalton participated. Dalton was a materialist and a realist, Davy a reductionist and an idealist. Where Dalton wished to weigh the ontological atoms of known elements, Davy assumed complexity in the provisional Lavoisierian elements and was committed to the search for the one or two "true elements" of nature. At the time Dalton and Thomson were first publicizing the atomic theory, Davy was busily engaged in discovering new metals in the alkalies and alkaline earths, and in detecting—or suggesting the presence of—hydrogen and oxygen in sulfur, phosphorus, nitrogen, ammonia, the metals, and all other oxidizable substances.[27]

Davy is usually portrayed as an implacable opponent, or at best a determined skeptic regarding the atomic theory.[28] Davy's idealism led him to view all subtle fluids and microscopic mechanisms as cancers to be excised from the body of positive science. Dalton's extended discussions of the caloric fluid enveloping ponderable, impenetrable atoms and of detailed molecular and crystalline architecture must have struck Davy as exactly the sort of speculation that a sophisticated scientist must sedulously avoid. Moreover, the supposed multiplicity of atoms violated Davy's belief in the unity of nature; Davy tended to prefer something similar to Prout's protyle hypothesis, where a single substratum is thought to underlie the apparent diversity of the elements.[29] Davy told Berzelius that he regarded Dalton's expansion of basic atomistic principles as "more ingenious than correct"; he told Dalton the same thing when the latter first spoke to him about the theory in December 1803.[30]

In a passage written at the end of his life and published posthumously, Davy expressed his view in more detail: "Indeed, in my opinion, Mr. Dalton is too much of an *Atomic Philosopher*; and in making atoms arrange themselves according to his own hypothesis, he has often indulged in vain speculation; and the essential and truly useful part of his doctrine, the expression of the quantities in which bodies combine, is perfectly independent of any views respecting the ultimate nature either of matter or its elements."[31] Similarly, when Davy, as president of the Royal Society, presented the first Royal Medal to Dalton in 1826, he noted carefully that his praise of his colleague's theory referred to "the fundamental principle, and not the details, as they are found in Mr. Dalton's system of chemical philosophy."[32]

Davy shared with many of his contemporaries the view that the "fundamental principle" of the atomic doctrine was the laws of stoichiometry, which he seemed to conflate with chemical atomism. The "details"—hard, ponderable, indivisible, spherical atoms surrounded by caloric—were thought to be expendable accretions to a very serviceable theory.[33] As Davy put it: "It is impossible not to admire the ingenuity and talent with which Mr. Dalton has arranged, combined, weighed, measured, and figured his atoms; but it is not, I conceive, on any speculations upon the ultimate particles of matter, that the true theory of definite proportions must ultimately rest."[34] Rather, Davy wrote, its "surer basis" lay in the stoichiometric relations themselves. Those who shared Davy's viewpoint did not hesitate to draw up equivalent weight tables for the known chemical elements and then use the weights in writing chemical formulas.

There can be little doubt that Davy became a resolute atomist in this sense—a chemical atomist—around 1808. Thomson certainly thought so in his *History*, as well as in a more contemporary article.[35] Dalton, Berzelius, and Davy's brother John all referred to him as an atomist.[36] Sometime after the spring of 1812 Coleridge added the marginal comment in one of his books: "Alas! . . . H. Davy is become Sir Humphry Davy and an *Atomist*!"; in a letter to Lord Liverpool, he bemoaned the "almost unanimous acceptance of Dalton's Theory in England."[37] Similar sentiments were expressed, though with no sense of distress, by Charles Sylvester.[38] Clearly Davy's contemporaries regarded him as *some* sort of atomist. Nor would Davy himself have repudiated that description. In a letter to Berzelius in March 1811, he wrote: "In the last two Bakerian lectures, those for 1809 and 1810, I have embraced the doctrine of definite proportions and our views on these subjects very nearly coincide. I believe that all the elements, i.e. the undecompounded bodies, are capable of being expressed by numbers which are invariable in all combinations, i.e. in every combination there is either one proportion or a multiple."[39]

It is indeed in Davy's Bakerian Lecture for 1809 that he first publicly discussed Dalton's ideas. Davy noted that a regularity that he had discerned in the oxygen content of the alkaline earths "must be considered merely as a consequence of Mr. Dalton's law of general proportions. Mr. Dalton had indeed, in the spring of 1808, communicated to me a series of proportions for the alkalies and alkaline earths; which in the case of the alkalies, were not very remote from what I had ascertained by direct experiments." Davy called Dalton's suggestion "ingenious" but regarded it merely as a supposition whose generality he left as an open question. His language indicates indecision as to whether Dalton's chemical atomism was a "law," a "theory," or a "doctrine" (we have noted a similar indecision in Thomson). Despite his reservations, however, Davy really was committing himself to chemical atomism here. He adduced a "strong argument in favour of the theory of chemical proportions," and stated, for instance, that sulfur could unite "with one proportion of oxygen, or a double proportion."[40]

In two papers read to the Royal Society in 1810, Davy proposed elemental weights for hydrogen, oxygen, nitrogen, sulfur, phosphorus, chlorine, potassium, and sodium.[41] In his 1811 letter to Berzelius, he repeated all of these except phosphorus and added weights for carbon, calcium, and iron. What Dalton called atoms and Wollaston would call equivalents Davy merely referred to as "proportions" or "proportional numbers."[42] Like Dalton, Davy set hydrogen equal to one as a basis for calculation and assumed the formula for water to be HO. But he disagreed with several other of Dalton's formulas, such as those for ammonia and the oxides of nitrogen; he also attempted, as we have seen, to draw a sharp distinction between his "proportions" and Dalton's "atoms."

Nonetheless, it is difficult to see where Davy's procedure differed from Dalton's. Both assumed the existence of chemically smallest parts for the elements and listed their weights. Both were compelled to assume formulas for the simple oxides and hydrides, though often their formulas for the same substance were different. As Dalton "arranged, combined, weighed, measured, and figured" his atoms, so Davy combined weighed, and measured his "proportions." Davy no doubt believed that he was dealing solely with the experimentally determined laws of chemical combination; however, those laws say nothing about chemical units with characteristic weights that combine with one another in simple multiples. In proposing a table of atomic weights, Davy was declaring his adherence to a theory of chemical atoms as I have defined it. He was wrong in identifying that theory with the laws of stoichiometry; unfortunately, many nineteenth-century chemists followed his lead.[43]

Davy's *Elements of Chemical Philosophy*, published in the summer of 1812, presented the proportional weights of thirty-seven elements for which good analytical data existed. Davy explained how he arrived at them:

> As in all well known compounds, the proportions of the elements are in certain definite ratios to each other; it is evident, that these ratios may be expressed by numbers; and if one number be employed to denote the smallest quantity in which a body combines, all other quantities of the same body will be multiples of this number; and the smallest proportions in which the undecomposed bodies enter into union being known, the constitution of the compounds they form may be learnt, and the element which unites chemically in the smallest quantity being expressed by unity, all the other elements may be represented by the relations of their quantities to unity.
>
> Hydrogene gas, or inflammable air is the substance of which the smallest weights seem to enter into combination; and it appears to exist in no definite compound in less proportion than water. The specific gravity of hydrogene is to that of oxygene as 15 to 1; and as 2 volumes of hydrogen to 1 of oxygen enter into the composition of water, the ratio of the hydrogene in water will be to the oxygene as 2 to 15; and it may be regarded as composed of two proportions of hydrogene and one of oxygene: and the number representing hydrogene will be 1, and that representing oxygene 15.[44]

Having thus derived the H_2O formula for water from the evidence of combining volumes, Davy was now compelled to double all of his previous weights as well as the number of hydrogen atoms in his hydride formulas.[45] He mentioned that William Higgins and Dalton had suggested a binary water formula, but he regarded this as a departure from "the doctrine of proportions derived from facts." He also rejected the HO formula because it would result in a fractional weight for oxygen: "The calculations are much expedited, and the formula rendered more simple, by considering the smallest proportion an integer."[46]

Here was where Davy quite properly noted the difference between his and Dalton's conceptions. "It is not necessary to consider the combining bodies, either as composed of indivisible particles, or even as always united, one and one, or one and two, or one and three proportions . . . at present, we have no means whatever of judging either of the relative numbers, figures, or weights, of those particles of bodies."[47] Such an argument indicated to Davy that assignment of elemental weights was merely a question of convention; this was one of his rationales for switching from the more cumbersome O = 7.5 to O = 15 (HO to H_2O). He noted that no future discovery, such as the fervently hoped-for fundamental protyle, could destroy the mathematical theory of "definite proportions."[48] The weights for the former "elements" would then simply represent the weights for the new compounds. Similarly, any future absolute determination of formulas could at most multiply or divide the old weights by small integers.

Davy's *Elements* received mixed reviews. Some were aghast at the author's bold speculations, such as the possible existence of water in nitrogen and the metals. Bostock wrote to Alexander Marcet regarding the book: "His discoveries are wonderful, his manner of operation perfection, but his hypotheses wild and his manner of reasoning loose."[49] Berzelius was similarly disappointed, as we shall see.

Thomas Young more charitably wrote that the book bears "no inconsiderable marks of haste."[50] Nevertheless, Young clearly perceived the importance of the "theory of the simplicity of the proportions in which all bodies combine with each other," and thought that "the work before us tends much more to its confirmation than any mass of evidence which has yet been collected on that subject"; indeed, "the proofs are so numerous and satisfactory, that there seems to be little room left for argument."[51] Young reproduced Davy's set of atomic weights in his review, making a few additions and modifications based on his own interpolations and on Richter's and Berzelius' research.[52] For Davy's term "proportion," Young substituted "combining weight."

WILLIAM WOLLASTON

Wollaston, a London physician who devoted his considerable talents to chemistry, introduced his 1808 paper on multiple proportions by reviewing the facts Thomson had recently communicated, and added:

> As I had observed the same laws to prevail in various other instances of super-acid salts, I thought it not unlikely that this law might obtain generally in such compounds, and it was my design to have pursued the subject with the hope of discovering the cause to which so regular a relation might be ascribed.
> But since the publication of Mr. Dalton's theory of chemical combination, as explained and illustrated by Dr. Thomson, the inquiry which I had designed appears to be superfluous, as all the facts that I had observed are but particular instances of the more general observation of Mr. Dalton.[53]

That Wollaston not only was a strong advocate, but also had been working on ideas similar to Dalton's was later asserted by Thomson, who stated that he "had the means of knowing" that Wollaston "had been struck with the proportions of oxygen in my table of metallic oxides,[54] and that he had begun to study the subject when my account of the Daltonian theory prevented him from proceeding with his investigations. . . . [M]y knowledge of Dr. Wollaston's sagacity prevents me from entertaining any doubt . . . that the result of his investigation would have been the discovery of the atomic theory."[55] A similar opinion was repeated by Andrew Ure, by William Whewell, and by George Peacock.[56]

Whether or not Wollaston was actually forestalled by Dalton, he was a strong adherent of the new theory as early as 1807.[57]

When Berzelius visited England in the summer of 1812, he and Wollaston talked at length about definite proportions and "the atomistic manner of considering the composition of bodies";[58] later he referred to his "many conversations" with Wollaston regarding the latter's "yet-unpublished atomic theory."[59] Greatly impressed by Wollaston, he wrote to Berthollet: "I am sure that among the chemists who are at present in the prime of life there is none that can be compared with Wollaston in mental depth and accuracy as well as in resourcefulness, and all this is combined in him with gentle manners and true modesty. I have profited more by an hour's conversation with him than frequently by the reading of large volumes. . . . Simplicity, clarity and the greatest appearance of truth are always the accompaniments of his reasoning."[60]

On 11 August 1812 Wollaston showed Berzelius a list of atomic weights, on the scale $H = 0.5$, $O = 8$. The next evening he sent Berzelius a slightly enlarged list, including a second column, consisting of the same weights converted to the scale $O = 10$; he pointed out that such a convention would greatly simplify the calculation of the weights of oxide series. In the letter of transmission, Wollaston appears as more of a chemical atomist even than Berzelius. He wrote that Berzelius' supposition of a nonintegral $4^{3}/_{4}$ units of base in one particular nitrate "offends against my prejudices . . . [but] it does not in the least shake my faith in the general truth of the doctrine, & I fully expect" that a systematic error will eventually be demonstrated.[61] This letter is interesting not only for clearly exhibiting Wollaston's atomistic "prejudice," but also in that it shows that he was the first adherent after Davy of the H_2O water formula and that it was he who proposed the $O = 10$ scale to Berzelius — not vice versa, as Berzelius later maintained.[62] Wollaston's particular weights also indicate that he was inclined to accept Davy's evidence for the existence of oxygen in sulfur and in nitrogen and that he was *not* inclined to accept the elementary nature of chlorine.[63]

Wollaston read his paper on chemical equivalents to the Royal Society in the autumn of 1813.[64] He listed the "equivalents" for a total of nineteen elements, thus becoming the fourth British chemist to publish a table of elemental weights. Since he relied heavily on the results of "the very industrious and ever accurate Berzelius," his values were notably more accurate than those of Dalton and Davy. He exhibited a sensitivity to significant figures that was rare then; moreover, like Thomson, he was always careful to specify the origin of the data that he used in his determinations.

Although Wollaston spoke highly of Dalton's ideas, he was eager to point out the differences in their methodologies:

> According to Mr. Dalton's theory, by which [multiple proportions] are best explained, chemical union in the state of neutralization takes place between single atoms of the substances combined; and in cases where there is a redundance of either ingredient, then two or more atoms of this kind are united to only one of the other.
> According to this view, when we estimate the relative weights of equivalents, Mr. Dalton conceives that we are estimating the aggregate weights of a given number of atoms, and consequently the proportion which the ultimate single atoms bear to each other. But since it is impossible in several instances, where only two combinations of the same ingredients are known, to discover which of the compounds is to be regarded as consisting of a pair of single atoms, and since the decision of these questions is purely theoretical, and by no means necessary to the formation of a table adapted to most practical purposes, I have not been desirous of warping my numbers according to an atomic theory, but have endeavored to make practical convenience my sole guide.[65]

Wollaston now took the formula for water as HO, and his weights resemble Davy's of 1810-11.

Wollaston's system of weights proved immensely popular, particularly in Britain. Here at last was atomism based exclusively on experiment and divorced from the conjectures about the nature of atoms found in Dalton's *New System*—or so at least it seemed. Davy thought that Wollaston's table of chemical equivalents "separates the practical part of the doctrine from the atomic or hypothetical part," and many British chemists concurred.[66] In Britain, Wollaston's weights were used almost exclusively until the 1860s. In Germany they were eclipsed for years by Berzelius' rival system; they finally came into vogue under the aegis of Leopold Gmelin in the mid 1840s and continued to be favored for fifteen or twenty years thereafter. French chemists followed either Wollaston or J. B. Dumas, whose system coincided with Wollaston's except that his weight for carbon was only half of Wollaston's. The justification often cited for adherence to Wollaston's table of weights was the increased level of empiricism.

However, as we suggested in chapter 1, the putative empirical origin of Wollaston's equivalents is largely illusory. A comparison of Wollaston's table with Dalton's atomic weights of 1810 shows that they are practically identical, notwithstanding the discrepancies arising from differences in the experimental data used. Of Wollaston's nineteen elemental weights only four—those of nitrogen, iron, copper, and phosphorus—differ by simple multiples from Dalton's values; that is, in only four cases did Wollaston disagree with Dalton regarding the formula to be used in calculating the "equivalent."[67] Indeed, Wollaston's weights have very little in common with modern chemical equivalents; they exhibit every essential characteristic of chemical atomic weights.[68] In spite of Wollaston's effort to produce an empirical system, his result turns out to be operationally identical to Dalton's theoretical one. Wollaston's numbers were closer to experiment only in the sense that the formulas from which they were derived were determined by the simplest a priori assumptions—*vide infra*—and that the numbers were explicitly regarded as possessing merely conventional status. To distinguish them from modern chemical equivalents, I will call them "conventional equivalents" and ask the reader to keep in mind that they are generically identical to atomic weights.

Wollaston is usually portrayed either as critical of the atomic theory or as indecisive about it.[69] However, his explicit support of the theory in the period 1807–12 and his implicit advocacy of chemical atoms in his paper on chemical "equivalents" contradict these interpretations. So do his repeated atomistic conjectures, which are often quite similar to Dalton's. In his 1808 paper on multiple proportions, for example, Wollaston was "inclined to think . . . that we shall be obliged to acquire a geometric conception of [the] relative arrangement [of the elementary atoms] in all the three dimensions of solid extension."[70] Wollaston intended to rationalize his finding that compounds of potash and oxalic acid could be prepared in the proportions 1:1, 1:2, and 1:4, but not 1:3. Assuming that the elementary particles are spherical in their "virtual extent" and that similar particles are mutually repellant, he argued that compounds of the form AB, AB_2, and AB_4 would be stable in a linear or tetrahedral arrangement, but that the trigonal planar AB_3 would be unstable.[71]

Wollaston expressed a similar idea in the Bakerian Lecture for 1812.[72] He suggested that close-packing of spherical atoms could explain the fact that different "primitive forms" could be produced from the same crystal by different methods of cleavage. Wollaston's espousal of this idea in his 1808 and 1812 papers—an idea Dalton had already sketched in his *New System*[73]—has been interpreted as advocacy of a point atomism similar to that of Bošković.[74]

In 1822 Wollaston was still thinking atomistically.[75] He argued that if the earth's atmosphere were composed of atoms, it would exhibit a distinct upper boundary where the repulsion of the atoms exactly balances gravitation. If, on the other hand, the atmosphere were composed of an infinite number of infinitely small particles, it would expand to fill all of space. The absence of any observable refraction near the sun or the planets indicated to Wollaston that space is a vacuum. It followed that the atmosphere has an upper limit, so that it must be composed of atoms. From the atomic nature of gases, Wollaston concluded that all matter is composed of atoms; therefore, "We may without hesitation conclude that these equivalent quantities which we have learned to appreciate by proportionate numbers, do really express the relative weights of elementary atoms, the ultimate objects of chemical research."[76]

To Marcet he wrote at this time: "The doctrine of ultimate atoms is so much more easy of conception, & thence at this time so commonly received, that it is possible I may have treated the notion of infinite divisibility with more respect than was absolutely necessary."[77] That Wollaston never committed the voltefaces regarding atomism that have been imputed to him was averred by Daubeny: "For after Dalton had proposed his new system of chemical philosophy and had presented a rather rough outline of it, [Wollaston] immediately felt that it was true and conformable to nature, nor did he ever cease confirming it by experiments or recommending it to others by his authority."

And Davy, who knew Wollaston well, thought that "his judgment was cool; his views sagacious; his inductions made with care, slowly formed, and seldom renounced."[78]

This evidence suggests that Wollaston was an early convert to the atomic theory and consistently defended it to the end of his life. What exactly did he mean, then, when he wrote: "I have not been desirous of warping my numbers according to any atomic theory, but have endeavored to make practical convenience my sole guide"? For one thing, he appears to have been troubled by the arbitrariness of the rules of procedure that chemists had to follow in the process of deciding on formulas. No doubt dismayed by the disregard for consistency with which Dalton and Davy used their own rules at times, Wollaston tried to overcome this arbitrariness by choosing the simplest rule and adhering to it scrupulously: he assumed that the lowest oxide of the element in question was the binary monoxide.[79] He applied this rule even to the oxides of iron, although the higher oxide then had to be formulated $FeO_{3/2}$. Only in one instance did he depart from his rule: he joined his predecessors in assuming that the two known oxides of sulfur were SO_2 and SO_3.[80]

Furthermore, Wollaston's purpose in publishing his 1814 paper was a practical one: he wished to describe his new "synoptic scale of chemical equivalents," a sort of slide rule for the rapid calculation of stoichiometric quantities. Marcet wrote to Berzelius the day after Wollaston read his paper: "Wollaston's memoir at first contained only an explanation of this instrument; but fortunately I was able to persude him to expand the paper a bit and to add, in the form of a preface, an historical sketch on definite proportions from Bergman to the present."[81] Interested only in "practical convenience" here, Wollaston did not intend the numbers on his scale to be absolute, nor did he intend to declare his acceptance or rejection of the atomic theory. Although he later asserted unequivocally that his equivalents corresponded to atomic weights, in 1814 he simply avoided the theoretical issue.

JACOB BERZELIUS

If John Dalton was the father of the atomic theory, then Jacob Berzelius was its midwife and fosterparent, playing perhaps the leading role in the development of both chemical and physical atomism during the second, third, and fourth decades of the nineteenth century. His importance for developments in Britain during the period 1808-15 justifies his inclusion in this chapter.

After defending his medical dissertation at Uppsala in 1802, Berzelius became an unpaid assistant to the professor of medicine and pharmacy at the School of Surgery in Stockholm and earned a meager living as physician to the poor. But Berzelius' rise in Swedish scientific circles was rapid. In 1807 he was appointed to the vacated chair of medicine and pharmacy and three years later served as president of the Swedish Academy of Sciences. In 1819 he was elected secretary of the Academy, a post that doubled his income, provided exceptional laboratory facilities, and released him from all unwanted teaching duties.[82]

From his earliest student days, Berzelius had taught himself antiphlogistic chemistry and was strongly influenced by Lavoisier's views on the nature of oxides, acids, bases, and salts. In 1800 the new chemistry was given a new instrumental paradigm with the discovery of current electricity and electrolysis. Berzelius, together with his friend and patron Wilhelm Hisinger, performed an elegant series of experiments on the "Voltaic pile."[83] From this early work derived many of Berzelius' most deeply held and characteristic beliefs, most notably the correlation between chemical and electrochemical properties, a theme that later developed into his theory of electrochemical dualism.

Humphry Davy carried out very similar experiments from 1806, experiments that led him to aver that the force of chemical affinity was nothing other than electricity. His most famous work was performed and reported in 1807, namely, the electrolysis of the caustic alkalies, resulting in the first isolation of sodium and potassium and the consequent demonstration of the oxide nature of soda and potash.[84] Berzelius lost no time in following up on this important discovery. In May 1808, using mercury as the negative electrode, Berzelius and his colleague M. M. Pontin succeeded in electrolyzing the alkaline earths as well as Davy's alkalies. In each case the electropositive component formed an amalgam with the mercury. What Lavoisier had only suspected now seemed to be demonstrated: bases, just as much as acids, are oxides of metals or of "radicals." And Berzelius and Pontin's success with the same procedure using an ammonia solution seemed to confirm Davy's speculation that ammonia must be an as yet undecomposed oxide of a highly electropositive (i.e., metallic) radical which he called "ammonium."[85] The Swedes' discoveries of the electrolysis of the earths and of ammonium amalgam proved as fruitful a source of ideas and experiments to Davy as Davy's work had proven (and would prove) for Berzelius.[86]

The year before (1807), while reviewing little known chemical works in preparation for writing his textbook,[87] Berzelius encountered J. B. Richter's *Ueber die neuern Gegenstände der Chymie*.[88] He was "astonished" at the "depth and importance" of Richter's law of equivalent proportions, and he immediately set out to confirm and extend it. "This investigation," he later wrote, "became the basis for the direction of my scientific endeavors during the greater part of my most active years of life, i.e., for my work on chemical proportions." Berzelius perceived that equivalent proportions could enormously simplify the proximate analysis of salts. It was only necessary to analyze, say, a series of sulfates and a series of salts of baryta to enable the calculation on paper of the compositions of all permutations of possible salts in the two series. But recently published analyses did not agree with the newly rediscovered law, so Berzelius undertook to repeat as many as seemed doubtful. It was at this time (early in 1808) that Berzelius learned of Davy's decomposition of the alkalies; he regarded this as opening a fertile new field for the study and application of the stoichiometric laws. Most important in Berzelius' mind loomed the intriguing possibility of *calculating* (since direct analysis never succeeded) the percentage of oxygen in ammonia.[89]

Such an indirect analysis proved (apparently) practicable when Berzelius independently rediscovered a "law" previously enunciated by Richter, Torbern Bergman, and Gay-Lussac. A recent scholar has referred to this as the "law of the constancy of basic oxygen"; Berzelius expressed it: "A given quantity of an acid always takes up the same quantity of oxygen with any oxide by which it is neutralized." For instance, 100 parts of sulfuric acid always combine with bases containing exactly 20.3 parts of oxygen; the numbers were different with different acids, but with a given acid the number was independent of the base forming the neutral salt.[90] Analysis of the oxygen content of various bases reacting with muriatic acid (HCl) provided the data indicating the amount of oxygen in the ammonia of sal ammoniac (NH_4Cl). By this reasoning, ammonia must contain no less than 47 percent oxygen.[91]

About the same time that Berzelius definitively verified the law of the constancy of basic oxygen he also encountered the law of multiple proportions—probably by recalculating previous analyses after reading Wollaston's 1808 paper.[92] We know that Berzelius first learned of Dalton's theory from reading that paper, sometime between November 1808 and June 1809. His earliest reference to the theory of which I am aware is in an undated letter to Wollaston that must have been written in late June 1809, in which he communicated some examples of multiple proportions he had encountered. He said he had only learned of Dalton's ideas from Wollaston's memoir, so the favorable results could not be a consequence of mere "prédilection" for the theory.[93] About the same time he professed to Davy "une grande vénération" for Dalton, and a year and a half later he told Berthollet that Dalton must be "un des physiciens les plus ingénieux de notre siècle."[94] He twice asked Davy (and both times in vain) to tell him more about the atomic theory, since: "There is every appearance that M. Dalton's ideas, of which I have a very high opinion, will help correct my own; furthermore, he must have done many experiments which will save me the trouble of doing them myself."[95] By the autumn of 1811 Berzelius still had not seen any detailed discussion of the new theory, so he was understandably thrilled when a new French correspondent, H. F. Gaultier de Claubry, offered to send him a copy of Dalton's *New System of Chemical Philosophy*. Berzelius gratefully replied: "I have long hoped to read this work, as it treats material with which I have been so involved these past four years."[96] But Gaultier failed to fulfill his promise.

This sudden outpouring of interest in Dalton's ideas in early or mid-1809 signaled an important transition in the development of Berzelius' chemistry. For two years he had been trying to establish and develop the laws of equivalent proportions and of the constancy of basic oxygen, to serve as controls on analysis and to enable the calculation of the composition of certain refractory substances. Now he had encountered both an important new stoichiometric relation—multiple proportions—and the theoretical key to all the phenomena he had been studying. Thereafter he seized on multiple proportions as the most critical law; however, lacking detailed knowledge of Dalton's already published theory, he properly avoided cloaking his investigation in any sort of corpuscular ontology. This understandable hesitation has been interpreted—I believe wrongly—as a lack of commitment to an atomist metaphysics.[97] In the introduction to a major work composed in the summer of 1809, Berzelius wrote that Dalton's hypothesis, if proven true, would be "the greatest step which chemistry has yet made toward its perfection as a science," and two pages later noted: "I will . . . completely refrain from theorizing. The extent that the results of the experiments confirm the theory will catch the eye of everyone, and the ideas to which they lead will surely occur to every attentive reader, with no guidance from me."[98] This is no positivist speaking.

Published in 1810-12, the paper consisted of a systematic survey of elemental combining proportions in a large number of inorganic oxides, acids, and salts.[99] This monumental work, more than any other, first established the experimental basis of chemical atomism; that is, it established equivalent and multiple proportions as general and exact scientific laws. But, true to his nontheoretical approach, Berzelius expressed all analyses either in percentages or in parts of one component per hundred parts of another; he proposed no formulas or atomic weights. He stated the law of multiple proportions as follows:

> *Whenever two bodies, A and B, combine together in various proportions, this always happens according to the following fixed and definite proportions*: $1A$ *with* $1B$ (combination in the *minimum*); $1A$ *with* $1\frac{1}{2}B$ (or perhaps more correctly $2A$ *with* $3B$); $1A$ *with* $2B$; $1A$ *with* $4B$.[100]

He illustrated the law by citing analyses of multiple oxides and sulfides of the same element, and of the basic and neutral salts of copper, iron, and lead.

After this paper was published in Swedish and was in press in German and French translations, Berzelius began to repeat certain of his analyses.[101] One of his principal goals here was to establish a point upon which he had already speculated: that the $1^1/_2$ relation merely indicated the existence of hitherto undiscovered lower oxides, so that the true relation would turn out to be some *even multiple* of $1^1/_2$. For instance, he was able to provide indirect evidence for the existence of two lower sulfur oxides (we would formulate them as S_2O and SO) in the muriates of sulfur, so the true multiple series is not 1 and $1^1/_2$, but rather 1, 2, 4, and 6.[102]

While pursuing this line of thought Berzelius discovered — probably in the late summer of 1810 — a new significant stoichiometric relation:

> Whenever two oxides saturate each other, they always contain oxygen in a proportion such that the amount in the substance that moves toward the positive pole of the electric pile is an integral multiple of the amount of oxygen in the other substance that moves toward the negative pole.

More simply, the acid always contains an integral multiple of the oxygen in the base — Berzelius' "oxide rule." First announced in letters to L. W. Gilbert and Berthollet in October 1810 that were subsequently printed, the rule underlay experiments published in German as three continuations of his stoichiometric "Versuch."[103] For Berzelius the rule formed the critical connecting link between first-order compounds (e.g., oxides) and higher-order compounds (combinations of oxides, e.g., salts); it proved central to his stoichiometric work throughout the next decade and motivated much of his research in mineralogy and organic chemistry.[104]

In addition to the oxide rule, the first continuation of the "Versuch" expanded Berzelius' calculation of the oxygen content of ammonia into a speculation regarding the presumed stages of oxidation of the "ammonium" radical. The first stage was thought to be hydrogen; the last stage — containing 72 times the oxygen — was water, and intermediate oxygen multiples produced ammonia, nitrogen, and the oxides of nitrogen. This clever scheme took account of the known stoichiometric relations of the compounds of nitrogen, maintained Berzelius' opinion of ammonia as a basic metal oxide analogous to the alkalies and alkaline earths, and, further, implied that ammonia could equivalently be viewed as a salt formed by the union of two different oxides, namely nitrogen and hydrogen![105] Though by this time Berzelius had read Gay-Lussac's memoir on gaseous combining proportions, he merely used the data for stoichiometric calculations and had not yet assumed any relation between atoms and volumes.[106]

Berzelius' second and third continuations consisted of an examination of nitric and nitrous acids and their salts. His purpose was to provide further support both for his oxide rule and his view of the complexity of nitrogen. Berzelius thought he was successful in these goals; he showed that nearly all nitrates and nitrites failed to obey the oxide rule if nitrogen were taken as elemental. If, however, nitrogen were viewed as the oxide of ammonium, all salts but two were subsumable under the oxide rule.[107]

As Berzelius' program continued to develop in 1811 and early 1812, certain ideas that Berzelius anticipated would eventually constitute general rules guided his work. Following the case of sulfur, he expected that for all cases of the $1^1/_2$ multiple: (a) the future detection of two lower oxides would transform the $1^1/_2$ multiple into the true multiple 6; (b) the highest oxide would provide to be the strongest acid; and (c) salts of the two acids (the 4 and 6 oxide multiples) would have oxide-rule multiples of 2 and 3.[108] Later translated into avowedly atomistic terms, this expectation would provide one of several compelling reasons for Berzelius to regard all strong bases as dioxides.

In the spring of 1812, Berzelius finally obtained the *New System of Chemical Philosophy* when Dalton sent him a complimentary copy. Never had a gift pleased Berzelius more; never had his expectations been more cruelly disappointed. He thought that Dalton had performed his experiments with fixed preconceptions, always seeking to confirm his favorite hypotheses, and that he had committed blunders that any competent chemist or pharmacist could easily have avoided.[109]

Shortly after receiving Dalton's book, Berzelius left for England where he spent the summer of 1812. He met many of the leading British scientists. He spoke with Wollaston about multiple proportions and about systems of atomic weights. Davy showed him the galley proofs of his *Elements of Chemical Philosophy*, just then being printed, and asked Berzelius for a thorough critique. Berzelius failed to meet Dalton, who was in Manchester, and Thomson, who coincidentally was travelling in Sweden that summer.[110]

Berzelius soon satisfied Davy's request for a critique with a thoroughness that did not please the Englishman. Berzelius thought Davy's work too esoteric to be a good elementary treatise and yet too simplistic to be a contribution to "chemical philosophy." He found that it contained far too many unsubstantiated hypotheses as well as numerous factual errors. He complained: "You have sought to introduce the doctrine of definite proportions, and that is good; but I do not at all like the way you have done it. I ask you to immediately banish from your writings, especially when they treat this doctrine, a word which you use all too often. This word is 'about'. To the extent that this word exists in chemistry, the doctrine in question would be an aborted foetus, sick and ready to die. Moreover, the use of this word makes all your numerical determinations, without exception, erroneous."[111]

It was around this time, early in 1813, that Berzelius began to compose articles in more distinctly atomistic terms. The direct and indirect contacts he had made during the preceding year with such chemical atomists as Dalton, Davy, Thomson, and Wollaston appear to have influenced his thinking profoundly. These new ideas were introduced and integrated into his ongoing research program in the first of these atomist papers, on compounds of nitrogen.[112] Berzelius revised his earlier work of the second and third continuations of the "Versuch," using new analyses he had made as well as a more precise figure for the composition of nitric acid borrowed from Dalton and Davy. He now rejected the ammonium radical, but still asserted — and for the first time with no troubling exceptions — that the oxide rule could only be maintained by assuming nitrogen to be the suboxide of a radical that he called "nitric" (Nt). His formulas for the various oxides and hydride of this radical were:[113]

NtO	=	nitrogen
NtO$_2$	=	nitrous oxide
NtO$_3$	=	nitrous gas
NtO$_4$	=	nitrous acid
NtO$_6$	=	nitric acid
NtOH$_3$	=	ammonia

The oxide formulas correspond to Davy's when the proper transformation is performed: nitrogen = N = NtO; they also correspond to modern anhydride formulas with a doubled atomic weight for nitrogen.[114]

In this article Berzelius introduced his famous symbolic atomic notation. He wrote:

> To avoid long circumlocutions, I shall here employ a simple and short method of expressing determinate combinations, which I always use in my annotations. [fn.: It] is founded on something very analogous to the corpuscular hypothesis of Dalton. It is known that bodies in their gaseous state either unite in equal volumes, or one volume of one combines with 2, 3, &c. volumes of the other. Let us express by the initial letters of the names of each substance a determinate quantity of that substance; and let us determine that quantity from its relation in weight to oxygen, both taken in the gaseous state, and in equal volumes; that is to say, the specific gravity of the substances in the gaseous state, that of oxygen being considered as unity. I have made choice of oxygen for unity, because it constitutes, as it were, the central point of chemistry, and because it enters more frequently than any other substance into compounds in various proportions, and ought therefore to be the easiest to add or subtract. It is obvious that this comes to the same thing as Mr. Dalton's weights of atoms; but I have the advantage over him, of not founding my numbers on an hypothesis, but upon a fact well known and proved.[115]

Of course, Berzelius was here assuming implicitly the equal volumes – equal numbers idea, conceived to be an unavoidable consequence of Gay-Lussac's law of combining volumes. Although cautiously phrased in the published paper, there is little doubt that Berzelius did in fact regard elementary vapor densities as proportional without exception to atomic weights.[116] Both Dalton and Davy had earlier used vapor densities in a desultory fashion to assist in the determination of elementary atomic weights. Berzelius was proceeding similarly, but in a more consistent and systematic way. Nonetheless, there is one significant inconsistency in this article. Berzelius' volumetric approach to the nitrogen oxide series only makes stoichiometric sense if he was taking NtO as *two* volumes of nitrogen gas combining with integral volumes of oxygen (i.e., NtO = 28). But that he meant NtO to stand for *one* volume is implied by his formulation of ammonia as $NtOH_3$.[117] Of course he shared this inconsistency with Davy, whose lead he was probably following but who never really claimed or even sought consistency in this question. Dalton, who happened to have viewed the nitrogen oxides correctly, had his own inconsistencies. All were the result of the unguessed existence of polyatomic elementary molecules.

Berzelius' first extended collation of atomic weights and formulas appeared in five installments published between December 1813 and May 1814.[118] He wrote:

> The most probable idea, and most conformable to our experience, is, that bodies are composed of atoms, or of molecules, which combine 1 with 1, 1 with 2, or 3, 4, &c.; and the laws of chemical proportions seem to result from this with such clearness and evidence, that it seems very singular that an idea so simple and so probable has not only not been adopted, but not even proposed before our own days. . . . As far as I know, the English chemists Dalton, Davy, and Young are the only persons who have yet attempted to make these [atomic weight] determinations. . . . But none of these philosophers have attempted to give any great degree of exactness to their determinations. They have frequently even omitted stating the experiments from which these determinations are derived. The method they have adopted of giving round numbers, though it facilitates the recollection and calculation, is scarcely consistent with the object of scientific researches, and ought to be rejected: for even supposing that perfect exactness could never be obtained, it is nevertheless the object towards which all our efforts should be directed.[119]

Perhaps the most skilled analyst of his day, Berzelius gloried in precision: he customarily used between four and seven significant figures. He was always careful to cite the experiments upon which he based his atomic weights; most of them were his own. He failed to mention Thomson's and Wollaston's papers on atomic weights for they had not yet reached him.[120]

Berzelius' atomic weights of 1813–14 are very similar to

Davy's of 1812; experimental differences aside, the two systems differ only in their values for a few of the metals.[121] The only systematic difference is that Berzelius assumed the generalized metal oxide formula MO_2, whereas Davy assigned the binary formula MO to every metal having only one known oxide.

Berzelius favored the general MO_2 formula for several reasons. It was a common axiom among the early atomists that in the simple oxides one element must be present as a single atom. Dalton regarded this idea as consonant with his rule of greatest simplicity. Thomson argued for it by pointing out that, since like atoms must repel each other, a serious mechanical instability would arise in such compounds as A_2B_2 or A_2B_3.[122] Berzelius agreed with this viewpoint. He also argued that in all the analyses he had made, he had never found empirical evidence to contradict the axiom:

> My idea, that in every inorganic combination one of the elements enters as unity, is founded on the circumstance that in all the inorganic bodies which I have analyzed, and I have analyzed a great number, I have found it to be so.[123]

Third, Berzelius stated that a composition such as A_2B_3 "would almost totally destroy chemical proportions." He worried that allowing such compounds would open the door to infinitely many formula assignments—for example, for the latter case, $A_{2n}B_{3m}$, where n and m could be any integers. Admitting such formulas, Berzelius added, would be "contrary to sound logic"; the single-atom axiom was one of the "necessary consequences or reflections inseparable from the theory of atoms."[124]

So Berzelius applied the formula MO_x to all metal oxides. Various facts led him to believe that for basic oxides x was generally 2. For instance, since hematite has 1½ times the oxygen of the basic black oxide of iron, these two substances were assigned the formulas FeO_3 and FeO_2. A similar case obtained with the base litharge (formulated by Berzelius as PbO_2) and the neutral minimum (PbO_3).[125] The two oxides of copper were related as CuO (neutral red cuprite) and CuO_2 (basic black oxide). Analyses of various sodium and potassium oxides by Gay-Lussac, Thenard, and Davy in 1811 seemed to Berzelius to strongly support dioxide formulas for the caustic alkalies. Specifically, sodium peroxide (modern Na_2O_2) was found to contain only fifty percent more oxygen than there is in soda, and potassium superoxide (modern KO_2) was found to contain only three times the oxygen of potash. Gay-Lussac and Thenard also found a (nonexistent) lower oxide of potassium, which Berzelius regarded as the monoxide.[126] Potash as well as soda thus acquired the dioxide formula.

So Berzelius had compiled five instances where the basic metal oxide seemed clearly to be a dioxide. Always on the watch for analogies to assist in the problematical process of assigning formulas, Berzelius decided to assume *generally* dioxide formulas for basic oxides, e.g., for the oxides of barium, strontium, calcium, magnesium, zinc, and silver.[127]

Berzelius determined the formulas for the nonmetals more systematically by using vapor densities. He now explicitly assumed that all atoms are spherical and of equal size and that elementary atoms and volumes are synonymous.[128] He wrote: "There is no other difference between the theory of atoms and that of volumes, than that the one represents bodies in a solid form, the other in a gaseous form. It is clear, that what in the one theory is called an *atom*, is in the other theory a *volume*. In the present state of our knowledge the theory of volumes has the advantage of being founded upon a well constituted fact, while the other has only a supposition for its foundation."[129] For Berzelius, then, the corpuscular theory is an example of what I call a physical atomic theory, hence is hypothetical, whereas the volume theory remains firmly on the empirical plane. Berzelius did not seem to realize that the volume theory also rested on an unproved physical assumption, that of equal volumes-equal numbers.

Furthermore, Berzelius was forced to deny the volume theory for compounds, since water, for instance, occupies twice the gaseous volume one would predict.[130] Even for elementary gases, he did not apply the theory consistently: he assumed that nitrogen atoms are heavier than oxygen atoms, even though nitrogen gas has a lower specific gravity.[131] Moreover, for substances of low volatility the method was inapplicable. Thus, like Dalton, Berzelius tried to construct his atomic theory on physical evidence but found that he had to fall back on purely chemical evidence, analogy, and semiempirical rules to arrive at probable chemical formulas.

Despite these similarities of method, Berzelius' atomism obviously differed in many respects from Dalton's, and Dalton was quick to send some "Remarks" for publication in the *Annals of Philosophy*.[132] Dalton took exception to Berzelius' seeming denigration of the atomic theory, and he criticized many of Berzelius' specific ideas. He opposed Berzelius' belief that all atoms must be spherical and of equal volume and denied that such compounds as A_2B_3 must be regarded as contradictions in terms. Berzelius soon penned a rebuttal.[133] He refused to argue further about the sizes and shapes of atoms, these questions not being subject to experimental test. He stoutly defended his rules of combination of oxides (such as the oxide rule) as having been empirically verified. And he firmly denied that he was in any way opposed to the atomic theory.

> When I endeavored to draw the attention of chemists to the difficulties in the atomic theory, it was not my intention to refute that hypothesis. . . . I wished the experiments to verify the theory; and I should have considered it as absurd, if I had taken the opposite road. I placed beside the corpuscular theory, a theory of volumes; because that theory is in some measure connected with facts which may be verified. To those who think that the theory of volumes may be fatal to the corpuscular theory, I would observe, that both are absolutely the same thing. . . . I am persuaded that both myself and Mr. Dalton will in time make use of these very difficulties to determine the true number of atoms in such and such compound bodies. . . . I suppose that Mr. Dalton will agree with me that by such researches we may render much more complete the beautiful theory for which he feels himself so much interested, and for which we are in a great measure indebted to him. . . . I differ, then, from Mr. Dalton, and must continue to consider the atomic theory as imperfect, and as clogged with difficulties, till it give us satisfactory explanations of all the phenomena relative to the chemical proportions. I do not think that we are very far from this explanation.[134]

There can be little doubt, then, that Berzelius regarded himself at this time as a true atomist in both the physical and chemical sense.[135]

The Early Reception in Great Britain: 1804-1815 79

RETROSPECT

Table 1 lists the formula or formulas that each chemist assumed as a basis for calculation of the respective combining weights. To conserve space, I have translated all formulas into Berzelius' notation—most were originally expressed verbally rather than symbolically. The chemical identities of the compounds will usually be unambiguous; for example, the three oxides of nitrogen are nitrous oxide, nitric oxide, and nitric acid,[136] the oxides of phosphorus are phosphorous and phosphoric acids, the oxides of sulfur are sulfurous and sulfuric acids. (Before Thomas Graham and Justus Liebig distinguished between anhydrous acids and acid anhydrides in the 1830s, all acids were formulated as the anhydrides.) Where ambiguity could arise—as in the alkali metal oxides—I give the formula for the most common oxide (potash, soda) in boldface. Where a chemist assumed inconsistent formulas for compounds of the same element, that is, formulas that should have led to two or more different atomic weight values, the single formula that was actually used in the calculation of the atomic weight is in italics. Occasionally an author stated an atomic weight without specifying a formula; in such cases the presumed formula is marked with an asterisk.[137] To facilitate comparison, I have converted Berzelius' formulas for the compounds of nitrogen from his "nitric" (Nt) to nitrogen (N = NtO). The different formulations of chlorine compounds reflect a difference of opinion on the nature of chlorine: Dalton, Wollaston, and Berzelius regarded chlorine as a higher oxide of the "murium" radical Mu;[138] Davy and Thomson thought of chlorine as an element and calculated its weight from the composition of Davy's "euchlorine," which was presumed to have the binary formula ClO.[139]

Table 2 is a collation of the atomic weights calculated by these chemists. I have included only nineteen elements; the selection was dictated by Wollaston's 1814 table. For each author, the first numbers are those that appear in the source cited; the second column lists the same figures recalculated to the standard $O = 8$ or $O = 16$, depending on the assumed formula for water. The last column gives modern atomic

TABLE 1
Assumed Formulas[a]

Element	Dalton 1810[b]	Davy 1810–11[c]	Davy 1812[d]	Thomson 1813[e]	Wollaston 1814[f]	Berzelius 1814[g]	Prout 1815[h]
H	—	—	—	—	—	—	—
O	HO	HO	H_2O	HO	HO	H_2O	HO
C	CO, CO_2	CO, CO_2*	CO, CO_2	CO, CO_2	CO, CO_2	CO, CO_2	CO, CO_2
N	N_2O, NO, NO_2, NH	NO, NO_2, NO_5, NH_3	NO, NO_2, NO_5, NH_6	N_2O, NO, NO_3, NH_3^i	NO, NO_2, NO_5, NH_3	NO, NO_2, NO_5, NH_6^j	NO, NO_2, NO_5, NH_3
S	SO_2, SO_3, SH	SO_2, SO_3, SH	SO_2, SO_3, SH_2	SO_2, SO_3, SH	SO_2, SO_3	SO_2, SO_3, SH_2	SO_2, SO_3, SH
P	PO, PO_2, PH^k	PO_3	PO, PO_2^m	PO, PO_2, PH_3	PO_2	PO, PO_2	PO_2
Cl	MuO_4^n	KCl	ClO	ClO	$(MuO_x)O^p$	MuO_3	ClO
Na[q]	—	NaO or NaO_2	NaO_2, NaO_3	NaO_2, NaO_3	NaO	NaO_2, NaO_3	NaO*
K[q]	—	KO	$K_2O?, KO, KO_3$	KO, KO_3	KO	KO, KO_2, KO_6	KO*
Ca	—	CaO*	CaO	CaO	CaO	CaO_2	CaO*
Mg	—		MgO	MgO	—	MgO_2	MgO*
Sr	—		SrO	SrO	—	SrO_2	SrO*

The Early Reception in Great Britain: 1804–1815

Ba	—	BaO	BaO	—	BaO₂	BaO*
Fe	FeO₂, FeO₃¹	FeO₂, FeO₃*	FeO₂, FeO₃	FeO, FeO_{3/2}	FeO₂, FeO₃	FeO₂,* FeO_{3/2}*
Cu	CuO, CuO₂⁵		CuO, CuO₂	CuO	CuO, CuO₂	CuO*
Zn	ZnO₂*		ZnO	ZnO	ZnO, **ZnO₂**	ZnO*
Ag	AgO		AgO	AgO	AgO₂	AgO*
Hg	HgO		HgO, HgO₂	HgO	HgO, HgO₂	HgO*
Pb	PbO		PbO₂, PbO₃, PbO₄	PbO	PbO₂, PbO₃, PbO₄	PbO*

NOTES: ᵃTypographical conventions are explained in the text.
ᵇDalton, *New System*, pt. 2, pp. 546–47, 560, passim; vol. 2, pt. 1, passim.
ᶜSee notes 39 and 41.
ᵈDavy, *Elements of Chemical Philosophy*, pp. 225–472 passim, 505; *Collected Works*, 4:167–352.
ᵉThomson, "Daltonian Theory," 2:42–49.
ᶠWollaston, "Scale," pp. 18–22.
ᵍBerzelius, "Essay on the Cause."
ʰProut, "On the Relation," pp. 326–29.
ⁱAltered to NO, NO₂, NO₅, and NH₃ in February 1814: *Ann. Phil.* 3:135; ibid., 7 (1816): 344–45.
ʲSee notes 114 and 131.
ᵏDalton, *New System*, pt. 2, pp. 414–15.
ᵐDavy, *Elements of Chemical Philosophy*, pp. 213–14.
ⁿDalton, *New System*, pt. 2, pp. 293, 308.
ᵖSee notes 121, 138, and 139.
ᵠSee note 126.
ʳDalton, "Remarks on the Essay," pp. 176–77.
ˢDalton, *New System*, 2, pt. 1, 27.
ᵗIn February 1814 Thomson changed his atomic weight for nitrogen to 1.803: "Daltonian Theory," 3:135. His first weight for chlorine dates from July of that year: ibid., 4:13. In August, he altered his values for phosphorus, magnesium, iron, zinc, and silver: ibid., 83–88. The following March he changed his weight for lead: ibid., 5:186.
ᵘSee note 141.

82 *Chemical Atomism in the Nineteenth Century*

TABLE 2
ATOMIC WEIGHTS*[a]

Element	Dalton 1810		Davy 1810–1811		Davy 1812		Thomson 1813		Wollaston 1814		Berzelius 1814		Prout 1815		IUPAC 1971
H	1	1	1	1	1	1	0.132	1.06	1.32	1.06	6.636	1.06	1	1	1.008
O	7	8	7.5	8.0	15	16	1.000	8.000	10.00	8.00	100.0	16.00	8	8	16.00
C	5.4	6.2	5.7	6.1	11.4	12.2	0.751	6.01	7.54	6.03	75.1	12.0	6	6	12.01
N	5	5.7	13.4	14.3	26	28	0.878	7.02	17.54	14.03	(179.5)	(28.73)	14	14	14.01
S	13	15	13.6	14.5	30	32	2.000	16.00	20.00	16.00	201.0	32.16	16	16	32.06
P	9	10	16.5	17.6	20	21	1.320	10.56	17.40	13.92	167.5	26.80	14	14	30.97
Cl	(29)	(33)	32.9	35.1	67	71	4.498	35.98	(44.1)	(35.3)	(439.6)	(70.33)	36	36	35.45
Na	(21)	(24)	22 or 44	23.5 or 47	88	94	5.882	47.06	29.1	23.3	579.3	92.69	24	24	22.99
K	(35)	(40)	40.5	43.2	75	80	5.000	40.00	49.1	39.3	978.0	156.5	40	40	39.10
Ca	(17)	(19)	20.8	22.2	40	43	2.620	20.96	25.46	20.36	510.2	81.63	20	20	40.08
Mg	(10)	(14)			28	30	1.368	10.94	(14.6)	(11.7)	315.5	50.47	12	12	24.31
Sr	(39)	(45)			90	96	5.900	47.20	(59)	(47)	1418	226.9	48	48	87.62
Ba	(61)	(70)			130	139	8.731	69.85	(87)	(70)	1709	273.5	70	70	137.3
Fe	50	57	50	53	103	110	6.666	53.33	34.5	27.6	693.6	111.0	28	28	55.85
Cu	56	64			120	128	8.000	64.00	40	32	806.5	129.1	32	32	63.55
Zn	56	64			66	70	4.315	34.52	41	33	806.5	129.0	32	32	65.38
Ag	100	114			205	219	12.62	100.9	135	108	2688	430.1	108	108	107.9
Hg	167	191			380	405	25.00	200.0	125.5	100.4	2532	405.1	100	100	200.6
Pb	95	109			398	425	25.97	207.8	129.5	103.6	2597	415.6	104	104	207.2

*Notes are provided on p. 81.

weights. I have retained significant figures except where they exceed four. Several substances now known to be elementary were represented as compound, or vice versa; I have adjusted these weights so that they correspond to the modern elements and have enclosed them in parentheses. For example, I have listed Berzelius' weight for his compound "nitric suboxide," NtO, which corresponds to nitrogen gas; I have given the respective "murium oxide" formula weights as the atomic weights of chlorine for Dalton, Wollaston, and Berzelius; and I have subtracted the weight of an atom of oxygen from the weights of metal oxides presumed by Dalton and Wollaston to be elements.[140]

Whether these chemists chose HO or H_2O as the formula for water was of course pivotal; when Davy changed his mind, about the summer of 1811, he was compelled to double every one of his atomic weights relative to hydrogen. Their systems of weights were also not strictly comparable since each chemist relied on different analytical data. But in the theoretical realm, there were large areas of agreement. All agreed on the formulas for the oxides of carbon and sulfur, so these atomic weights seemed well established. All but Wollaston agreed on the formulas for the oxides of iron. Where only one oxide of a metal was known, most assumed it to be binary. Most also thought that the two oxides, that is, acids, of phosphorus were PO and PO_2.[141] Finally, by 1814 all but Dalton regarded the lowest oxide of nitrogen as binary, nitric acid being NO_5.[142] Thus, the chemists who thought water was H_2O got an atomic weight for nitrogen that is roughly double the modern value; the HO advocates got a value roughly equal to the modern figure.

All of these areas of agreement may be regarded as consequences of two implicit assumptions: first, that oxide molecules contain only a single atom of the oxidized element — what I will call the "single-atom" axiom; and, second, that molecules of the lowest oxide in a series have one oxygen atom — the "binary" axiom. Wollaston was most consistent in basing his formulas on these assumptions: the first he never contradicted, the second he contradicted only once[143] (Prout followed Wollaston fully). Davy's and Berzelius' formula for water and Dalton's formula for nitrous oxide contradicted the single-atom rule; these assignments were justified on the basis of vapor densities. Dalton, Wollaston, Thomson, and Berzelius all arrived at this rule by picturing gases as composed of mutually repulsive particles. Where the oxygen content of two known oxides of an element stood in the ratio of two to three, these chemists rarely hesitated to contradict the binary rule, suggesting that the protoxide was not yet known. Berzelius was the only one of the six who consistently disregarded this second rule; we have discussed his probable reasons for adopting the general metal oxide formula, MO_2.

Dalton, the founder of chemical atomism, was well aware that his rules were to some extent arbitrary, that they merely expressed probabilities. He pointed out the problematical nature of formula assignments at least four times between 1810 and 1831.[144] All of the other chemical atomists we have treated in this chapter expressed similar reservations at one time or another. However, Dalton, Thomson, and Berzelius regarded their formulas, and thus their weights, as at least highly probable.

Wollaston appears to have been equally confident, asserting in 1822 that, "We may without hesitation conclude that [my] equivalent quantities . . . do really express the relative weights of elementary atoms." Despite his emphasis on "practical convenience" and on an empirically derived system, Wollaston here leaves little room for doubt regarding his opinion of the correctness of his 1814 weights and formulas. The description of Wollaston as the conscious founder of an empirical system of chemical formulation is, therefore, misconceived. Wollaston's system was neither more nor less empirical than the systems of his contemporaries; he was as much a chemical atomist as Dalton or Berzelius. He was indeed more consistent than his fellows in following conventions, but consistency has nothing to do with empiricism.

Much of the confusion is semantic. Today we distinguish between the experimental determination of chemical equivalents and the theoretical task of deducing atomic weights. Although many contemporary chemists made a similar distinction between Wollaston's equivalents and Dalton's atomic weights, many others used the two terms as synonyms, and I suggest that the latter were the more perspicacious. Dalton, for instance, clearly thought that Davy and Wollaston both were advocating chemical atomism.[145]

Thomson is another good example of this group. His physical — or metaphysical — atomism was very similar to Dalton's, yet by 1814 he had settled on a table of atomic weights that was virtually identical to Wollaston's equivalents.[146] In 1815 Thomson compared his weights with Wollaston's and Berzelius' and remarked: "The weights given in these three different tables do not always coincide with each other; but in general a very near approach to coincidence will be perceived. In some cases the weights that I have assigned are half those given by Berzelius. The reason of this is obvious; and the circumstance can occasion no difficulty or ambiguity."[147] In 1831 Thomson explained why he could refer to Wollaston as a "convert" to the atomic theory and to Davy as a "strenuous supporter" of it:

> The only alteration which [Davy] made was to substitute *proportion* for Dalton's word *atom*. Dr. Wollaston substituted for it the term *equivalent*. The object of these substitutions was to avoid all theoretical annunciations. But, in fact, these terms, *proportion*, *equivalent*, are neither of them so convenient as the term *atom*; and, unless we adopt the hypothesis with which Dalton set out, namely, that the ultimate particles of bodies are *atoms* incapable of further division, and that chemical combination consists in the union of these atoms with each other, we lose all the new light which the atomic theory throws upon chemistry, and bring our notion back to the obscurity of the days of Bergman and Berthollet.[148]

Many of Thomson's contemporaries agreed with this viewpoint. In Andrew Ure's *Dictionary of Chemistry*, the entry "Atomic Theory" is merely a cross-reference to the article "Equivalents (Chemical)." Ure made no distinction between the various denotations for elemental weights and discussed the easiest methods of interconverting the systems of Dalton, Davy, Thomson, and Wollaston.[149] William Henry wrote in 1829: "*Chemical Equivalents* . . . may be considered as denoting the *relative weights of the atoms of bodies*", and shortly thereafter he appended a "Table of Chemical Equivalents or Atomic Weights."[150] Charles Sylvester equated "Definite Proportions" with atoms and regarded Davy as a true atomist.[151] William

Brande, Davy's successor at the Royal Institution, conflated chemical and physical atomism in an 1826 lecture, but correctly identified the atomic theory in chemistry with what he called the *"theory of . . . chemical equivalents."*[152] Examples could easily be multiplied.[153]

Thomson was not the only contemporary historian to discount the significance of positivistic terminology not founded in an empirical methodology. William Whewell noted that Wollaston and Davy "objected, indeed, to Dalton's assumption of atoms; and, to avoid this hypothetical step, Wollaston used the phrase *chemical equivalents*, and Davy the word *proportions*, for the numbers which expressed Dalton's atomic weights. We may, however, venture to say that the term 'atom' is most convenient, and it need not be understood as claiming our assent to the hypothesis of indivisible molecules."[154] Hermann Kopp's treatment is similar; he made no distinction between Wollaston's "atomic weights, or as he called them, equivalent weights," and Berzelius' system.[155]

This characteristic lack of concern among contemporary chemists and historians regarding the particular multiples or submultiples chosen by early compilers of atomic weights supports the thesis I have sought to develop here. It cannot be decided from a chemist's terminology, from his system of weights, or even from his statements on the subject whether or not he adhered to the doctrine of chemical atomism. Rather, a chemist's advocacy of any fixed system of elemental weights is necessary and sufficient cause to claim that he advocated a chemical atomic theory.

Dalton, of course, went far beyond simple chemical atomism. In accordance with the etymology of the word, he regarded atoms as indivisible particles: in short, he identified chemical with physical atoms. It was this point that was disputed throughout the nineteenth century.[156] The many scientists who disagreed with this identification have too facilely been regarded as being indifferent or even opposed to atomism because their position has erroneously been interpreted as a reaction against chemical atomism. We have already argued for the falsity of the nearly universal judgment that Davy and

Wollaston were antiatomists. A similar corrective must also be applied, admittedly with greater caution, to the French chemists J. L. Gay-Lussac and J. B. Dumas, to whom we now turn.

1. Thomson, *History of Chemistry*, 2 vols. (London, 1830-31), 2:289-91; [R. D. Thomson,] "Biographical Notice of the Late Thomas Thomson," *Glasgow Med. J.* [3] 5 (1857): 69-80, 121-53, 379-80 (131).

2. Thomson, *System of Chemistry*, 1st ed., 4 vols. (Edinburgh, 1802), 3:196-203, 270-71; 2nd ed. (1804), 3:440-50, 316ff.; cf. 5th ed. (1817), 3:174; *History*, 2:287. All but the last are cited by Partington, *History of Chemistry*, 4 vols. (London: Macmillan & Co., 1961-70), 3:776-77, and by S. Mauskopf, "Thomson Before Dalton," *Ann. Sci.* 25 (1969): 229-42, who has a thorough discussion of this subject.

3. Thomson's statement that his account of the atomic theory was published with Dalton's permission (*System*, 3d ed., 3:425) implies such a correspondence, though apparently no relevant letters survive. Dalton revised his weight for oxygen upward from $O = 5.66$ after reading Gay-Lussac and Humboldt's 1805 memoir (implying $O = 6.9$). See H. E. Roscoe and A. Harden, *A New View of the Origin of Dalton's Atomic Theory* (London, 1896), pp. 83-85, and A. Thackray, "The Origin of Dalton's Chemical Atomic Theory," *Isis* 57 (1966): 35-55 (43, 55).

4. Thomson, *System*, 3d ed., 3:424.

5. Ibid., p. 622.

6. $H = 1$, $O = 6$, $C = 4.4$, $N = 5$, $P = 8$, $S = 15$; ibid., pp. 429, 451-53, 472, 478, 514-33, 539-40, 612-28.

7. H. Debus, "Die Genesis von Daltons Atomtheorie," *Z. physik. Chem.* 20 (1896): 359-76 (373); Dalton, "Remarks on the Essay of Dr. Berzelius on the Cause of Chemical Proportions," *Ann. Phil.* 3 (1814): 174-80 (175).

8. Thomson, "On Oxalic Acid," *Phil. Trans. Roy. Soc.* 98 (1808): 63-95; Wollaston, "On Super-Acid and Sub-Acid Salts," *Phil. Trans. Roy. Soc.* 98 (1808): 96-102. The two papers describe multiple proportions in the oxalates and carbonates of strontium and potassium. Dalton had recognized the phenomenon in the carbonates nearly a year earlier: A. Thackray, *John Dalton: Critical Assessments of His Life and Science* (Cambridge, Mass.: Harvard University Press, 1972), p. 87.

9. A. Ure, "Equivalents (Chemical)," *Dictionary of Chemistry* (London, 1821), n.p.; Thackray, "Origins," p. 43.

10. Thomson, "Oxalic Acid," pp. 85-86. Thomson's conflation of the laws of definite and multiple proportions with chemical atomism was one of the sources of a confusion that lasted the entire century and still exists today.

11. Thomson, "On the Daltonian Theory of Definite Proportions in Chemical Combinations," *Ann. Phil.* 2 (1813): 32-52, 109-15, 167-71, 293-301; 3 (1814): 134-40, 375-78; 4 (1814): 11-18, 83-89.

12. Ibid., 2: 33-34.

13. Ibid., pp. 35-37.

14. Ibid., pp. 33, 36, 38, 41, 43, 113; cf. letters from Thomson to Dalton of 13 November 1809 and 13 August 1818, in Roscoe and Harden, *New View*, pp. 148 and 172; and Thomson, *History* 2: 299-300.

15. Thomson, "Daltonian Theory", p. 36; from O ≅ 7 to O ≅ 7.5. The methods and data derived from the work of Gay-Lussac ("Mémoire sur la combinaison des substances gazeuses, les unes avec les autres," *Mém. soc. Arcueil* 2 [1809]: 207-34); for a detailed exposition, see M. P. Crosland, *Gay-Lussac: Scientist and Bourgeois* (Cambridge: Cambridge University Press, 1978), chapter 5.

16. Thomson's weights and formulas for nineteen of the more important elements and their compounds are summarized in Tables 1 and 2.

17. On Prout, see W. H. Brock, "William Prout," in *Dictionary of Scientific Biography* 11 (1975): 172-74, and works cited therein, esp. by Brock and Benfey; also see R. Siegfried, "The Chemical Basis for Prout's Hypothesis," *J. Chem. Educ.* 33 (1956): 263-66.

18. [W. Prout,] "On the Relation between the Specific Gravities of Bodies in their Gaseous State and the Weights of their Atoms," *Ann. Phil.* 6 (1815): 321-30, and "Correction of a Mistake in the Essay on the Relation . . .," *Ann. Phil.* 7 (1816): 111-13. See tables 1 and 2.

19. Prout, "Reply," *Phil. Mag.* [3] 5 (1834): 132-33.

20. Prout, "Relation," pp. 330, 113. For the distinction between Prout's two hypotheses and a discussion of some of the nineteenth-century Proutian debates, see W. Brock, "Dalton versus Prout: The Problem of Prout's Hypotheses," in D. S. L. Cardwell, *John Dalton and the Progress of Science* (Manchester: Manchester University Press, 1968), pp. 240-58; also Partington, *History*, 4: 222-32, and D. M. Knight, *The Transcendental Part of Chemistry* (Folkestone, Kent: Dawson, 1978).

21. Thomson, "Some Observations on the Relations between the Specific Gravity of Gaseous Bodies and the Weights of their Atoms," *Ann. Phil.* 7 (1816): 343-46.

22. Berzelius needed four-volume weights to harmonize inorganic with organic chemistry: see chapter 6.

23. Thomson, *An Attempt to Establish the First Principles of Chemistry by Experiment*, 2 vols. (London, 1825); Thomson to Dalton, 19 April 1825, in Roscoe and Harden, *New View*, p. 179.

24. Brock, "Dalton versus Prout," pp. 249-55; Berzelius, *Jahresbericht* for 1825 6 (1827): 77-78; Thomson, *History*, 2: 306-8; Partington, *History*, 4: 225-26.

25. Thomson, *History*, 2: 293-94.

26. T. H. Levere, "Coleridge, Chemistry, and the Philosophy of Nature," *Stud. Romant.* 16 (1977): 349-79; *Affinity and Matter: Elements of Chemical Philosophy* (Oxford: Oxford University Press, 1971).

27. H. Davy, ed. J. Davy, *Collected Works of Sir Humphry Davy*, 9 vols. (London, 1839-40), 5: 61, 67-73, 89-92, 131-32, 160-63, 168-70; 4: 132, 358-60 (publications from 1807-12); Davy to Berzelius, 20 March 1809, in H. G. Söderbaum, ed., *Jac. Berzelius Bref* (Stockholm and Uppsala, 1912-32), vol. 1, pt. 2, p. 11. See R. Siegfried, "The Discovery of Potassium and

Sodium, and the Problem of the Chemical Elements," *Isis* 54 (1963): 247-58; R. Siegfried, "The Phlogistic Conjectures of Humphry Davy," *Chymia* 9 (1964): 117-24; Knight, *Transcendental Part*.

28. For instance, by Knight and Siegfried, two of the most eminent Davy scholars; these historians completely discount Thomson's testimony.

29. Knight, *Transcendental Part*; Knight, *Atoms and Elements: A Study of Theories of Matter in England in the Nineteenth Century* (London: Hutchinson, 1967); Siegfried, "The Chemical Philosophy of Humphry Davy," *Chymia* 5 (1959): 193-201; Siegfried, "Boscovich and Davy: Some Cautionary Remarks," *Isis* 58 (1967): 236-38; L. P. Williams, *Michael Faraday* (London; Chapman & Hall, 1965), pp. 66-80.

30. *Berzelius Bref*, 1:ii 23 (letter of 24 March 1811); Thackray, *Critical Assessments*, p. 92.

31. Davy, *Collected Works*, 5: 330n.; from a fragment originally intended for inclusion in his *Consolations in Travel, or The Last Days of a Philosopher* (London, 1830).

32. Ibid., 6: 97. He compared Dalton to Kepler and expressed the hope that another Newton might soon emerge to provide a general theoretical explanation for the laws of chemical combination.

33. Davy seems to have been more sympathetic to point atomism; see his *Consolations in Travel*, where point atoms are described as "not mere supposition unsupported by experiments" (ibid., 9: 388).

34. Davy, "On Some of the Combinations of Oxymuriatic Gas and Oxygene," *Phil. Trans. Roy. Soc.* 101 (1811): 1-35 (17n.).

35. Thomson, *History*, 2:293-94; "Daltonian Theory," pp. 33, 41-43 (1813).

36. Dalton, cited in Thackray, *Critical Assessments*, pp. 93-94, 100-101 (Ms from 1830); Berzelius, *Ann. chim.* 89 (1814): 81; J. Davy, *Memoirs of the Life of Sir Humphry Davy* (London, 1836), 1: 439.

37. Both cited by Levere, "Coleridge," pp. 349, 357, 361.

38. "Proportions, Definite," A. Rees' *Cyclopaedia*. I thank June Fullmer for this reference.

39. *Berzelius Bref*, 1, ii, 22.

40. Davy, "On Some New Electrochemical Researches," *Phil. Trans. Roy. Soc.* 100 (1810): 16-74 (69, 70-71n.); *Collected Works*, 5: 278, 279n.

41. Davy, "Researches on the Oxymuriatic Acid, its Nature and Combinations . . .," *Phil. Trans. Roy. Soc.* 100 (1810): 231-57 (245, 254-55); Davy, "On Some of the Combinations," pp. 15-18n.; *Collected Works*, 5: 298, 308-9, 326-28n. A collated summary of the values given in the 1811 letter to Berzelius and in the two papers just cited, constituting a set of weights for a total of eleven elements, is provided in table 2. The only inconsistency in the three sources is in the value for sodium; in the 1811 letter he suggested that its "proportion" was 22, corresponding to a formula for soda of NaO, whereas in the Bakerian Lecture for 1810 he stated that it was "probable" that soda is a "deutoxide," hence the proportion for sodium would be double this figure, or 44.

42. Thomson had used the word "proportion" to denote atomic weight in 1807: *System*, 3d ed., 3: 621.

43. We have seen that Thomson had been the first to commit this error (above, n. 10). Sir Harold Hartley, in *Humphry Davy* (London: Nelson, 1966), pp. 77-84, has noted that Davy's "proportions" were, in effect, atomic weights.

44. Davy, *Elements of Chemical Philosophy* (London, 1812), pp. 112-13; *Collected Works*, 4: 80-82.

45. See tables 1 and 2 for his new weights and formulas. He also assumed a different formula for phosphoric acid and revised most of his previous determinations. Davy retained this second set of atomic weights to the end of his life; see his *Elements of Agricultural Chemistry*, 2d ed. (London, 1827); *Collected Works*, 7: 169-391.

46. Davy, *Elements of Chemical Philosophy*, pp. 114-15; *Collected Works*, 4: 82-83; 5: 342; *Phil. Trans. Roy. Soc.* 102 (1812): 409.

47. Davy, *Elements*, pp. 114-15.

48. Ibid., pp. 480-82. Cf. passage cited above, n. 31.

49. The previous year he had written to the same correspondent: "Entre nous, I have been concerned to observe Davy become of late so very prone to hypothesyze, & to draw hasty conclusions" (Transcribed letters dated 25 February 1811 and 5 December 1812 in L. F. Gilbert Papers, D. M. S. Watson Library, University College, London [hereafter cited as "Gilbert Papers"]).

50. Young, "Elements of Chemical Philosophy," *Quart. Rev.* 8 (1812): 65-86 (85); rpt. G. Peacock, ed., *Miscellaneous Works of the Late Thomas Young* (London, 1855), 1: 575-600.

51. Young, "Elements of Chemical Philosophy," *Quart. Rev.* 8 (1812): 67, 76-77.

52. Ibid., pp. 78-79. This table was reprinted the following year in Young's *Introduction to Medical Literature* (London, 1813), pp. 58-65.

53. Wollaston, "Salts," p. 96.

54. Thomson, *System*, 2d ed., 1: 272-73.

55. Thomson, "On the Discovery of the Atomic Theory," *Ann. Phil.* 3 (1814): 329-38 (337-38).

56. Ure, "Equivalents (Chemical)"; Whewell, *History of the Inductive Sciences*, 3d ed. (London, 1857), 3: 131; Peacock, *Thomas Young*, p. 469.

57. Thomson, *History*, 2: 293-94, 305-6.

58. *Berzelius Bref*, 1, ii, 32 (Berzelius to Davy, August or September 1812).

59. Berzelius to J. Schweigger, 4 December 1812, rpt. in *J. Chem. Phys.* 9 (1813): 296n.

60. *Berzelius Bref*, 1, i, 41-42 (October 1812); trans. M. E. Weeks, *Discovery of the Elements*, 7th ed. (Easton, Pa: Journal of Chemical Education, 1968), p. 411; see also n. 58. Berzelius cited with approval the "common proverb" that "whoever argues with Wollaston is wrong."

61. Letter headed "Wedn^y Ev^g 12 Augt 1812," Gilbert Papers. The list contains 26 elements and compounds.

62. *Berzelius Bref*, 1, iii, 51, 69. It is however possible that both came to the same conclusion independently. Thomson was the first to publish a weight table on the basis of $O = 1$ in July 1813; he acknowledged that Wollaston and Berzelius were the originators of the idea. Wollaston's paper setting $O = 10$

and Berzelius' setting O = 100 both appeared within the next six months. Wollaston said he regretted he had not mentioned the idea in his 1808 paper. See also ibid., ii, p. 32, and Berzelius, trans. E. Wöhler, "Aus Berzelius' Tagebuch während seines Aufenthaltes in London im Sommer 1812," *Z. angew. Chem.* 18 (1905): 1946-48; 19 (1906): 187-90, 571-76 (572). Thomson agreed that Wollaston assumed water to be H_2O: "Daltonian Theory," 4: 84n.

63. He remained unconverted as late as 1816, as is revealed by another letter to Berzelius (16 January 1816, Gilbert Papers).

64. Wollaston, "A Synoptic Scale of Chemical Equivalents," *Phil. Trans. Roy. Soc.* 104 (1814): 1-22. The word had been used by Henry Cavendish, *Phil. Trans. Roy. Soc.* 57 (1767): 102, and by Wollaston himself, *Phil. Trans. Roy. Soc.* 98 (1808): 100-101, as applied to acids and bases.

65. Ibid., p. 7

66. Davy, *Collected Works*, 7: 96-97. For an excellent review of the reception of Wollaston's paper, see D. C. Goodman, "Wollaston and the Atomic Theory of Dalton," *Hist. Stud. Phys. Sci.* 1 (1969): 37-59.

67. See tables 1 and 2. For the first three of these elements, Wollaston assumed that the lowest oxide was binary, whereas Dalton assumed ternary combinations. Dalton calculated his value for phosphorus on the basis of phosphine as PH, which yields a value equal to one-third of the modern figure, whereas Wollaston chose the acid formula PO_2, which results in a value equal to two-fifths of the modern figure.

68. Goodman, "Wollaston and Dalton," pp. 44-50, has argued similarly regarding Wollaston's "equivalents." See also Cannizzaro, "Considerations on Some Points of the Theoretic Teaching of Chemistry," *Faraday Lectures*, ed. C. S. Gibson and A. J. Greenaway (London, 1928), pp. 17-43, esp. 19-22.

69. For an example of the former viewpoint, see P. T. Hinde, "William Hyde Wollaston: The Man and His Equivalents," *J. Chem. Educ.* 43 (1966): 673-76; Goodman suggests the latter interpretation. T. I. Krasovitskaia and S. IA. Plotkin, "Vollaston i atomnaia teoriia Daltona," *Vop. ist. est. tekh.* 45 (1973): 41-44, have argued that Wollaston was a consistent atomist.

70. Wollaston, "Salts," p. 101.

71. Ibid., pp. 101-2. Cf. Dalton's justification of his simplicity rule.

72. Wollaston, "On the Elementary Particles of Certain Crystals," *Phil. Trans. Roy. Soc.* 103 (1813): 51-63.

73. Dalton, *New System of Chemical Philosophy*, pt. 1 (Manchester, 1808), pp. 210-11; Charles Daubeny, *Introduction to the Atomic Theory* (Oxford, 1831), p. 137. Wollaston claimed that the idea came to him independently in 1809.

74. Goodman, "Wollaston and Dalton," pp. 41-42.

75. Wollaston, "On the Finite Extent of the Atmosphere," *Phil. Trans. Roy. Soc.* 112 (1822): 89-98.

76. Ibid., p. 94. For discussions of Wollaston's astronomical argument, see George Wilson, *Religio Chemici* (London, 1862), pp. 259-63, and Goodman, "Wollaston and Dalton," pp. 53-58.

77. Wollaston to Marcet, 26 May 1822, Gilbert Papers.

78. Daubeny, cited in W. Munk, *Roll of the Royal College of Physicians of*

London (London, 1878), 2:442n. (my trans. from the Latin); Daubeny was F. R. S. and professor of chemistry at Oxford from 1822. Davy, cited in J. Davy, *Memoirs*, 1:258n.

79. Wollaston, "Scale," p. 7. His rule is nowhere explicitly stated, but can be inferred from his table of formulas and weights.

80. He probably supposed that the monoxide had not yet been discovered. If he had taken the lower oxide to be SO, consistent with his rule, then his values for sulfur and oxygen would have coincided, clearly an uncomfortable state of affairs.

81. *Berzelius Bref*, 1, iii, 83 (letter of 12 November 1813).

82. H. G. Söderbaum, *Jac. Berzelius: Levnadsteckning*, 3 vols. (Uppsala, 1929-31); Söderbaum, *Berzelius' Werden und Wachsen* (Leipzig, 1899); J. E. Jorpes, *Jac. Berzelius: His Life and Work* (Stockholm: Almqvist & Wiksell, 1966).

83. J. Berzelius and W. Hisinger, "Versuch, betreffend die Wirkung der elektrischen Säule auf Salze und auf einige von ihren Basen," *Neues allg. J. Chem.* 1 (1803): 115-49.

84. H. Davy, "On Some Chemical Agencies of Electricity," *Phil. Trans. Roy. Soc.* 97 (1807): 1-56; Davy, "On Some New Phenomena of Chemical Changes Produced by Electricity, particularly the decomposition of the Fixed Alkalies . . .," *Phil. Trans. Roy. Soc.* 98 (1808): 1-44; *Collected Works*, 5:1-101.

85. Berzelius and M. M. Pontin, "Elektrisch-chemische Versuche mit der Zerlegung der Alkalien und Erden," *Ann. Phys.* 36 (1810): 247-80. The authors completed the experiments, published them in Swedish, and reported them in a letter to Davy all in May 1808: Söderbaum, *Werden und Wachsen*, pp. 49-52; *Berzelius Bref*, 1, ii, 7, and 4, ii, 36-39.

86. *Berzelius Bref*, 1, ii, 7-12. Berzelius' research stimulated much of the work reported in Davy's 1808 Bakerian Lecture, "An Account of Some New Analytical Researches . . .," *Phil. Trans. Roy. Soc.* 99 (1809): 39-104.

87. Berzelius, *Lärbok i kemien*, 1st ed., 6 vols. (Stockholm, 1808-30).

88. J. B. Richter, *Ueber die neuern Gegenstände der Chymie*, 11 pts. (Breslau, 1791-1802); Berzelius summarized Richter's work in *Lärbok* 1 (1808): 398-99.

89. Berzelius, *Autobiographical Notes* (Baltimore: Williams & Wilkins, 1934), pp. 62-65; "Versuch, die bestimmten und einfachen Verhältnisse aufzufinden, nach welchen die Bestandtheile der unorganischen Natur mit einander verbunden sind" *Ann. Phys.* 37 (1811): 249-334, 415-72 (249-54) (hereafter abbreviated "Versuch"); *Essai sur la théorie des proportions chimiques* (Paris, 1819), pp. 16-18; *Berzelius Bref*, 1, i, 6-7; A. Lundgren, *Berzelius och den kemiska atomteorin* (Uppsala: Almqvist & Wiksell, 1979), pp. 76-80.

90. Berzelius, "Ueber das Verhältniss der Sauerstoffmengen etc. zu einander . . .," *J. Chem. Phys.* 1 (1811): 257-62; Berzelius, "Versuch," pp. 250-51. See Evan M. Melhado, *Jacob Berzelius: The Emergence of His Chemical System* (Madison: University of Wisconsin Press, 1981), p. 170, for a full discussion.

91. Berzelius, "Verhältniss der Sauerstoffmengen," p. 260; Berzelius, "Versuch," p. 449.

92. Berzelius' own testimony on this point is inconsistent. See *Berzelius Bref*, 1, i, 7; *Essai*, p. 17; "Verhältniss der Sauerstoffmengen," p. 257; "Versuch," pp. 251, 415-16; *Autobiographical Notes*, pp. 172-73.

93. Berzelius, *Berzelius Bref*, 3, ii, 286-88. Wollaston's paper is "On Superacid and Sub-acid Salts," as reprinted in *Nicholson's J.* 21 (Nov. 1808): 164-69, which provides a terminus a quo. In this letter Berzelius mentioned a letter he had received from A. G. Ekeberg that Söderbaum dates 16 June 1809 (*Berzelius Bref*, 3, ii, 316), and Berzelius also cited some "just completed" analyses of the alkalies discussed more fully in a letter to Davy of 30 June 1809 (*Berzelius Bref*, 3, ii, 12). The letter to Wollaston was thus surely written in the latter half of June 1809, our terminus ad quem. Sometime between then and June 1811, Berzelius also read John Murray's account of Thomson's summary of the Daltonian theory, in *A Supplement to the First Edition of a System of Chemistry* (Edinburgh, 1809). For further information regarding this sequence of events, see Berzelius, *Autobiographical Notes*, pp. 172-73; Berzelius, "Versuch", p. 251; Berzelius, *Essai*, p. 17; Berzelius, *Berzelius Bref*, 1, ii, 29. Berzelius had read Dalton's 1803/1805 paper on the solubilities of gases in water, which contained the first atomic weight table, but Lundgren (*Berzelius*, pp. 89-95) has shown that he paid no attention to the table.

94. Berzelius to Davy, 30 June 1809, *Berzelius Bref*, 1, ii, 15; Berzelius to Berthollet, undated but probably February 1811, *Berzelius Bref*, 1, i, 16.

95. Berzelius to Davy, 22 July [1810] and 10 June 1811, *Berzelius Bref*, 1, ii, 19 and 29.

96. *Berzelius Bref*, 3, ii, 101-2.

97. E.g., by Melhado, *Berzelius*, and esp. by Lundgren, *Berzelius*.

98. Berzelius, "Versuch," pp. 252, 254. Melhado (p. 168n.) notes that in a letter dated 8 October 1809 published in Gehlen's journal, Berzelius mentioned that the work was in press, hence my estimate of summer 1809 as the date of composition: "Vermischte Notizen," *Neues allg. J. Chem.* 9 (1810): 585-89 (587); see also corroborating evidence in "Versuch," pp. 415-16.

99. Berzelius, "Versuch," originally published as "Försök rörande de bestämde proportioner, hvari den oorganiska naturens beståndsdelar finnas förenade," *Afhandlingar i fysik, kemi och mineralogi* 3 (1810): 162-275. Translations of the paper also appeared in *Ann. chim.* 78-83 passim (1811-12), and *Phil. Mag.* 41-43 passim (1813-14). Collected in *Ostwalds Klassiker der exakten Wissenschaften*, nr. 35 (Leipzig, 1892). Melhado provides a painstaking and insightful study of this epochal work: *Berzelius*, pp. 165-241 passim.

100. Berzelius, "Versuch," p. 252.

101. Ibid., pp. 324-34, 465-72; these represent passages added to the German translation in press, and are not found in the Swedish original. For the chronology of publication of the various parts of the German "edition," see Melhado, *Berzelius*, p. 181n.

102. Berzelius, "Versuch," pp. 332-34, 467-71.

103. Berzelius, "Schreiben des Herrn Prof. Berzelius . . . über ein zweites neues Gesetz . . . Stockholm, d. 1. Oct. 1810," *Ann. Phys.* 37 (1811): 208-20; *Berzelius Bref*, 1, i, 10; "Lettre de M. Berzelius à M. Berthollet," *Ann. chim.* 77 (1811): 63-84 (75); "Fortsetzungen" to the "Versuch," *Ann. Phys.* 38

(1811): 161–226; 40 (1812): 162–208, 235–330. The passage cited is from *Ann. Phys.* 38 (1811): 163.

104. This is one of the principal themes of Melhado's monograph: *Berzelius*, pp. 145–48, 193–202 and passim.

105. Berzelius, "Lettre," pp. 81–82; "Versuch," 38: 186.

106. Berzelius, "Lettre," pp. 76–77.

107. Berzelius, "Versuch," 40: 162–208, 235–330. For a detailed discussion, see Melhado, *Berzelius*, pp. 203–10, 228–31.

108. Berzelius, "Versuch, die chemischen Ansichten, welche die systematische Aufstellung der chemischen Nomenclatur bergründen, zu rechtfertigen," *J. Chem. Phys.* 6 (1812): 119–44, 284–322; 7 (1813): 43–78; "Zwei Schreiben des Hrn. Prof. Berzelius . . .," *Ann. Phys.* 42 (1812): 276–98; *Essai*, pp. 114–16. For a discussion, see Melhado, *Berzelius*, pp. 239–41.

109. Berzelius to Gaultier de Claubry, 27 April 1812, *Berzelius Bref*, 3, ii, 104–5.

110. "Aus Berzelius' Tagebuch," p. 188; *Berzelius Bref*, 1, ii, 32–33 and 3, i, 7; *Autobiographical Notes*, pp. 74–89; and letters between Berzelius and Dalton, August to October 1812, printed in Roscoe and Harden, *New View*, pp. 156–62.

111. *Berzelius Bref*, 1, ii, 35–59 (37). Davy was hurt and angered by Berzelius' critique: *Autobiographical Notes*, pp. 83–89.

112. Berzelius, "Experiments on the Nature of Azote, of Hydrogen, and of Ammonia, and of the Degrees of Oxidation of Which Azote is Susceptible," *Ann. Phil.* 2 (1813): 276–84, 357–68. The paper was dated 22 April 1813. The misadventurous story of the transmission to Thomson, editor of the *Annals of Philosophy*, both of this article and of the nearly concurrent "Essay on the Cause of Chemical Proportions" (discussed below) can be followed in Berzelius' correspondence with Marcet and Thomson: *Bref*, 1, iii, 38–39, 57, 60, 64, 69–70, 72, 82; 3, i, 7–13.

113. For the sake of clarity, I have slightly modified the formulas as they appear in the article. Berzelius wrote N for "nitric" and Az for azote (nitrogen gas); his formula for nitric acid was "N + 6O," and the other formulas were written in an analogous way.

114. Berzelius' "nitric" was assigned the weight 79.64, where O = 100 (or 12.74, where O = 16); nitrogen gas was then 179.64 (or 28.74); "Experiments," p. 284. Davy's weight for nitrogen was 27.7, also about twice the modern value. This was, of course, a direct result of the formulas chosen for the oxides. $2HNO_3 - H_2O = N_2O_5$ is the modern formula for anhydrous nitric acid; $2HNO_2 - H_2O = N_2O_3$ is anhydrous nitrous acid.

115. Berzelius, "Experiments," p. 359 and 359n.

116. *Berzelius Bref*, 1, iii, 60; 3, i, 11.

117. Corrected, in German translation, *Ann. Phys.* 46 (1814): 172, to $NtOH_6$.

118. Berzelius, "Essay on the Cause of Chemical Proportions, and on Some Circumstances Relating to Them; Together with a Short and Easy Method of Expressing Them," *Ann. Phil.* 2 (1813): 443–54; 3 (1814): 51–62, 93–106, 244–57, 353–64. See especially the table of forty-seven elements in the

last installment, pp. 362-63, but note that the values listed in this table do not always correspond exactly to the values in the body of the paper. Berzelius published a mineralogical treatise early in 1814 containing a similar table: *Försök att . . .grundlägga ett rent vettenskapligt system för mineralogien* (Stockholm, 1814); H. Black, trans., *An Attempt to Establish a Pure Scientific System of Mineralogy* (London, 1814), pp. 117-18. The table contains some revised figures, but the assumed molecular formulas are all the same.

119. Berzelius, "Essay on the Cause," 2: 445, 453-54.

120. *Berzelius Bref*, 1, iii, 83-93; 3, i, 11.

121. That is, they assumed the same formulas for the oxides of the nonmetals and also for several of the metals. Berzelius believed that chlorine was the trioxide of the muriatic radical, atomic weight 139.56 on the scale oxygen = 100 (22.33 for O = 16). Thus the weight of chlorine was Cl = MuO_3 = 439.56 (70.33). Davy's weight was 71.5.

122. Thomson, "On the Daltonian Theory," 2 (1813):35.

123. Berzelius, "An Address to those Chemists who wish to examine the Laws of Chemical Proportions, and the Theory of Chemistry in general," *Ann. Phil.* 5 (1815): 122-31 (124). Berzelius does not seem to have recognized that one cannot apply empirical tests to axioms. Dalton correctly responded that it all hinges on the particular atomic weights chosen. See below for a discussion of this exchange.

124. Ibid., pp. 124 and 128; "Experiments on Azote," p. 358; "Essay on the Cause," 2:447, 450-51; "Ueber die bestimmten und einfachen Verhältnisse . . .," *J. Chem. Phys.* 2 (1811): 297-326 (318n).

125. Here the true relation is not $1:1^1/_2$, but $1:1^1/_3$, since the two substances are properly formulated PbO and Pb_3O_4. Berzelius' analytical data (1:1.43) was in fact closer to the $1:1^1/_2$ ratio.

126. Gay-Lussac and Thenard, *Recherches physico-chimiques* (Paris, 1811), 1: 124-75, esp. 126-28; Davy, "On Some of the Combinations of Oxymuriatic Gas and Oxygene," *Phil. Trans. Roy. Soc.* 101 (1811): 1-35, on p. 4; Davy, *Elements of Chemical Philosophy*, p. 326. Thomson expressed doubt regarding the existence of the suboxide of potassium: "On the Daltonian Theory," pp. 45-46n.

127. Berzelius, "Essay on the Cause," 3:360.

128. Ibid., 2:446, 450.

129. Ibid., p. 450.

130. Ibid., 3:361.

131. Berzelius had to assign nitrogen gas a heavier weight than oxygen because he believed it to be an oxide. This meant that he had to accept the higher formulas for the oxides of nitrogen, even though these formulas were also inconsistent with relative vapor densities. Dalton pointed out the inconsistency in 1830: see the MS cited by Thackray, in *Critical Assessments*, p. 100.

132. Dalton, "Remarks on the Essay of Dr. Berzelius on the Cause of Chemical Proportions," *Ann. Phil.* 3 (1814): 174-80.

133. Berzelius, "Address."

134. Ibid., pp. 124-27.

135. Some recent scholars have doubted this; see for example G. Eriksson, "Berzelius och atomteorin," *Lychnos*, 1965, pp. 1–37, on pp. 30, 36.

136. Dalton long referred to nitrogen dioxide gas as "nitric acid"; by 1830 he had accepted the common opinion that this substance was "nitrous acid gas," and reserved the higher oxide N_2O_5 for the nitric acid formula.

137. Since we are dealing with integral multiples, inferring the assumed formulas from the stated atomic weight is normally unambiguous and straightforward.

138. Wollaston did not specify the degree of oxidation of muriatic acid, merely suggesting that chlorine has one additional oxygen "equivalent."

139. Davy, "On a Combination of Oxymuriatic Gas and Oxygene Gas," *Phil. Trans. Roy. Soc.* 101 (1811): 155–62. "Euchlorine" has the modern composition Cl_2O and is actually a mixture of chlorine and chlorine dioxide.

140. Except for a period of time around 1810, Dalton generally accepted Davy's view that potash and soda are oxides, and there is little doubt that Dalton would have chosen binary formulas for these compounds. The same conclusion also applies to Wollaston, who did represent the alkalies as binary molecules but declined to speculate on the possible compound nature of the alkaline earths.

141. The modern formulas for phosphorous and phosphoric acid anhydrides are P_2O_3 and P_2O_5. The chemists who regarded phosphoric acid as PO_2 thus calculated an atomic weight for phosphorus of about four-fifths of the modern value (or two-fifths for $O = 8$). Davy initially assumed the formula PO_3, which converts to three-fifths of the modern value ($O = 8$); in 1812 he derived the "proportion" of phosphorus from phophorous acid = PO, hence two-third of the modern figure ($O = 16$). Dalton derived his weight from phosphine = PH, hence one-third the modern value.

142. Thomson in July 1813 wrote nitrous oxide as N_2O, but in February 1814 had changed his mind and agreed with the others ("Daltonian Theory," 2:42; 3:135).

143. This was in the case of the oxides of sulfur, where I have suggested he had good reason.

144. Dalton, *New System*, pt. 2 (1810), p. 276; "Remarks on the Essay," p. 178; Thackray, *Critical Assessments*, pp. 98–102; Daubeny, *Introduction*, pp. 134–37 (letter from 1831).

145. Dalton, cited in Thackray, *Critical Assessments*, pp. 93–94, 100–101.

146. They were identical except for doubled weights for sodium and iron. On the other hand, Thomson disagreed with Dalton on the weights for sodium, nitrogen, phosphorus, lead, and zinc.

147. Thomson, "A Sketch of the Latest Improvements in the Physical Sciences," *Ann. Phil.* 5 (1815): 1–53 (12).

148. Thomson, *History*, 2:293–94.

149. Ure, "Equivalents (Chemical)"; 2d ed. (1823), pp. 420–30.

150. W. Henry, *Elements of Experimental Chemistry*, 11th ed. (London, 1829), 2:654.

151. [Sylvester,] "Proportions, Definite," in Rees' *Cyclopaedia*. I thank June Fullmer for identification of the passage and of the anonymous author.

152. Cited in C. A. Russell and D. C. Goodman, *Atoms and Electricity* (Bletchley: Open University Press, 1973), p. 29, who unjustly fault Brande for using the word "theory" rather than "law." See also Brande's *Manual of Chemistry*, 6th ed. (London, 1848), 1:139-44.

153. E. Turner, *Elements of Chemistry*, 3d ed. (London, 1831), pp. 887-91; T. Graham, *Elements of Chemistry* (London, 1843), pp. 726-27.

154. Whewell, *History*, 3d ed., 3:131.

155. H. Kopp, *Geschichte der Chemie*, 4 vols. (Brunswick, 1843-47), 2 (1844): 370-76.

156. Knight, *Atoms and Elements*, pp. 3, 32-34, draws attention to the distinction between chemical and physical atomism. One of Knight's major concerns is to demonstrate the surprising infertility of the atomic theory until mid-century, and he argues the point convincingly. However, he goes on to suggest that "Dalton had little success in persuading the positivists that his theory swept away mystery" (p. 32). This may be true as regards Dalton's physical theory of matter, but it does not apply to his theory of chemical atoms. Also in his more recent *Transcendental Part* (pp. i-ii), Knight fails to see the implications of his quite proper distinction between physical and chemical atomism.

4

The French Response, 1808–1836

BERTHOLLET

Although Adolphe Wurtz' dictum that "la chimie est une science française"[1] betrays more than a little chauvinism, it is reasonable to describe the chemistry of France as preeminent at the beginning of the nineteenth century. Lavoisier was executed in 1794, but such disciples and colleagues as L. B. Guyton de Morveau, A. F. Fourcroy, and especially C. L. Berthollet continued to utilize and expand upon the new gravimetric and oxygen-based chemistry promulgated by their deceased friend. In 1801 Berthollet bought a country house at the suburban village of Arcueil and fitted there a laboratory for scientific work. From 1807 to 1813 this house was the scene of regular meetings of the informal Société d'Arcueil, where such prominent young scientists as Alexander von Humboldt, J. B. Biot, L. J. Thenard, and P. L. Dulong read and subsequently published their work in the Society's own *Mémoires*. Particularly prominent in the new generation was J. L. Gay-Lussac, who had been Berthollet's handpicked protégé since 1800 and on whom his master's mantle fell when Berthollet died in 1822.[2]

In chapter 1 we examined Berthollet's philosophical outlook and his unfavorable response to Proust's enunciation of the law of definite proportions. Berthollet was a corpuscularian, like most chemists of his day, but he held to a dynamic and continuous conception of chemical reaction and composition that he found difficult to reconcile with stoichiometry and chemical atomism. His first reaction to Dalton's theory was critical but not uniformly unfavorable. By the late summer of 1808, it appears that he had not only been able to obtain, but also to carefully study Thomson's 1807 *System of Chemistry* (containing the first published description of Dalton's theory), Wollaston's and Thomson's 1808 papers on multiple proportions, and the first part of Dalton's *New System of Chemical Philosophy*.

Berthollet had the opportunity to comment on the new theory almost immediately by writing an introduction to the French translation of Thomson's *System*, published in 1809.[3] He admitted the accuracy of Thomson's and Wollaston's examples of multiple proportions and acknowledged Dalton's ingenuity. Nonetheless, he thought that the generality of chemical atomism was as yet unproven and took decided exception to certain aspects of Dalton's physical atomism. This response is remarkably similar to Davy's in the Bakerian Lectures for 1809 and 1810. Berthollet also expressed some discomfort with the arbitrary nature of the rule of greatest simplicity, a criticism J. Bostock would repeat two years later.[4]

At the same time that Berthollet was composing his critique, Gay-Lussac was busy discovering his law of combining volumes of gases. We have seen in chapter 2 that in his memoir announcing the discovery, Gay-Lussac remained faithful to his mentor's ideas while according Daltonian atomism a good deal of respect. His final conclusion seems to have been that the new law was consistent with both doctrines.

Responses to Gay-Lussac's discovery were varied. Probably the first thought in more than one reader's mind was that it supported the idea that equal volumes of gases contain equal numbers of particles, i.e., that the molecules of all gases have the same volume. However, this conception is difficult to reconcile with the many reactions where the volume of product exceeds the volume of at least one reactant. For instance, if water is formed from two atoms (two volumes) of hydrogen and one atom (one volume) of oxygen, one would expect exactly one volume of water vapor to result, not two as is the case. Such apparent anomalies led most chemists instinctively to reject the equal volumes-equal numbers (EVEN) hypothesis, even given Gay-Lussac's newly discovered relation.

Dalton, for instance, regarded atoms and molecules as having all *approximately* the same volume, but adamantly refused for many years to credit Gay-Lussac's law as exact. Such men as Thomson and Prout, on the other hand, accepted the law as exact but concluded from it not the equality of atomic volume, but merely that volumes were related by simple integral multiples. Yet a third response was devised by an Italian chemist, who accepted EVEN and managed to escape the anomalies by means of an important and ingenious auxiliary hypothesis.

THE THEORY OF AVOGADRO AND AMPÈRE

When Amedeo Avogadro (1776-1856) began his professional career as a lawyer in Turin—the capital of Piedmont in the Kingdom of Sardinia—a French army under the command of Napoleon Bonaparte had just overrun the northern Italian states. Even after the years of Napoleonic domination, Sardinia and Turin retained a strong French influence, particularly in the arts and sciences.[5] Consequently, when Avogadro turned his attention to physics and chemistry after 1800, his position in the social hierarchy of science was similar to scholars in the French provincial universities. All his early papers were written in French and published in French journals; but he had no direct personal access to the Parisian chemists or institutions, and this was one reason why his work remained little known throughout his life.

Avogadro read the description of Dalton's theory in the French translation of the third edition of Thomson's *System*—though probably not Dalton's *New System* itself—and was very favorably impressed by Gay-Lussac's 1809 memoir. In 1811 he published a landmark article that he appears to have regarded as an extension of Gay-Lussac's work.[6] After mentioning the law of combining volumes, he wrote that: "The first hypothesis to present itself in this connection and apparently even the only admissible one, is the supposition that the number of integral molecules in all gases is always the same for equal volumes, or always proportional to the volumes."[7] This is of course the EVEN assumption, which, we have seen, was made here neither for the first nor the only time. Avogadro then drew the ready conclusions that relative vapor densities would be directly as the relative weights of the "integral molecules" and combining volumes would give directly the formulas of compounds. For instance, since oxygen is about 15 times as dense as hydrogen, and since twice the volume of hydrogen combines with oxygen to form water, the weight of an oxygen molecule is 15 relative to hydrogen as unity, and the formula for water is H_2O.[8]

But, as noted, EVEN leads to anomalies for many compounds formed from the elements: two volumes of water are formed, not just one. Avogadro's signal contribution in this paper was to suggest submolecularity in elementary molecules, a second hypothesis that was capable of obviating all of these vapor-density anomalies. If each oxygen molecule splits in two during combination with a molecule of hydrogen, the EVEN principle would suggest the observed two volumes of water produced:[9]

$$2H_2 + O_2 = 2H_2O;$$

or, to write the reaction in a more direct and accurate representation of Avogadro's ideas:

$$2H + O = 2HO_{1/2}$$

Just as Prout's dual hypotheses were until recently referred to ambiguously in the singular number, so "Avogadro's hypotheses" of EVEN and submolecularity have too often been considered as signifying EVEN alone, whereas it was only this second suggestion that proved to be truly epochal and unique.

Closely following Gay-Lussac's methodology, Avogadro went on to discuss the use of his gas theory in determining weights and formulas for a wide variety of gaseous and (using appropriate assumptions) nongaseous compounds. He was vague as to the precise degree of submolecularity of elementary molecules, usually assuming that it was normally two, but at other times implying four or some other even integer.[10] He was much closer in spirit to the French physicalist tradition represented by Berthollet and Gay-Lussac than the nonreductionist Daltonian school.[11] Like Gay-Lussac, he recognized the conformity of his hypotheses with "Dalton's system," but also asserted their reconcilability with Berthollet's ideas.[12]

In stressing the very real differences between the British chemical-atomist and the the French physical-molecularist approaches, some recent historians have gone so far as to assert the absolute incommensurability of the two traditions. But Avogadro's avoidance of such words as "atom" or "diatomic" and his uncertainty—indeed, his unconcern—regarding the degree of divisibility of his molecules do not constitute sufficient grounds for the claim that "[t]here was no well developed ontological distinction between his molecules and submolecules," hence "the category of chemical atom was absent from" his ontology.[13] To be sure, Avogadro's reductionist bias led to a certain conceptual isolation that paralleled his geographical isolation and was one of many reasons for his obscurity.[14] But virtually all atomists of the day—including, on rare occasions, Dalton himself—admitted the possibility or even probability of subatomic/submolecular structure, and we have noted that few took the word "atom" at literal etymological face value. Historically sensitive philosophers of science are beginning to create models of scientific research traditions that are able to assimilate such apparently inconsistent approaches as those of Dalton and Avogadro into a single meaningful category.[15]

Three years after Avogadro's paper appeared in J. C. de Lamétherie's *Journal de physique*, André Marie Ampère published a similar theory in the much more prestigious *Annales de chimie*.[16] It appeared in the form of a long letter to Berthollet, in which Ampère related ideas he said he had been working on for a long time; he intended soon to publish them in a more detailed and formal way. Like Avogadro, he viewed his work as an extension of Gay-Lussac's law of combining volumes. Also like Avogadro, he began by enunciating both the EVEN and the submolecularity hypotheses; he cited the Italian in a footnote, adding that he had read the earlier paper only after his own memoir was already written.[17]

But Ampère introduced new and distinct aspects to this conception of gases by also adopting some elements of R. J. Haüy's theory of crystal structure. Haüy had claimed to discern that every crystal exhibits one of a small number (three to six) of "primitive forms," revealed by appropriate cleavage. These macroscopic forms reflect, though are not identical to, the shape of the microscopic polyhedral "integrant molecules" of which the crystals are composed.[18] Ampère took he step that Haüy did not, asserting that the geometry of the integrant molecule was identical to that of the primitive form.

Now the simplest substances must be in the simplest polyhedral form, the tetrahedron, with one point atom at each of the four vertices. Thus for Ampère such gases as hydrogen, oxygen, and nitrogen were composed of *particules* (molecules) consisting of four *molécules* (atoms) each. In combination these particles split in half, so that in modern dress Ampère was suggesting the following sorts of schematic reactions:

$$2H_4 + O_4 = 2H_4O_2$$
$$N_4 + O_4 = 2N_2O_2$$

Ampère represented chemical combination as the interpenetration of the molecular polyhedra until the two centers of gravity coincide.[19]

The rest of Ampère's paper explored the consequences of this theory and is in most respects parallel to Avogadro's earlier work. Like Avogadro, he had succeeded in applying essential elements of Dalton's atomic theory to a notion of gases derived from Gay-Lussac's law; and he had incorporated structural, geometric, and crystallographic ideas as well. Although the promised detailed memoir never appeared, and although Ampère's sketch seems to have made as little immediate impression on his colleagues as Avogadro's paper had, Ampère held to the theory for the rest of his life, publishing a reaffirmation of it in 1835.[20]

The historical record reveals little evidence that there was in the scientific community any approval, or even so much as awareness, of the Avogadro-Ampère gas theory for a number of years after its publication. Still, it would be wrong to assume, as many historians have, that its very existence was unknown to all but a few. Ampère, at least, became known as a world-class scientist and had published his version in the premier chemical journal of the day; and any attentive reader, of which there must have been legion, would have noticed Ampère's citation of his Italian predecessor. The lack of published comment is, I think, due largely to the frankly speculative and ad hoc nature of the theory. As probable as some aspects may have appeared, there was in this period not a scrap of corroborative evidence. The initial position of the theory was analogous to speculations early in this century on the existence of the neutron, later of the neutrino, and currently of the "intermediate vector boson." In such cases there is often little published comment, and a good deal of oral discussion, until some clever investigator finally adduces empirical evidence to corroborate probable notions.

Abruptly in 1826 we begin to see some sympathetic vibrations in the French scientific community. J. B. Dumas asserted in that year that "all physical scientists" subscribe to the EVEN assumption, although he added that the submolecularity hypothesis was still very little applied.[21] Dumas was probably strongly influenced by Ampère, to whom he owed his position at the Athénée de Paris; he also cited Avogadro, but only secondarily.[22]

In the following year, Ampère's lectures at the Collège de France inspired a student there, M. A. Gaudin (1804-80), to devote much of his professional career to elaborating a version of Ampère's theory of matter.[23] The first product of Gaudin's research was sent to the Académie des Sciences in 1831, and a revised version was published in the *Annales de chimie* two years later.[24] Gaudin was following self-consciously in Haüy's and Ampère's footsteps, but he modified the latter's theory by denying that all molecules need be polyhedral. Instead, he always assumed the smallest degree of submolecularity that would accord with the EVEN assumption: the permanent elementary gases and the halogens were diatomic, mercury was monatomic, and sulfur was hexatomic. This somewhat ad hoc procedure led to formulas and atomic weights that happen to accord exceedingly well with modern ones.[25] From symmetry considerations he concluded that triatomic water molecules were linear and tetratomic ammonia molecules were planar. Chemical stability was determined by the degree of symmetry in the atomic arrangements; chemical reactions result in a reshuffling of the atoms to the most symmetrical form. Hence Gaudin rejected all attempts to identify proximate molecular groups that persist unchanged through a series of chemical reactions. Mauskopf has called Gaudin's work of 1831-33 "the clearest exposition of the gas hypothesis prior to Cannizzaro"; Graebe has shown that Cannizzaro was in fact influenced by his reading of Gaudin's memoirs.[26] Gaudin cited Ampère often, but seems never to have mentioned Avogadro's name.

A. E. Baudrimont (1806-80) was a close contemporary of Gaudin, and his concerns were similar. He, too, accepted and elaborated the Avogadro-Ampère gas theory (acknowledging Avogadro's priority), distinguished between atoms and molecules, and speculated on atomic arrangements guided by considerations of symmetry.[27] An examination of Baudrimont's writings, as well as those of Dumas and Gaudin, indicates that the Avogadro-Ampère theory was more widely known and more favorably regarded than most current assessments would admit. In addition to these three chemists, F. Choron, M. Donovan, W. Prout, and A. Laurent all discussed the theory before the Karlsruhe Congress of 1860.[28] Others, such as Berzelius and C. Gerhardt, used elements of the theory without acknowledging explicitly either the assumptions themselves or their provenance.

Most historians of chemistry have asserted that conceptual confusion helped to delay acceptance of the theory, especially the failure clearly to distinguish between the atoms and molecules of elementary gases. Avogadro himself has been vindicated of this charge by recent writers,[29] but Dumas continues to draw criticism for obscurity. However, much of this purported obscurity is removed simply by close reading of the various memoirs. As we have remarked, such words as atom and molecule had no fixed or standardized denotations, much less the particular ones we use today. It would consequently be ahistorical for us to praise only those writers who happen to approach the modern usage. Even where an author ambiguously used the same word (such as "molécule") to denote both atoms and molecules, we must still ask whether contexts always provide clear distinctions; if they do, there is no basis for the charge of obscurity. In short, we must sharply distinguish between conceptual and definitional confusion. It is my judgment that all of the chemists treated in this chapter clearly distinguished *concep-*

tually between atom and molecule. Consequently, we must seek other reasons for the delay in acceptance of the Avogadro-Ampère theory.[30] Some of these, including the absence of convincing corroborative evidence and Avogadro's association with an obsolescent approach, have been mentioned. Another important reason, the dominance of the theory of electrochemical dualism, will be treated in the following chapters.

THE LAW OF DULONG AND PETIT

The Avogadro-Ampère gas theory could usefully be applied to the determination of molecular weights and formulas, but it had limited applicability due to the relatively small number of elements that happen to exist in the gaseous state. Consequently, the discovery of a method for the determination of approximate atomic weights of solids constituted another landmark in the early history of the chemical atomic theory. This also occurred in the French scientific community at the hands of two young Parisian chemists, P. L. Dulong and A. T. Petit. Petit, the younger man, had had a brilliant student career at the Ecole Polytechnique and began teaching there in 1815 when he was twenty-three years old.

Dulong, also a Polytechnicien, began his professional life as a physician, subsequently becoming assistant to Thenard; in 1810 he attracted Berthollet's notice and became a member of the exclusive Arcueil circle. In 1816 Berthollet virtually retired, and the editorship of the influential *Annales de chimie* reverted to Gay-Lussac and Arago; Dulong was made a member of the editorial board.[31] In that year Dulong revealed his commitment to chemical atomism by writing for the *Annales* a largely favorable review, with translated excerpts, of Prout's paper on atomic weights. He later told Berzelius that he was the only one of the editorial board to be favorably impressed by the article. In the same year, he read an atomistic memoir on the oxides of phosphorus to the Académie des Sciences. He calculated the atomic weight of phosphorus from the correct formula of the acid anhydride, P_2O_5. This memoir betrays the strong influence of Davy and Berzelius; he used Wollaston's and Gay-Lussac's convention of $O = 10$.[32] Dulong's commitment to atomism is also revealed in a letter to Berzelius, where he refers to the atomic theory as "the most important idea of the century."[33]

The famous collaboration between Dulong and Petit began in 1815 and ended only with Petit's tragically early death in 1820.[34] Their association appears to have been formed in response to a prize competition announced by the Institut de France (from 1816 renamed the Académie des Sciences), and it culminated in two excellent joint memoirs in the field of thermometry, coefficients of expansion, and specific heats. The discovery of the law of atomic heat capacities seems to have been an artifact of this purely physical research program. Many years later Dumas related how the law was indeed discovered fortuitously and was published only as the result of a "calculated indiscretion" of Arago, to whom Petit had confided.[35]

Dulong read the joint paper to the Académie on 12 April 1819. Their discovery amounted to the fact that the specific heats of thirteen elemental solids were apparently precisely inversely proportional to their atomic weights. Since specific heats are taken at equal weights, and atomic weight is clearly inversely as the number of atoms in such a standard mass, what this law seemed to imply was that the specific heat of all atoms is precisely the same. Conversely, if the equality of atomic heat capacities is assumed, then the sort of relation discovered by Dulong and Petit ought to result. The authors explicitly stated this atomistic conclusion in their paper.[36]

Dulong and Petit used the highly reliable experimental results of Berzelius as the basis for their atomic weights. However, they were forced to use simple submultiples of most of Berzelius' weights to achieve their inverse proportionality. The values for sulfur and platinum could remain unchanged, but eight other weights had to be halved; they took two-third of Berzelius' figure for bismuth, one-third that for cobalt, and one-fourth that for silver. We have noted more than once that there was never much sense of concern when early chemical atomists disagreed over particular multiples or submultiples of atomic weights. Halving most of Berzelius' metallic atomic weights merely implied that most metal oxides were assumed by Dulong and Petit to be of the form MO, rather than Berzelius' preferred standard MO_2 oxide formula. In fact, this modification accorded better with the assumptions of most other chemical atomists of the day. But Dulong and Petit did not make this point explicit, possibly in an attempt at tact.

Coincidentally, Berzelius was in Paris at the time this discovery was made. He and Dulong had carried out new determinations of several important atomic weights at Arcueil between late January and early March 1819. Dulong had long been impressed by Berzelius' researches, possibly even converting to atomism under the Swede's influence. Dulong's influence on Berzelius is documented by the fact that when compiling a revised set of atomic weights in 1818, Berzelius altered the multiple of only one, phosphorus, and that as the result of Dulong's 1816 research.[37]

However, the discovery of the Dulong-Petit law seems to date from the beginning of April, after the period of close collaboration between Dulong and Berzelius, and the latter played no role at all in that discovery. In fact, two years later he published a mild criticism of the conclusions of the two Frenchmen. He thought that the generality and precision of the law was unproven, at least for the present, and was unwilling to alter his atomic weight multiples on this evidence alone. However, he admitted the importance of Dulong and Petit's research, and expressed the hope that Dulong might continue to pursue it.[38]

The law of Dulong and Petit was in fact soon found to be only an approximate and not an exact law of nature.[39] Nonetheless, however useless an approximate relation of this sort is for the *precise* determination of atomic weights, it can indicate which of the various multiples or submultiples is the correct one; it can, in other words, provide evidence as to the true formulas of simple compounds of the elements, and the intrinsically much more precise measurement of combining proportions can then be used to calculate exact weights. The law of Dulong and Petit was in fact used in just this way as early as 1826, and by none other than Berzelius, who by then had taken a more favorable stance toward the new law.

GAY-LUSSAC

Just about the time that Dalton's atomic theory became known to the chemical world at large, Gay-Lussac, one of the rising stars of French chemistry, announced the discovery of the law of gaseous combining volumes, which many interpreted in the light of chemical atomism. However, he maintained an equivocal stance toward this interpretation, and indeed toward most of the critical problems in atomic theory from 1809 until the end of his life. This was at least partially due to the prevailing positivistic, mathematical, and physicalist bias prevalent in France at this time, and to Berthollet's distaste for the atomic theory.

But atomistic ideas inexorably penetrated the French scientific establishment, as they had in Britain. Translations or translated excerpts of many atomistic memoirs appeared regularly in French chemical journals after 1809: Davy's Bakerian lectures, Wollaston's scale of "equivalents," Ampère's theory, and, above all, Berzelius' numerous influential papers.[40]

Gay-Lussac began implicitly to adopt certain aspects of chemical atomism. Proceeding logically from his law of combining volumes, he chose to express himself nearly exclusively in terms of "volume formulas," stating that water, for example, was composed of one volume of hydrogen united to half a volume of oxygen. This could be written symbolically in modern (atomistic) notation as $HO_{1/2}$.

In the 1809 paper and in subsequent work, Gay-Lussac frequently discovered or otherwise encountered examples of the law of multiple proportions in the vapor phase. All oxide series, especially those with gaseous reactants such as nitrogen, could be so expressed; Wollaston's 1808 work with the carbonates, where carbonic acid gas was the multiple, is another obvious example. Gay-Lussac justified his emphasis on the volumetric approach to the study of chemical composition on several bases: it was quite accurate (especially if the law of combining volumes was used and *assumed* to be exact); it seemed to reveal simpler mathematical relations than weight proportions did; and eudiometric measurements may have seemed metaphysically more natural than gravimetric measurements. Using the law of combining volumes, gravimetric composition could be *deduced* by simply measuring the vapor densities of the reactants. The percentage of hydrogen in hydrogen chloride gas, for example, could be calculated by dividing the density of hydrogen into the sum of the densities of the two gases.[41]

Gay-Lussac was even able to apply his volumetric approach to nonvolatile solids, such as carbon. For instance, in his 1809 work he determined that one volume of oxygen on combining with carbon expands to two volumes of carbonic oxide. Now all other equations of formation of compounds resulted in a *contraction* of the sum of the volumes of reactants. Gay-Lussac maintained this regularity by assuming that if a proportional amount of carbon could be vaporized, it would occupy *two* volumes.[42] Accordingly, French chemists began to write the formulas of carbonic oxide and carbonic acid as C_2O and CO, which put the atomic weight of carbon at 6 on the scale of $H = 1$, $O = 16$.

From 1814 Gay-Lussac pursued chemical atomism much more obviously, though still adhering to "volume" formulas. In three major articles published between 1814 and 1816, he verbally specified volume formulas for many common substances, such as ammonia (NH_3), nitric acid (N_2O_5), sulfuric acid ($SO_{3/2}$), hydrochloric acid (HCl), carbonic oxide (C_2O), carbonic acid (CO), cyanogen (C_2N), prussic acid ($H_{1/2}CN_{1/2}$), sugar ($CHO_{1/2}$ or $C_3H_3O_{3/2}$), alcohol ($C_2H_3O_{1/2}$), and ether ($C_4H_5O_{1/2}$).[43] It should be noted that the use of volumes rather than atoms permits the presence of fractions in these formulas. From a modern perspective they are all correct, using the approximate relative weights $H = 1$, $C = 6$, $O = 16$, $N = 14$, $Cl = 35$, and $S = 16$.

During this period Berzelius also spoke not of atomic weights but of "volume" weights and formulas. He stated that the "volume theory" and the "corpuscular theory" are identical, but he preferred the former denotation because of its empirical derivation.[44] In other words, Berzelius concluded from Gay-Lussac's law that equal volumes of elementary gases under similar conditions contain equal numbers of particles (EVEN), so that relative volume weights—vapor densities—indicate relative atomic weights. When Gay-Lussac subsequently began to use volume weights and formulas it was natural for the readers of his papers to infer that the discoverer of the law of combining volumes was reasoning in a similar fashion and could therefore be counted among the adherents of the chemical atomic theory. Other French chemists, notably Dumas, interpreted Gay-Lussac's volume formulas in an unambiguously atomistic fashion and employed the system of atomic weights implicit within them.[45] We have previously mentioned Dumas' 1826 assertion that everyone at that time accepted the EVEN assumption.[46]

However, throughout the 1810s Gay-Lussac studiously avoided any theoretical discussions, or even suggestive terminology, on the topic of chemical or physical atomism. In this he was probably influenced by Berthollet's disfavor regarding the atomic theory. In at least one other instance—recognition of the elementary nature of chlorine—Berthollet managed to dissuade his protégé from advocating views that appeared to him to be too hypothetical.[47]

After Berthollet's death Gay-Lussac became bolder in his advocacy of atomistic concepts and terminology. In March 1823 he published a paper on ferrocyanic acid, in which his (correct) gravimetric formulation employs the term "atome" for the first time in his career.[48] A year later he presented a paper on silver fulminate written jointly with an unknown young German chemist, Justus Liebig, again using the word "atome" in describing the formula of the substance.[49] When Friedrich Wöhler's analysis of silver cyanate, indicating an identical elementary composition, was published in the *Annales de chimie*, Gay-Lussac commented that "to explain their difference, one must assume a different mode of combination between their elements."[50]

Early in 1827 Gay-Lussac and Dulong reported favorably to the Académie des Sciences on Dumas' memoir on the atomic theory and prefaced their report with a general discussion of chemical atomism.[51] They drew a clear distinction between the physical atomic theory, which they regarded as speculative, and chemical atomism, which they thought had been adequately verified. They particularly recommended two methods for the problematical task of determining chemical formulas: the law of combining volumes (which they called a "confirmation évidente" of Dalton's hypothesis) when the constituent elements are gases, and the law of Dulong and Petit for solids.

Gay-Lussac's favorable attitude toward chemical atomism in the 1820s is also documented in his *Cours de chimie*, where he stated: "This theory has been confirmed since [Dalton's and Wollaston's work of 1808] by a very large number of facts."[52] In 1832 Gay-Lussac cosigned (with A. C. Becquerel) another referee report on an atomistic paper, Gaudin's "Recherches sur le groupement des atomes."[53] Again, the referees were impressed and pleased with the author's efforts. They stopped short of endorsing his views, however, urging that he be encouraged to seek "new facts" that could transform "what can only be considered probable conjectures" into well-established truth. Clearly, Gaudin's extensive speculations on actual arrangements of physical atoms within molecules hindered acceptance of his ideas.

Gay-Lussac's apparent change of heart regarding the atomic theory around 1823 may have represented nothing more than the revelation for the first time of convictions held much earlier. In 1814 he published a masterly investigation of the newly discovered substance iodine, in which he calculated weights for six elements.[54] He referred to his chemical atomic weights as "nombres proportionnels," "proportions," or "rapports," a terminology that seems to have been modeled on Davy's. He even went to so far in this paper as to offer the same atomistic explanation for the phenomenon of isomerism that he repeated ten years later in connection with Liebig's and Wöhler's analyses of silver fulminate and silver cyanate.[55] I suggest that the restraining influence of the doyen of Arcueil may have inhibited Gay-Lussac from further development of this atomistic proclivity, an influence that was removed with Berthollet's death.

By that time also the atomic theory was gaining rapidly in respectability. The law of Dulong and Petit, and isolated examples of such newly discovered phenomena as isomorphism, polymorphism, isomerism, polymerism, and allotropy all seemed to call for atomistic explanations. Dulong had been an ardent atomist since 1816. Gay-Lussac's close collaborator Thenard had presumably initially been as skeptical of Dalton's theory as most French chemists, but by 1813 he was beginning to sound like a chemical atomist.[56] In his textbook he incorporated chemical atomism as early as 1816; in the second edition he spoke of this theory as acquiring "every day a greater probability by the numerous applications of which it is susceptible."[57] Personal contact with Dulong and Thenard, as well as with other atomists such as Petit, Ampère, Berzelius, M. E. Chevreul, and Dumas may have encouraged Gay-Lussac to commit himself publicly to the atomic theory.[58]

An anomaly remains to be discussed. Although in 1827 he stressed the importance of the law of combining volumes as an empirical confirmation of the atomic theory, the formulas Gay-Lussac actually used in the 1820s were not in accord with his volume determinations of 1808–1816; rather, they were based on the gravimetric "equivalents" of Wollaston (H = 1, C = 6, O = 8, N = 14). Hence he regarded the atomistic formula for carbonic acid as CO_2 rather than his earlier volumetric CO; silver oxide as AgO; anhydrous cyanic acid as C_2NO; acetic acid as $C_4H_3O_3$.[59] This would imply HO or H_2O_2 as the formula for water, though to my knowledge he never discussed this topic in print. Paradoxically, throughout this period Gay-Lussac's earlier volume weights (H = 1, C = 6, O = 16, N = 14) were gaining in popularity until by the early 1830s most French chemists had adopted them.[60]

I suggest that Gay-Lussac's inconsistency may be only apparent and can be resolved if he is regarded as an adherent of the views of Avogadro and Ampère on the nature of gases. Let us suppose Gay-Lussac regarded the submolecularity of such elements as hydrogen, carbon, nitrogen, and chlorine as two and the submolecularity of oxygen as four. Using gravimetric atomic weights of H = 1, C = 6, O = 8, N = 14, and Cl = 35 (Wollaston's "equivalents"), Gay-Lussac could then conceive chemical reactions that would be consistent both with the law of combining volumes and with the Avogadro-Ampère hypotheses. For instance, using modern notation and Wollaston's atomic weights:[61]

$$H_2 + Cl_2 = 2HCl$$
$$2H_2 + O_4 = 2H_2O_2$$
$$2C_2 + 2H_2 = C_4H_4$$
$$2C_2 + O_4 = 2C_2O_2$$

In short, I am suggesting that Gay-Lussac may have provisionally regarded his volume weights as *molecular* weights in the modern sense and used Wollaston's figures for the actual *atomic* weights. He would not have wanted to make such a convention explicit, at least not in print, due to the highly speculative nature of his assumptions.

A statement by Dumas lends support to this conjecture. After asserting in 1826 that he and all of his contemporaries accept the EVEN assumption, he went on to say that an immediate and necessary consequence of this view—the divisibility of elementary molecules—had, as far as he knew, been applied by only one chemist, Gay-Lussac.[62] Dumas was then a struggling young chemist relatively new to the Parisian scientific scene; surely the last thing he would want to do was antagonize the premier chemist of France by making an unsupported or incorrect statement of this type. Dumas may well have known more about Gay-Lussac's theoretical views than is available in published form to twentieth-century historians. In the referee report on Dumas' paper, Gay-Lussac and Dulong made no disclaimer regarding Dumas' assertions; their tone is highly approving. They also offered no criticism of Dumas' use of the Gay-Lussac volume weights as true atomic weights.

DUMAS

Strongly influenced by the work of Berzelius, Dumas had been an atomist since his student days in Geneva; he used Gay-Lussac's volume weights consistently after his arrival in Paris in 1823.[63] His 1826 statement on the Avogadro-Ampère theory of gases indicates that he regarded both chemical and physical atomism as not only respectable, but as no longer even very controversial. Again in 1828, in the first volume of his *Traité de chimie appliquée aux arts*, Dumas explicitly professed his adherence to the Avogadro-Ampère hypotheses. Here for the first time he introduced—though he did not always use them consistently—the terms "molécule chimique" and "molécule physique," corresponding approximately to the modern distinction between atoms and molecules.[64]

According to an historical interpretation first given by

Hermann Kopp and followed by most historians to the present day,[65] Dumas' determinations of anomalous vapor densities of such elements as phosphorus, arsenic, and sulfur led by 1832 to disintegration of his belief in the EVEN hypothesis. There is no indication of this in the papers cited by these historians;[66] all Dumas says is that "we have been wrong to suppose that the vapors of nonvolatile compounds must resemble the well known permanent gases with respect to their mode of division."[67] I construe this to mean only that Dumas no longer believed that the submolecularity of molecules is the same for all elements.

Indeed, four years later Dumas asserted just as firmly as ever the tenets of the Avogadro-Ampère gas theory.[68] Again he suggested that "chemical molecules" are subunits of "physical molecules"; expressed in a memorable phrase, this thought becomes: "la chimie coupait les atomes que la physique ne pouvait pas couper."[69] He stated: "The gases, even elementary gases, do not contain in equal volumes the same number of atoms, at least [not] the same number of *chemical atoms*."[70] This, Dumas explained, applies particularly to phosphorus and arsenic, which have twice the number of "atomes chimiques" per "particule gazeuse" as nitrogen, and to sulfur, which has three times as many atoms per molecule as oxygen. It is clear that far from denying EVEN, he has shown here that the Avogadro-Ampère hypotheses can be brought into complete agreement with all the empirical data available.[71]

Another supposed indication of Dumas' rejection of EVEN and of his growing skepticism about atomism in the early 1830s is his increasing use of four-volume formulas. Thus the formula for "hydrogène bi-carboné" (ethylene), which in the 1820s Dumas wrote C_2H_2, was by 1832 changed to C_8H_8, where "each atom [i.e., molecule] is equivalent to four volumes."[72] But this statement involved no denial of the Avogadro-Ampère gas theory. What it represented was an attempt to reconcile volume formulas with the gravimetric equivalent formulas that were then proving so useful in England and Germany.

Dumas thought that taking formulas at four volumes would be most simple and consistent. For instance, assuming that basic lead oxide was the binary PbO, elemental analysis of lead acetate resulted in calculation of a formula for acetic acid of $C_8H_6O_3$. This is actually twice the modern formula, minus water, a consequence of assuming that a single molecule of acetic acid combines with PbO. It also involves a halved weight for carbon—doubling the number of carbon atoms in the formula—which was the result of Gay-Lussac's 1809 speculation on the nature of carbon vapor. Alcohol and ethylene, both directly related to acetic acid by oxidation and hydration, received the analogous formulas $C_8H_{12}O_2$ and C_8H_8.[73] By taking formulas at four volumes Dumas was in effect using an atomic weight scale of H = 0.5, C = 3, O = 8, and Pb = 207, different from the modern one due to the different theoretical assumptions. But there was no contradiction with the ideas of Avogadro and Ampère.

Four-volume formulas simply amount to the use of halved modern atomic weights. In this sense modern formulas would be referred to as two-volume formulas, this a result of the fact that the simple elementary gases are mostly diatomic, so that one molecule of reactant creates two molecules (two volumes) of product. Dumas pleaded ignorance of the actual number of atoms per molecule at least three times in print between 1826 and 1836—merely implying that it was probably a small even integer.[74]

A discussion of this question also was the occasion for a statement that has been much misunderstood. When he averred that if he were in charge, he would erase the word "atom" from science, he prefaced this by the forthright affirmation: "It is my conviction that the chemists' equivalents . . . which we call *atoms*, are nothing but molecular groups."[75] His dissatisfaction with the word amounted to the firm belief that the physical "atoms" of many elements are, in accordance with the ideas of Avogadro and Ampère, chemically divisible.

We have argued here that not only Dumas, but also his older contemporary Gay-Lussac, adopted such a viewpoint. However, both men maintained a cautious attitude toward speculation. Both avoided theoretical discussions on this topic, except in rare, explicitly philosophical contexts. Neither was completely consistent.[76] These tendencies may have been at least partially due to the unresolved nature of two critical questions:

How many atoms are there in a molecule? and Is this number the same for every element?[77] One might be fully persuaded of the probability of the views expressed by Avogadro and Ampère, but without unequivocal answers to these questions the application of those views was fraught with uncertainty. Nor were the methods for atomic weight determinations derived from the laws of isomorphism and atomic heats totally free from contradictions, exceptions, and uncertainties;[78] indeed, Dumas' discomfort with the anomalies led him to make statements that have been understandably interpreted as at least anti-Avogadrian, if not antiatomistic.[79] For these reasons, and others, a clear consensus in favor of a single system of atomic weights did not emerge until the 1860s.

    1. A. Wurtz, *Histoire des doctrines chimiques* (Paris, 1869), p. 1.

    2. M. P. Crosland, *The Society of Arcueil: A View of French Science at the Time of Napoleon I* (London: Heinemann, 1967); M. P. Crosland, *Gay-Lussac: Scientist and Bourgeois* (Cambridge University Press, 1978). Also see Crosland's article "The First Reception of Dalton's Atomic Theory in France," in D. S. L. Cardwell, ed., *John Dalton and the Progress of Science* (Manchester: Manchester University Press, 1968), pp. 274–87.

    3. T. Thomson, *Système de chimie . . . traduit de l'anglais sur le dernier edition de 1807 par M. Jean Riffault, précédé d'une introduction de M. C. L. Berthollet* 1 (Paris, 1809):1–170.

    4. Ibid., pp. 21–27. Nonetheless, by 1815 Berthollet had arrived at a limited rapprochement with chemical atomism (M. P. Crosland, "First Reception," pp. 280, 283; Michelle Sadoun-Goupil, *Le chimiste Claude-Louis Berthollet (1748–1822): Sa vie, son oeuvre* [Paris: Vrin, 1977], pp. 228–33).

    5. It will be recalled that the mathematical physicist Lagrange, and Napoleon himself, were born Italians.

    6. A. Avogadro, "Essai d'une manière de déterminer les masses relatives des molécules élémentaires des corps . . .," *J. phys.* 73 (1811): 58–76; trans. in *Foundations of the Molecular Theory* (Edinburgh: Alembic Club Rpt. no. 4, 1923), pp. 28–51. All citations will be from this translation.

    7. Ibid., p. 29.

    8. Ibid., p. 30.

    9. Ibid., pp. 31–33. I have represented Avogadro's verbal description in modern (Berzelian) symbols.

    10. Ibid., pp. 31–32, 36. On this important point see B. W. Mundy, "Avogadro on the Degree of Submolecularity of Molecules," *Chymia* 12 (1967): 151–55; M. A. Morselli, "The Manuscript of Avogadro's 'Essai . . .'," *Ambix* 27 (1980): 147–72; N. Fisher, "Avogadro, the Chemists, and Historians of

Chemistry," *History of Science* 20 (1982): 77-102, 212-31 (84); J. K. Bonner, "Amedeo Avogadro: A Reassessment of His Research and its Place in Early Nineteenth Century Science," (Ph.D. diss. Johns Hopkins University, 1974), esp. pp. 212-14. These authors also discuss Avogadro's later papers, which are too often ignored.

11. Bonner, "Amedeo Avogadro," p. 57; J. H. Brooke, "Avogadro's Hypothesis and Its Fate," *History of Science* 19 (1981): 235-71.

12. Avogadro, "Essai," rpt. p. 51.

13. Brooke, "Avogadro's Hypothesis," pp. 243-44. For similar arguments see also Fisher, "Avogadro," p. 84, and Bonner, "Avogadro."

14. Brooke rightly stresses this point, relying in part on Robert Fox's classic article, "The Rise and Fall of Laplacian Physics," *Hist. Stud. Phys. Sci.* 4 (1974): 89-136. Fox demonstrates that the physicalist Laplacian style of physics was replaced around 1815-20 by a more dynamical, empirical, and atomistic tradition.

15. I am thinking especially of Larry Laudan, *Progress and Its Problems: Towards a Theory of Scientific Growth* (Berkeley: University of California Press, 1977), p. 72 and passim.

16. A. M. Ampère, "Lettre . . . sur la détermination des proportions dans lesquelles les corps se combinent . . .," *Ann. chim.* 90 (1814): 43-86.

17. Ibid., pp. 43-47, 47n.

18. R. J. Haüy, *Essai d'une théorie sur la structure des cristaux* (Paris, 1784). See S. H. Mauskopf, "The Atomic Structural Theories of Ampère and Gaudin: Molecular Speculation and Avogadro's Hypothesis," *Isis* 60 (1969): 61-74; *Crystals and Compounds: Molecular Structure and Composition in Nineteenth Century French Science*, n. s., 66:3 (Philadelphia: *Trans. Amer. Phil. Soc.*, 1976), pp. 11-14, 35-37.

19. Ampère, "Lettre," pp. 45, 48-49, 55.

20. Ampère, "Note . . . sur la chaleur et sur la lumière considerée comme resultant de mouvemens vibratoires," *Ann. chim. phys.* [2] 58 (1835): 432-44 (434-36).

21. J. B. Dumas, "Sur quelques points de la théorie atomistique," *Ann. chim. phys.* [2] 33 (1826): 337-91 (337-38).

22. C. Graebe, "Der Entwicklungsgang der Avogadroschen Theorie," *J. prakt. Chem.* 87 (1913): 145-208 (159-61).

23. M. A. Gaudin, *L'architecture du monde des atomes* (Paris, 1873), pp. xii-xiii. See discussions in Graebe, "Avogadroschen Theorie," pp. 165-72; Mauskopf, "Molecular Speculation," pp. 65-72; T. M. Cole, Jr., "Early Atomic Speculations of Marc Antoine Gaudin: Avogadro's Hypothesis and the Periodic System," *Isis* 66 (1975): 334-60 (341-58).

24. Gaudin, "Recherches sur la structure intime des corps inorganiques définis . . .," *Ann. chim. phys.* [2] 53 (1833): 113-33. For details on this paper, and for citations pertaining to the following discussion, see Mauskopf, "Molecular Speculations," and Cole, "Early Speculations."

25. Cole, "Early Atomic Speculations," pp. 356-57.

26. Mauskopf, "Molecular Speculation," p. 66; Graebe, "Avogadroschen Theorie," p. 172.

27. A. E. Baudrimont, *Introduction à l'étude de la chimie par la théorie atomique* (Paris, 1833); *Traité de chimie générale et expérimentale*, 2 vols. (Paris, 1844–46), 1: 115. See Mauskopf, "Molecular Speculation," pp. 72–73, and J. R. Partington, *A History of Chemistry*, 4 vols. (London: Macmillan & Co., 1961–70), 4 (1964):393–95.

28. F. Choron, *Théorie des atomes et des équivalents chimiques*, 2d ed. (Paris, 1839), p. 3; M. Donovan, *Treatise on Chemistry* (London, 1832), pp. 379–81, 399–400; W. Prout, *Chemistry, Meteorology and the Function of Digestion considered with reference to Natural Theology* (London, 1834), pp. 144–46; A. Laurent, *Méthode de chimie* (Paris, 1854), pp. 82, 85. Fisher ("Avogadro," p. 85) also names Thomson, Persoz, and T. S. Hunt.

29. Brooke, "Avogadro's Hypothesis," p. 256; Cole, "Early Atomic Speculations," p. 338; Mauskopf, "Molecular Speculation," pp. 73–74; and Partington, *History*, 4: 213–14, 218.

30. For excellent analyses of this subject, see Brooke, "Avogadro's Hypothesis," and Fisher, "Avogadro."

31. Crosland, *Gay-Lussac*, pp. 166–68.

32. [W. Prout, ed. P. L. Dulong,] *Ann. chim. phys.* [2] 1 (1816):411–16; Dulong, "Mémoire sur les combinaisons du phosphore avec l'oxigène," *Mém. Soc. Arcueil* 3 (1817): 405–52; J. Berzelius, ed. H. Söderbaum, *Jac. Berzelius Bref* (Stockholm and Uppsala, 1912–32), 2, iv, 36 (Dulong to Berzelius, 8 January 1822).

33. *Berzelius Bref*, 2, iv, 12 (Dulong to Berzelius, 15 January 1820).

34. On this collaboration, see Robert Fox, "The Background to the Discovery of Dulong and Petit's Law," *Brit. J. Hist. Sci.* 4 (1968): 1–22.

35. Dumas, "Eloge historique de Henri-Victor Regnault," *Mém. Acad. Sci.* 42 (1883): xlviii. Dumas most likely had the story from Arago: Fox, "Background," pp. 2–3.

36. A. T. Petit and P. L. Dulong, "Recherches sur quelques points importants de la théorie de la chaleur," *Ann. chim. phys.* [2] 10 (1819): 395–413.

37. On Dulong and Berzelius, see Fox, "Background," pp. 16–18, and Crosland, *Arcueil*, pp. 411–16.

38. *Berzelius Bref*, 1, iii, p. 193 (Berzelius to A. Marcet, 27 April 1819); Berzelius, *Jahresbericht* for 1820, 1 (1822), pp. 18–19.

39. Ida Freund, *The Study of Chemical Composition: An Account of its Method and Historical Development* (Cambridge: Cambridge University Press, 1904; rpt. New York: Dover, 1968), pp. 361–84.

40. *Ann. chim.* [*phys.*] 75 (1810): 155; 77 (1811): 63; 78 (1811): 5, 105, 217; 79 (1811): 113, 233; 80 (1811): 5, 171, 225; 81 (1812):5, 278; 82 (1812): 5, 113, 225; 83 (1812): 5, 117; 85 (1813): 326; 89 (1814): 67; 90 (1814): 43, 138; 92 (1814): 89, 141; 94 (1815): 5, 170, 296.

41. Crosland, *Gay-Lussac*, pp. 92–114, 129–31, 136–40.

42. Gay-Lussac, "Mémoire sur la combinaison des substances gazeuses, les unes avec les autres," *Mém. Soc. Arcueil* 2 (1809): 207–34 (8–9); Crosland, *Gay-Lussac*, p. 102.

43. I have transcribed Gay-Lussac's verbal volume formulas into modern symbolic notation for the sake of clarity and economy. The first six of these

are found in his "Mémoire sur l'iode," *Ann. chim.* 91 (1814): 5-160 (16-17, 25, 30, 48, 76, 102-5, 132-36); for cyanogen and prussic acid see his "Recherches sur l'acide prussique," *Ann. chim.* 95 (1815): 136-231 (149-51, 162, 182); for alcohol, ether, and sugar see his "Lettre . . . à M. Clément, sur l'analyse de l'alcool et de l'éther sulfurique, et sur les produits de la fermentation," *Ann. chim.* 95 (1815): 311-18. Also compare his "Pesanteurs spécifiques des fluides elastiques," *Ann. chim. phys.* [2] 1 (1816): 218-21.

44. Berzelius, "Essay on the Cause of Chemical Proportions," *Ann. Phil.* 2 (1813): 443-54 and later installments, on p. 450; *Lärbok i kemien*, 3 (Stockholm, 1818): 42-44; *Essai sur la théorie des proportions chimiques* (Paris, 1819), pp. 48-49.

45. J. B. Dumas and P. J. Pelletier, "Recherches sur la composition élémentaire et sur quelques propriétés caracteristiques des bases salifiables organiques," *Ann. chim. phys.* [2] 24 (1823): 163-91. This work on alkaloids was Dumas' first publication in the field of organic chemistry. Though incorrect, the formulas imply the use of Gay-Lussac's volume weights, to which Dumas thereafter remained faithful. Cole, in "Early Atomic Speculations," has drawn attention to an earlier atomistic memoir by the pharmacist A. LeRoyer "and his student, J.-A. Dumas," "Essai sur le volume de l'atome des corps," *J. phys.* 92 (1821): 401-11, where Berzelius' atomic weights are employed.

46. Dumas, "Théorie atomistique," pp. 337-38.

47. Crosland, "First Reception," pp. 278-83; *Arcueil*, pp. 309-10. For the relation between Gay-Lussac and Berthollet, see also Crosland, "The Origins of Gay-Lussac's Law of Combining Volumes of Gases," *Ann. Sci.* 17 (1961): 1-26, and S. Pierson, "Gay-Lussac and Berthollet's Theory," *Actes XII*ᵉ *Congrès Int. d'Hist. Sci.* 6 (Paris: Blanchard, 1971): 83-86.

48. Gay-Lussac, "Sur l'acide des prussiates triples," *Ann. chim. phys.* [2] 22 (1823): 320-23. This was his first paper published after Berthollet's death (on 6 November 1822).

49. J. Liebig and Gay-Lussac, "Analyse du fulminate d'argent," *Ann. chim. phys.* [2] 25 (1824): 285-311.

50. F. Wöhler, "Recherches analytiques sur l'acide cyanique," *Ann. chim. phys.* [2] 27 (1824): 196-99; editorial footnote by Gay-Lussac, pp. 199-200.

51. Gay-Lussac and Dulong, "Rapport . . . sur un mémoire de M. Dumas," *Ann. chim. phys.* [2] 34 (1827): 326-31.

52. Gay-Lussac, *Cours de chimie* (Paris, 1828), Leçon 1, p. 12.

53. *Procès-verbaux des séances de l'Académie* 10 (Hendaye, 1922): 142-45. This paper by Gaudin was never printed.

54. Gay-Lussac, "Mémoire sur l'iode." The elements are hydrogen, oxygen, chlorine, sulfur, nitrogen, and iodine. His weights are similar to those of Wollaston, except for the value of nitrogen, which was one-third Wollaston's (corresponding to a binary formula for ammonia).

55. He suggested that the identity in composition of acetic acid and cellulose ("matière ligneuse"), and of sugar, starch, and gum arabic, were "new proof[s] that the arrangement of the molecules in a compound" is of the greatest importance in determining its properties (Ibid., p. 149n).

56. L. J. Thenard, "Observations sur le phosphore," *Ann. chim.* 85 (1813): 326-29.

57. Thenard, *Traité de chimie élémentaire, théorique et pratique*, 1st ed., 4 vols. (Paris, 1813-16), 4:247-54; 2d ed., 4 vols. (Paris, 1817-18), 1:21-24; 4:214-27.

58. Berzelius was in Paris from August 1818 to July 1819. On Chevreul's atomism see Partington, *History*, 4:256-57. For Dumas see below. Fox has a perceptive discussion of this topic: "Background," pp. 16-18; and in his "Laplacian Physics" he relates the rise of French chemical atomism to the fall of the Laplacian program.

59. Gay-Lussac, *Cours de chimie*, Leçon 3, p. 2; Leçon 23, p. 16; Liebig and Gay-Lussac, "Analyse," p. 296; Gay-Lussac, fn. to Wöhler, "Recherches," p. 200. Again, the formulas were discussed by Gay-Lussac only verbally, as was usually at this time. I have transcribed them into modern symbolic notation for the sake of clarity.

60. In 1833 Dumas commented that "the ablest chemists of France" all considered Gay-Lussac's value for the atomic weight of carbon (6) to be "more probable" than Berzelius' (12); "Sur la composition de l'acide pyrocitrique," *Ann. chim. phys.* [2] 52 (1833): 295-303 (299n).

61. These schematic equations are intended to indicate composition, not synthetic reactions. Gay-Lussac discussed the composition of ethylene in his "Lettre à M. Clément," pp. 313 and 317.

62. ". . . il ne parait avoir encore été admis dans la pratique par aucun chimiste, si ce n'est par M. Gay-Lussac," "Théorie atomistique," p. 338. Fox is, I believe, the only non-French historian to have correctly understood this phrase (*The Caloric Theory of Gases* [Oxford University Press, 1971], p. 284). Berzelius applied the volume theory only to permanent elementary gases, since he did not accept the divisibility of elementary molecules.

63. See citations above, n. 45; also see L. J. Klosterman, "Studies in the Life and Works of J. B. A. Dumas (1800-1884): The Period up to 1850" (University of Kent at Canterbury, 1976), p. 133.

64. Dumas, *Traité de chimie appliquée aux arts* 1 (Paris, 1828): xxxiii-xl. The celebrated passage on p. xxxviii regarding the formation of hydrogen chloride from its elements is, contrary to previous opinions, a confirmation of Dumas' advocacy of EVEN, if it is noted that Dumas' "atome" is often our concept of molecule.

65. H. Kopp. *Geschichte der Chemie*, 4 vols. (Brunswick, 1843-47), 2 (1844):397-98.

66. Dumas, "Sur la densité de la vapeur du phosphore," *Ann. chim. phys.* [2] 49 (1832): 210-14; "Dissertation sur la densité de la vapeur de quelques corps simples," *Ann. chim. phys.* [2] 50 (1832): 170-78.

67. Dumas, "Vapeur du phosphore," p. 214.

68. Dumas, *Leçons sur la philosophie chimique* (Paris, 1837), pp. 262-70. This lecture was delivered on 28 May 1836.

69. Ibid., p. 265.

70. Ibid., p. 268.

71. Gaudin, in "Recherches sur la structure," had delineated very similar

ideas, and Dumas may have been influenced by them; but Dumas' 1836 lecture is consistent with his earlier writings, and it is also possible that the influence went in the other direction.

72. Dumas, *Traité* 1 (1828):483; "Recherches sur les combinaisons de l'hydrogène et du carbone," *Ann. chim. phys.* [2] 50 (1832): 182–97 (184): *Traité* 5 (1835):59, 452.

73. *Traité* 5 (1835):30–72, esp. 54–59; 452, 466, 494. I thank T. M. Cole, whose correspondence helped to clarify this issue.

74. Dumas, "Théorie atomistique," p. 338; *Traité* 1 (1828):lx; *Leçons*, pp. 268–70. Some historians ascribe to Dumas the idea of diatomic gases. I know of no statement by Dumas to that effect.

75. Dumas, *Leçons*, p. 290.

76. Although Dumas asserted that "each atom of ether represents four volumes" according to the formula C_8H_8, H_2O (*Traité*, 5:494), it is clear that this is in fact a two-volume formula; this error enabled Dumas to advocate a theory of etherification derived from Gay-Lussac, against the theories of Liebig and Berzelius.

77. Gay-Lussac and Becquerel regarded Gaudin's attempted resolution of these two questions to be the "interesting part" of his 1832 paper (*Procès-verbaux*, p. 144). Gaudin used the Dulong-Petit law as well as vapor densities in his determinations; see Cole, "Early Atomic Speculations."

78. A. Ladenburg, trans. L. Dobbin, *Lectures on the History of the Development of Chemistry since the Time of Lavoisier* (rev. ed., Edinburgh: Alembic Club, 1906), pp. 105–7.

79. Fisher, "Avogadro," pp. 88 and 91.

5

Romantic Interlude: Atomism Encounters German Idealism

ROMANTICISM AND NATURPHILOSOPHIE

When chemical atomism was first emerging in Great Britain, the Germanic lands were in flux. A patchwork quilt of sovereign duchies and grand duchies, bishoprics and archbishoprics, kingdoms and free cities, all loosely federated under the aegis of the obsolescent Holy Roman Empire, "Germany" was only an abstract — or at most an ethnographic — unity. Ironically, it was the Napoleonic wars that helped simultaneously to destroy the last vestiges of the Empire and to create the beginnings of a pan-German sentiment.

The desire for national unification was also stimulated by the growth of a new intellectual and artistic movement, romanticism. Influenced by such writers as J. W. von Goethe, J. G. von Herder, J. G. Fichte, and G. W. F. Hegel, German romantics exhibited such characteristics as rejection of classical rationalism, exaltation of the emotions, philosophical idealism, political conservatism, and an intense nationalism.

Idealism and antirationalism might seem inimical to the successful pursuit of science, and this impression would be reinforced by noting that the health of chemistry in the homeland of the early romantic movement has generally been viewed as moribund. German chemistry in the first quarter of the nineteenth century was certainly less prestigious than the chemistry of England or France. Hufbauer counts a total of only thirty "active chemists" (i.e., those who published papers and had "at least a modicum of contemporary renown") in all the Germanic lands at the turn of the century, scattered among a similar number of small universities.[1] Most of these chemists had been trained and continued to teach in ancillary disciplines such as medicine or pharmacy. Furthermore, the Napoleonic incursions severely hampered scientific investigation and communication until about 1814. Only eighteen German universities survived. Justus Liebig, who was ten years old at the time of the liberation, later stated flatly, "At the end of the wars there were no longer any scientists in Germany."[2]

But German chemistry at this time was by no means as bleak as Liebig portrayed it. At the beginning of the second decade of the century, most of Germany was still occupied or controlled by the French, yet many chemists were publishing good papers and treatises. For example, Friedrich Stromeyer was professor of chemistry at the justly famed University of Göttingen. At the new University of Berlin, Heinrich Klaproth and Sigismund Hermbstaedt were active in pharmacy, mineralogy, and technical chemistry; also in Prussia J. B Trommsdorff and C. F. Bucholz were doing significant work at the University of Erfurt. In Bavaria, J. S. C. Schweigger was teaching at the Nuremberg Realschule, J. N. Fuchs was professor at the University of Landshut, and A. F. Gehlen was active in the Munich Academy of Sciences. Young J. W. Döbereiner had recently replaced J. F. A. Göttling at the University of Jena. K. W. G. Kastner was then teaching at the University of Heidelberg; later he would serve at Halle, Bonn, and Erlangen. L. W. Gilbert was transferring the site of his activities from Halle to Leipzig. In sum, German chemistry at this time was practiced by an identifiable community of mature scholars, which, although admittedly less prominent than those of England or France, was by no means negligible in importance.

German philosophy, in contrast to the sciences, reached something of a zenith in the late eighteenth and early nineteenth centuries, best represented by the work of Immanuel Kant of Königsberg. Kant's treatment of the sciences was one of the sources of that set of ideas known as *Naturphilosophie*, first given detailed expression by Friedrich Schelling. Strongly influenced by the romantic movement, Schelling and other *Naturphilosophen* invoked an antimaterialist ontology of active forces and polarities, and an organismic and unified view of nature. Elements of this philosophy were derived from Kantian idealist epistemology, which in turn was indebted to Humean skepticism.

Naturphilosophie had considerable resonance with the Newtonian physicalists' emphasis on forces and dynamical mechanisms. It is significant, and not coincidental, that this physicalist or dynamical scientific metaphysics experienced a Europe-wide resurgence during the period when *Naturphilosophie* was emerging. Phlogiston, a materialist explanation for combustion, was essentially abandoned by the turn of the century. Shortly thereafter such scientists as Humphry Davy, Thomas Young, Benjamin Thompson, and Augustin Fresnel urged dynamical explanations to replace the materialist theories of the caloric fluid of heat and the corpuscular view of light. The invention of the Voltaic pile seemed to some to provide an experimental paradigm for the essential polarity in nature, and its application to chemistry (in electrolysis) and to magnetism (in H. C. Oersted's and Michael Faraday's discoveries) seemed dramatically to illustrate both the ubiquity and the unity of natural forces. The surprising heuristic utility of romantic and *naturphilosophische* concepts in the early nineteenth century have been fully recognized and explored only in recent years.[3]

Naturphilosophie intersected with chemistry in interesting and complex ways. There would seem to be a fundamental irreconcilability between late eighteenth-century materialist atomistic ideas and the emerging idealist dynamical trend. This conflict was indeed stressed by J. J. Winterl—one of the few really unambiguous *Naturphilosoph*-chemists—who attempted to construct a purely dynamical chemistry. However, his purported discoveries were decisively refuted by Gilbert, R. Chenevix, and four of the most prominent Parisian chemists.[4]

One of the intents of this chapter will be to show that when chemical atomism traveled across the North Sea and the Rhine it encountered a German scientific community that was conditioned by *naturphilosophische* ideas, but one that was also willing to strike a compromise between atomism and dynamism, between classical and romantic *Weltanschauungen*.[5] Thoroughgoing dynamical chemistry was discredited after the Winterl episode, and the atomistic explanations were too ready to hand once the stoichiometric relations were established. To be sure, physical atomism was generally shunned and denigrated as materialistic, but chemical atomism was immediately and enthusiastically received by the majority of active German chemists. Most advocated some compromise position between atomism and dynamism — one might say between materialism and idealism. One such model already existed in the dynamical point-atomism of Rudjer Bošković.[6]

LUDWIG WILHELM GILBERT

Berzelius served the same role of advocate and elaborator of chemical atomism for the Germans as he did for the British and French. He corresponded with the two leading German chemical editors, Gilbert and Schweigger, and it was their journals that gave German chemists their first view of the new ideas then emerging in Britain, Sweden, and France. Gilbert printed the first translation into a major language of Berzelius' classic stoichiometric paper in the March and April 1811 numbers of his *Annalen der Physik*; the German translation of a Swedish summary appeared in Schweigger's *Journal für Chemie und Physik* in July of the same year.[7]

Gilbert and Berzelius regarded chemical atomism from similar standpoints. As early as 1804, Gilbert indicated his sympathy with atomistic ideas, broadly defined. After suggesting that neither pure (Cartesian) corpuscularianism nor pure (Kantian) dynamism is fruitful alone, he added: "Most physicists imagine all bodies to consist of smallest parts (moleculae) which have certain forces, and leave undetermined whether the differences between substances are a result of the compositions of these smallest parts or of a specific difference in their forces."[8] Thus Gilbert may be seen as following the Newtonian tradition of dynamical atomism, modified by a certain positivistic caution. That he scorned unsupported speculation and sought to follow an inductive-experimental methodology is clear from his own and other chemists' papers that he published around this time. For instance, during an extended visit to Germany, the Irish chemist Richard Chenevix wrote a vitriolic letter to Gilbert protesting against *Naturphilosophie* in general and the ideas of Winterl in particular. Gilbert printed the letter along with four short critical papers by Chenevix, with the editorial comment that it was time for "the voice of earnest truth" to be heard.[9] Soon thereafter Gilbert himself wrote a long critique in the same vein. Chenevix, Gilbert thought, was "a true critical scientist"; physicists, he added, "should question nature scientifically," that is, experimentally.[10]

Gilbert published the first detailed historical and critical review of the recently discovered stoichiometric laws in September 1811.[11] J. B. Richter was depicted in this essay as the true founder of the new mathematical branch of chemistry, and Berzelius as the leader in the important work of verification and extension of Richter's ideas. But Gilbert by no means ignored the implications of those laws for the theoretical side of the science; he regarded the laws of stoichiometry as leading directly to a "purified form of corpuscular philosophy."[12] Given Gilbert's sympathy with corpuscular ideas and his empirical bias, it is understandable that he found himself favorably disposed toward Berzelius' 1810-11 paper and perceived the implicit support of atomism within it. It is somewhat surprising, however, that Gilbert also mentioned Dalton's much more speculative theories in the same tone of respect that he used in describing the experimental work of the Swedish chemist. He wrote that he intended to discuss at some later date "the corpuscular hypothesis proposed and developed by the ingenious and bold English scientist Dalton, which seems to explain [simple and definite proportions] in the simplest way."[13]

This promise was never kept. Like Berzelius, Gilbert could not have been very successful in learning the details of Dalton's theory. Napoleon's blockade of Britain, known as the Continental System, was still in effect, severely hindering scientific as well as economic intercourse between Britain and Germany. The only German-language descriptions of Dalton's theory published to that time were a few brief and unilluminating references in translated French and English papers.[14] In the same month in which he read his inaugural dissertations at Leipzig, Gilbert was forced to add an editorial footnote to a passage in his journal where Davy gave "proportions" (atomic weights) for four elements, admitting that "this will be incomprehensible to most of my readers." Unfortunately, Gilbert's note was just as cryptic as Davy's "proportions" were obscure.[15]

Even a year and a half later (April 1813), Schweigger complained, likewise in a footnote to a passage in one of Davy's translated articles, that "Dalton's theory, due to the current conditions, can yet be little known in Germany."[16] Later that summer Schweigger attempted, using Berzelius as intermediary, to exchange back issues of Thomas Thomson's new *Annals of Philosophy* for his own *Journal für Chemie und Physik* — unsuccessfully, since all channels of communication between London and Nuremberg were still closed.[17] Thomson's celebrated 1807 description of Dalton's theory was available in French translation by 1809; it appeared in German translation in 1811, but was not yet seen by Gilbert in late September of that year.[18] This secondhand account served as the principal source of information on the new theory in Germany until the same translator — Friedrich Benjamin Wolff, a Berlin *Gymnasium* teacher — completed a German edition of Dalton's *New System of Chemical Philosophy* in 1812-13.[19] Even then, it was not until early in 1814 that Schweigger was able to see the second volume of Wolff's translation — the earliest detailed firsthand exposition of Dalton's atomic theory published in Germany.[20]

JOHANN S. C. SCHWEIGGER

But Schweigger's opinion of Daltonian atomism was already formed by that time. In May 1812 he printed a paper he had read five months earlier to the Erlangen Physikalisch-Medicinische Gesellschaft as the "introduction to a more detailed investigation on crystal electricity."[21] Schweigger had seen initial reports of Berzelian stoichiometry and Daltonian atomism and was familiar with Davy's work; this paper shows the influence of all three chemists. According to Schweigger, many hitherto mysterious chemical phenomena could be explained by assuming that the ultimate particles (*Körpertheile*) possessed crystalline forms and electrical poles. He elaborated and defended this theory to the end of his life; although similar to other, more highly regarded electrochemical theories (particularly that of Berzelius), it remained without appreciable influence.[22]

Schweigger wanted to dissociate himself from Daltonian

atomism. The presumption of electrical poles in his *Körpertheile*, he argued, prevented them from being atoms in Dalton's sense; Dalton's hypothesis was "very artificial," whereas his own "leads to a simpler view." A year later Schweigger compared Dalton's views on chemical atomism with Davy's; he did not find much difference between them.[23]

Finally, after examining the German translation of the important second volume of Dalton's *New System*, Schweigger wrote an extended review. By this time (March 1814) Schweigger had discovered Richter and, like Gilbert, he invoked Richter's name as the true founder of the science of chemical "metrology" (*Messkunst*, a term that had been used by Richter as synonymous with *Stöchiometrie*). According to Schweigger, Dalton's accomplishment lay in elaborating the fundamental ideas first introduced by Richter. But in Schweigger's eyes this was no small merit; he referred to Dalton and Berzelius as "the most ingenious chemists of our time," and metrology (by which he meant chemical atomism) he considered to be "the most important doctrine in modern chemistry," and indeed "the greatest advance that both theoretical and practical chemistry have ever made together."[24]

Schweigger was attempting to buttress his own atomistic theory of crystal electricity while showing the disadvantages of Dalton's ideas. He had nothing against the concept of "atom" but thought that the word was ill-chosen because of the physical connotation of indivisibility; he preferred such expressions as Davy's "proportion" or "proportional number" (*Verhältnisszahl*).[25] Schweigger himself proposed the term *Körperdifferential* and used it habitually thereafter, to emphasize that metrology need not make physical statements concerning the mechanical structure of ultimate particles, but rather was designed as a mathematical scheme for the calculation and explanation of combining weight relations.[26]

Nonetheless, here as well as in his 1812 paper, Schweigger was advocating a form of physical as well as chemical atomism. He took issue with Dalton's implicit suggestion of *spherical* atoms (his own assumption being that they possessed crystalline forms) and pointed out that his definition of *Körperdifferential*—"infinitely small parts of bodies"—could be considered either absolute, as in the dynamical viewpoint, or relative, as in the atomistic viewpoint. "In any case," he wrote, clearly inclining to the latter, "even Leibniz, the inventor of differential calculus, interpreted the expression 'infinitely small quantities' merely in a relative sense."[27]

Thus Schweigger properly distinguished between chemical and physical atomic theories and showed that advocacy of the former need not imply anything to do with the latter. But he did not perceive that there was also an important distinction between the purely stoichiometric work of such men as Richter, J. L. Proust, and Berzelius (before 1813), and the chemical atomic theory of Dalton and his followers. His repeatedly expressed view was that Dalton altered Richter's ideas "only semantically."[28]

The publication in German of Wollaston's influential paper on chemical "equivalents" did nothing but reinforce this error;[29] the work was celebrated on the Continent as well as in Britain as a purging of all theoretical and hypothetical assertions from Dalton's atomism. Schweigger regarded it as a complete return to and vindication of Richter's original views. Wollaston's new term "equivalent" was, in Schweigger's opinion, an "excellent choice," though he regarded it as completely synonymous with Richter's *Massentheile*, Dalton's "atom," and his own *Körperdifferential*.[30] Thereafter Schweigger became much more critical of Dalton's physical atomism, asserting that "these atomic fictions, which one can easily and arbitrarily twist and turn and alter however one likes," are entirely dispensable; Wollaston's paper was above all notable for its "avoidance of all atomistic sophistry." However, he never relinquished his own physical-atomistic theory of crystal electricity.[31]

Schweigger played an important role in spreading the new ideas on chemical atomism to Germany; he was, in fact, the first enthusiastic German adherent of the doctrine.[32] Furthermore, as editor of one of the country's leading chemical journals, he was able to publish original contributions on stoichiometry and atomic theory by a variety of German chemists; examples include papers by F. C. Vogel, A. F. Gehlen, G. H. Schubert, J. N. Fuchs, J. W. Döbereiner, and J. L. G. Meinecke.[33] The first four of these chemists were countrymen of

Schweigger (Bavarians) and did relatively little work in the new field; the important contributions of Döbereiner and Meinecke will be discussed below.

KARL W. G. KASTNER

Karl Kastner (1783–1857) also dealt with chemical atomism at an early stage. After receiving the Ph.D. degree in chemistry under Göttling at the University of Jena—one of the first such degrees ever awarded—Kastner taught successively at the universities in Heidelberg (1805), Halle (1812), Bonn (1818), and Erlangen (from 1821). He was a prolific author and a popular teacher, and fame came early; between the death of Klaproth in 1817 and the rise of his student Justus Liebig around 1830 he was widely regarded as the most preeminent chemist in Germany. During his student days he was an ardent disciple of Winterl, who strongly influenced Kastner's first book, *Materialien zur Erweiterung der Naturkunde* (Jena, 1805). The association of Kastner with *Naturphilosophie* stuck, even though he later repudiated Winterl. This partially explains why Kastner is now almost completely forgotten, even by German historians—his name is usually mentioned only in connection with his famous protégé.[34] Liebig himself may have had much to do with his former teacher's eclipse, since he became very critical of most of the older German chemists, and especially of Kastner, after his more thorough chemical education in Paris.[35]

Kastner's *Einleitung in die neuere Chemie* (Halle, 1814) was a popular work intended for students and intelligent laymen. As the title implies, it was quite up to date, reflecting the literature available in Germany through the end of 1812. A lengthy section (pp. 396–466) dealt with the laws of chemical combination discovered by Richter and extended and perfected by Berzelius. Kaster summarized Dalton's ideas, then appended a short commentary (pp. 501–9). He did not attempt to equate the work of Dalton and Richter, as Schweigger did. He pointed out that strict atomistic mechanisms in the Cartesian sense cannot account for chemical phenomena—even Dalton ascribed attractive and repulsive forces to his atoms. Nor are the radical dynamists fully correct: "The two scientists will meet each other in their efforts, if they proceed on the path of observation and experiment." In the same section, Kastner described in favorable terms the dynamical atomism of Bošković. In contrast, the ideas of Kant and Schelling were characterized (p. 265) as interesting but inconclusive; scientists, Kastner argued, should leave such investigations to metaphysicians. In summary, Kastner may perhaps be regarded as a romantic, but by no means a *Naturphilosoph*. Like many of his contemporaries (Gilbert, Schweigger, Davy) he was critical of the speculations of the *Naturphilosophen* and seemed to be more attracted to dynamical atomism than to either Daltonian physico-chemical atomism or pure Kantian dynamism.

This philosophical position was also characteristic of G. F. Hildebrandt, who taught at Erlangen from 1793 until his death in 1816. Influenced greatly by Kant's and Schelling's dynamical views, Hildebrandt nonetheless clearly recognized the heuristic utility—and even the necessity—of atomistic notions for the explanation of chemical phenomena.[36]

JOHANN WOLFGANG DÖBEREINER

One of the most prominent representatives of the romantic style of chemistry in Germany during the 1810s was J. W. Döbereiner (1780–1849).[37] Döbereiner had little systematic training in chemistry. Between 1794 and 1802 he served as an apprentice and journeyman pharmacist in four German cities; he then made three attempts at various sorts of chemical manufacturing, all of them unsuccessful. Duke Carl August of Saxony-Weimar and his minister, J. W. von Goethe, rescued the young chemist from poverty and obscurity by appointing him successor to Göttling at the University of Jena. After receiving *pro forma* the necessary doctoral degree in 1810, Döbereiner assumed his duties as professor of chemistry, pharmacy, and technology. The call to Jena was the "greatest event" in Döbereiner's life, and he repaid the debt by selfless devotion to his work, his university, and his noble patrons.[38]

Döbereiner had become acquainted with Schweigger during one of his abortive business ventures, a brewery-distillery near Bayreuth, and they became good friends.[39] It may well have been Schweigger's enthusiastic advocacy of chemical atomism that led Döbereiner to investigate the same subject. In any case,

Döbereiner's first essays on "stoichiometry" date from the summer of 1815. Following in the steps of earlier chemists, Döbereiner calculated atomic weights for several important elements: H = 1, O = 7.5, S = 15, N = 4.5, C = 5.7, and Fe = 25.[40] Wollaston's influence was especially strong; Döbereiner used his term *Aequivalent* (as well as Davy's *Verhältnisszahl*), and his numbers were calculated on the basis of similar implicitly assumed formulas.[41]

By 1816 Döbereiner was an enthusiastic chemical atomist. In the course of that year, he published three works on stoichiometry: *Grundriss der allgemeinen Chemie: Anfangsgründe der Chemie und Stöchiometrie*, a general introduction for the use of his students; *Darstellung der Verhältnisszahlen der irdischen Elemente zu chemischen Verbindungen*, a collection of atomic weight tables intended as an appendix to his *Grundriss*; and *Neueste stöchiometrische Untersuchungen: Beiträge zur chemischen Proportionslehre*, an eighty-six-page update of the preceding works.[42] "Stoichiometry" (i.e., chemical atomism) became his favorite field. He taught the subject in his classes, instructed Goethe in it, experimented and speculated actively on it. He investigated numerous organic substances and expressed the results atomistically.[43] His organic analyses were not as precise as Berzelius' work two years earlier, but he was able to correct one of Berzelius' formulas, that for oxalic acid.

After puzzling over the various weights for the known elements, Döbereiner perceived several interesting regularities. He noted that the "proportion" for strontia (50) is precisely the arithmetical mean of the values for the chemically analogous earths lime (27.5) and baryta (72.5). Döbereiner disclosed several more examples of such "triads" twelve years later.[44] He claimed to have shown that mineral waters were always composed of stoichiometric proportions of the various constituent salts.[45] He pointed out that common elements tend to have small numbers and uncommon elements large ones, and that many chemically analogous elements have nearly the same proportional values (such as iron, chromium, cobalt, nickel, and manganese). Finally, he attempted to correlate similar "chemical constitutions" of certain compounds with similar stoichiometric numbers. For instance, according to his formulation of acetic acid, $(C_3H_3O) + (CO_2)$, there was an apparent similarity in both constitution and stoichiometric number to limestone, $(CaO) + (CO_2)$.[46] These observations and speculation find significance as the first attempts to perceive relationships between atomic weights and properties of the elements, an endeavor that culminated in the discovery of the periodic law in the 1860s.

In assessing the effect of romantic ideas on Döbereiner's chemistry, it is important to recall that Jena was one of the wellsprings of the romantic movement in Germany. Hegel, Schelling, Fichte, Goethe, Ritter, and the Schlegel brothers all were active in Jena around 1800, and together they created or propagated many of the most characteristic ideas of German romantic literature, philosophy, and science. These men had all departed by the time Döbereiner arrived in 1810, but their influence remained strong. Thus it is not surprising to see romantic and sometimes even *naturphilosophische* themes in Döbereiner's scientific writings. Early in his career, Döbereiner was strongly attracted by the work of such *Naturphilosoph* scientists as Winterl, Oersted, and Ritter. He was intensely chauvinistic and given to the romantic cult of friendship.[47] He sought to reveal a fundamental underlying unity in nature and speculated much more than would have been respectable for a classical scientist. Like Schweigger, his predilection was toward a Pythagorean synthesis, the mathematization of nature according to dynamical principles. For instance, Döbereiner viewed chemical reactions as caused by the mutual interpenetration of two bodies, which, however, retained their physical individualities.[48]

But Döbereiner's views were also essentially atomistic. Although he did, in the romantic style, stress the importance of forces and polarities in chemistry, he was also well aware of the theoretical implications of "stoichiometry" or chemical atomism. Against his nationalistic inclinations, he felt compelled after 1814 to accept the French-English-Swedish theoretical developments and their implications. After proposing a theory of electrolysis involving a material role for light, heat, and "the two electricities," Döbereiner added: "One may call this view raw or atomistic, but I will not relinquish it, for it corresponds to the laws of chemical analysis."[49]

So Döbereiner's study of chemical atomism led to intimations of further belief in some sort of *physical* atomism, as with Gilbert, Schweigger, and Kastner. In this respect historians might well have been able to view him as a progressive forerunner of modern chemistry; furthermore, much of his experimental work was of great value. And yet, like Kastner, his associations with romantics and *Naturphilosophen* led to neglect of his work by non-German contemporaries and denigration of his contributions after his death.[50] Berzelius, a good contemporary model of the classical style of chemistry, regarded Döbereiner as one of the two worst chemists in Europe.[51]

JOHANN L. G. MEINECKE

Another chemist who enjoyed a modicum of respect in Germany during his lifetime but who later dropped into nearly complete obscurity was J. L. G. Meinecke (1781–1823). He studied at Erlangen, taught at a primary school in Halle, then held professorships at Kassel and Halle (from 1814), publishing books and papers on mineralogy, botany, anatomy, and chemistry. Beyond this, very little is known about his life.[52] From his places of residence we may guess that by 1814 he had come into personal contact with Hildebrandt, Schweigger, Gilbert, and Kastner; he knew Döbereiner by early 1816 at the latest.[53] This list of associates would suggest that Meinecke might share a common interest in chemical atomism. Such an interest was first revealed by the publication in 1815 of *Die chemische Messkunst, oder Anleitung, die chemischen Verbindungen nach Maass und Gewicht auf eine einfache Weise zu bestimmen und zu berechnen* . . . [54]

The purpose of this book was to describe a "stereometrische [volumetric] theorie" independent of the ideas of Dalton and Berzelius. Meinecke began by recalling the remarkable relations governing the combining volumes of gases discovered by Gay-Lussac in 1809. Seeing no reason why similar relations should not also apply to liquids and solids, he proposed to examine all elementary substances considered in the solid state, relating their densities to their combining proportions. He established to his own satisfaction that oxygen in the solid state would be about as dense as ice, which he took as unity. He was then able to determine volume weights for all the known elements, primarily from density measurements. For instance, the density of the purest form of carbon, diamond, was known to be 3.176 times that of ice, so Meinecke took this figure as the weight of a unit volume of carbon. He then had to formulate carbonic acid as composed of one volume of carbon and eight volumes of oxygen. Translated into atomistic terms, Meinecke was suggesting an atomic weight for carbon about four times that determined by Berzelius. Few of his weights came very close to those accepted by most of his contemporaries—not surprising considering how they were derived.

But Meinecke protested strongly against any possible atomistic interpretation of his theory. He criticized "Dalton's doctrine of atoms" in much the same terms that Schweigger had. However, like his colleagues, he did not want to be considered as a pure "*Dynamiker*": "The author considers atomism and dynamism to be merely opposing views of the same thing. . . . There is a third view of nature uniting both of these, but which belongs more to the field of philosophy than that of experimental science."[55]

Despite Meinecke's protests, his procedure involved only a minor modification of that of most early nineteenth-century atomists. For instance, Meinecke followed Berzelius in using gaseous combining volumes, wherever possible, to determine formulas and hence certain elemental weights. Berzelius then used a set of a priori and semiempirical rules to calculate, from stoichiometric data, the density of all other elements imagined to be in the gaseous state; Meinecke merely used a different set of rules and pictured all elements in the solid state. In general, Meinecke's book was either denigrated, or, more frequently, simply ignored.[56]

According to his later testimony, Meinecke wrote the *Chemische Messkunst* and submitted it to the publisher in 1814.[57] By early 1816 Meinecke had shifted ground, now agreeing much more closely with the thinking of colleagues such as Schweigger and Döbereiner. For his stereometric theory derived from solid-state densities he now substituted the more conventional stoichiometric determinations from combining weights, allied with consideration of vapor densities wherever possible, and simplifying assumptions for solid substances. Meinecke's new views were expressed, beginning in June 1816, in articles in Trommsdorff's, Schweigger's, and Gilbert's journals,[58] and in 1817 he published a book-length exposition.[59] Like Schweigger, Meinecke credited the new science almost exclusively to Richter, and still disparaged Dalton's "modification."[60] His atomic weights, now referred to as "portions" (*Antheile*), were modeled on those of Davy, Thomson, Wollaston, Berzelius, Gay-Lussac, and Prout.[61]

Like his compatriot Döbereiner, Meinecke was eager to discern mathematical regularities among the *Antheile* and to discover laws correlating stoichiometric values with chemical and physical properties.[62] He appears to have come across examples of the triad relationship at about the same time as Döbereiner.[63] Most of the "laws" that Meinecke claimed to have discovered have proven illusory, merely a product of patient juggling and clever manipulation of data. A recent claim that Meinecke anticipated the law of Dulong and Petit is without merit.[64]

One of the difficulties encountered in studying Meinecke is his disconcerting inconsistencies. For instance, he never clearly decided between the atomic weight systems advocated by Berzelius ($H = 1$, $C = 12$, $O = 16$, $N = 14$) and Wollaston ($H = 1$, $C = 6$, $O = 8$, $N = 14$).[65] Attracted initially by the former, he seemed to decide by late 1816 in favor of the latter, yet returned at least once thereafter to Berzelian atomic weights.[66]

Meinecke is best known in the annals of science for having espoused the hypotheses published somewhat earlier by William Prout. In his first stoichiometric paper, Meinecke merely asserted that a "round number"—an integral multiple of the value for hydrogen—is preferable to a simple arithmetic average from several determinations.[67] In subsequent writings Meinecke gave several justifications for his exclusive use of integral atomic weights after 1815: (1) empirically, all *Antheile* were found to be either exactly or approximately integral multiples of the weight of hydrogen; (2) this is consonant with Dalton's atomic theory (though he still decried that theory as hypothetical); (3) the law of multiple proportions in compounds would then have a counterpart for elements; and (4) the "protyle" hypothesis, inferred from the integral multiples hypothesis, would enable a revitalization, in somewhat altered form, of the phlogiston theory.[68] He did not mention Prout, whose anonymous paper appeared in Thomson's *Annals of Philosophy* for November 1815 and February 1816 and whose identity was revealed by Thomson in May 1816. Oddly, Meinecke did mention Dalton as originator of the integral multiples hypothesis, a notion that Brock and others have shown is almost certainly wrong.

A circumstantial case can be made that Meinecke had in fact read Prout's paper before writing his own. Certainly something persuaded him in the winter of 1815-16 that his previous work on the "stereometric" theory was ill-conceived. To be sure, it is quite possible that he could have come upon the hypotheses independently; in fact, Thomson and Davy had anticipated certain aspects of Prout's work.[69] However, examination of the 1816 volumes of the *Journal für Chemie und Physik*, of which Meinecke was then coeditor, leads to the impression that Meinecke must have been *au courant* with English chemistry in general and Thomson's *Annals* in specific — articles from this periodical were translated for the *Journal*, and by this time there was little delay in the mails. It would be surprising if Meinecke had not read an article that treated a subject so close to his heart.[70] Even if he had not seen the original English paper, he could have read a French translation of it that appeared in April 1816 (two months before any of Meinecke's new papers appeared) in the *Annales de chimie*, another journal that Meinecke was obviously reading in connection with his editorial duties.[71]

Still more incriminating is the manner in which Meinecke developed his case for the integral multiples hypothesis. The single most important atomic weight determination was that of oxygen. Most chemical atomists at that time relied on Gay-Lussac's measurements of gaseous combining proportions for water, and on J. B. Biot and F. Arago's determination of the vapor densities of the two gases,[72] which led to a value for oxygen of 15.07 (or 7.04 for those who regarded the formula for water as HO). The reader will recall Prout's more indirect approach, which enabled him to avoid the problematical density measurement of hydrogen, and to arrive at the value of O = 8 exactly. This new determination was crucial to Prout's case for the integral multiples hypothesis. In his *Erläuterungen*, published about a year after this paper,[73] Meinecke first asserted—unjustifiably—that Biot and Arago's results actually lead to a value of O = 8; he then reproduced Prout's new argument, without acknowledgment.[74]

These coincidences of content and method did not pass unnoticed by Meinecke's contemporaries. Thomson printed a translated excerpt of Meinecke's *Annalen der Physik* paper, adding: "I . . . suspect that Meinecke has been influenced by Dr. Prout's paper, though he has taken no notice of it." Bischof also drew attention to Prout's priority and to Meinecke's failure to cite his predecessor. Bischof doubted the validity of the multiples hypothesis, quite justifiably pointing out that for most atomic weights the experimental uncertainty was at least as large as the atomic weight of hydrogen.[75] Schweigger wrote a very complimentary review of Bischof's book, and quoted extensive passages relating to the Prout-Meinecke question.[76]

Meinecke replied to these attacks.[77] He claimed that his earliest paper on the subject (the "Anleitung"), although appearing only in late 1816, was submitted in the spring of that year, and composed during the previous winter, that is, at the same time that Prout was publishing his papers. He also suggested that there was never any need to cite Prout, since both he and Prout derived their ideas from the work of other well known chemists, especially Dalton and Gay-Lussac. The tone is unmistakably bitter. Meinecke never published another article on stoichiometry. Four years later, at the age of forty-two, he took his own life.[78]

RETROSPECT

A few summarizing comments are now in order. In Germany, just as in England and France, few perceived the real distinction between empirical stoichiometry and theoretical atomism, leading many to the inconsistent position of denigrating "Dalton's atomic theory" and simultaneously advocating what was generally known as the "doctrine of definite proportions"—chemical atomism. This error enabled Schweigger and others to credit

Richter with the origination of "stoichiometry," a term the Germans used to denote the chemical atomic theory as well as the laws of combining proportions. Dalton's work could then be viewed as a mere modification of Richter's, a deleterious one since it went beyond empirical fact into the realm of hypothesis. The physical atomism so prominently explicit in both volumes of Dalton's *New System* was repellent to the sensibilities of many Germans, highly respectful—if not blind followers—of Kant's philosophy. When William Higgins wrote one of several vitriolic priority claims against Dalton, Meinecke contributed a brief commentary. He asserted—incorrectly—that Richter had published his works on stoichiometry before *both* Higgins and Dalton, then added that no German will contest Higgins' style of atomism anyway; even though "we sometimes play with chemical atoms and particles, nevertheless science in Germany has taken too serious and sound a course to give any significance to such hyperhypothetical trivialities."[79] This historiographic tradition was perpetuated for years thereafter in Germany,[80] though by about 1830 more balanced assessments began to appear.[81]

A similar confusion led to an entirely spurious distinction between Wollaston's conventional equivalents and Berzelius' atomic weights, a distinction that has survived to the present day. The German "stoichiometers," like their counterparts in England and France, were on the whole initially more attracted to conventional equivalents because of their apparent empirical certification; Döbereiner is the most unambiguous example. But as such chemists as C. G. Gmelin, H. and G. Rose, G. Magnus, E. Mitscherlich, and F. Wöhler took pilgrimages to Stockholm and returned to spread their master's doctrines, the Berzelian atomic weights began to acquire the stamp of dogma in Germany. Liebig's conversion in 1830 made opinion nearly unanimous. The most significant exception was Leopold Gmelin at Heidelberg, who retained his *Mischungsgewichte* (patterned after Wollaston's conventional equivalents) from the first edition of his famous *Handbuch der theoretischen Chemie* (Frankfurt, 1817-1819). This work, however, exercised little influence on his colleagues until the 1840s, when a general return to the Wollaston-Gmelin system took place, not only in

Germany, but throughout Europe. These events will be described in detail in chapter 6.

The German chemists discussed here were generally young; Schweigger, Kastner, Döbereiner, and Meinecke were all around thirty years of age in 1811, when the earliest reports of the new theory of chemical composition were printed in Germany. They were among the progressives who accepted the developing French-British theoretical chemistry, with its oxygen theory of combustion, operational definition of the elements, and emphasis on precise quantitative methods. They provided important elaboration of chemical atomism, tilling soil that would ultimately prove immensely fertile. Several of their speculative notions later led to influential and fruitful concepts in chemistry.[82] Nevertheless, much of their work in this field was of poorer quality than that being done in other parts of Europe; specifically, it was too often desultory and speculative. Germany had neither a master systematizing theoretician such as Dalton or Berzelius nor a consummate experimentalist of the caliber of Berzelius or Gay-Lussac. Consequently, the lack of attention by contemporary foreign chemists was not entirely undeserved.

It seems clear that the romanticism prevalent in Germany at this time was conducive to the expression of speculative ideas. Speculation in science is a notoriously two-edged sword. On the one hand, the romantic *Weltanschauung* to some extent favorably conditioned the response to Dalton's theoretical ideas. Schweigger, Döbereiner, and Meinecke were receptive to such synthetic scientific hypotheses; they regarded as outmoded the more descriptive style of older German chemists represented in the analytical field by the eminent Klaproth at Berlin and in pharmaceutical chemistry by Trommsdorff at Erfurt. The latter chemists were primarily oriented toward practical and empirical investigations and had virtually nothing to say regarding the atomistic ideas filtering into Germany from the *Ausland*. But the speculations of the German stoichiometers often went too far afield, leading to ideas insufficiently supported by the available data. In such cases the criticisms of the classical scientists were reasonable.

The reputations of these men also suffered by their association, however tenuous, with *Naturphilosophie*. After the second decade of the nineteenth century there was a marked intensification of the distaste for Schelling's philosophy as applied to chemistry. During a visit to Tübingen in the summer of 1819, Berzelius remarked in his diary on the unkempt appearance of the university students:

> The reason for this intentionally barbaric and dissipated appearance is the philosophical school which is known in Sweden as phosphorism and in Germany as *Naturphilosophie*. Its basis is ignorance of everything real, love of poetry and the fine arts, and a trusting uncritical devotion to the views of such persons who through incomprehensibility have acquired a reputation for profundity, especially when their foolishness goes so far that the government find it necessary in the name of sound reason to remove the idiots to where they can do no more damage.[83]

As a student Liebig was initially attracted by the *Naturphilosophen*, but, he later wrote, he "awoke from this delusion" with "fear and horror," and thereafter regarded *Naturphilosophie* as insidious and destructive, the "Black Death of our century."[84]

But, typically, Liebig's depiction of the situation is overdrawn. Several of the German stoichiometers were indeed sympathetic with Schelling's philosophy, particularly in their early years. However, this did not, as is generally believed, preclude a commitment to some kind of atomism.[85] Schelling himself described his *Naturphilosophie* as containing both dynamical and atomistic elements.[86] Similarly, the German stoichiometers all advocated some sort of compromise between dynamism and atomism. Their espousal of the chemical atomic theory left little room for doubt regarding the particulate fine structure of matter, though they seemed always to suggest that the ultimate particles were endowed with dynamical qualities, and all except Schweigger thought that atomic and molecular magnitudes were as yet inaccessible to experimental science. Nonreductionist materialism had proven the most fruitful assumption for the founders of chemical atomism, but these German dynamical physicalists seemed to have little trouble in incorporating the new discipline into their own metaphysics.

At any rate, by 1810 orthodox Kantian dynamism was no longer being advocated by competent German chemists. In the second decade of the century, most were careful to avoid the stigma attached to the more radical followers of the new philosophy. It is probably a mistake to describe any of the German stoichiometers as *Naturphilosophen* at all; the term "romantic," though also slippery and ill defined, is more fitting. Though often given to speculation, these chemists paid at least lip service to the empirical ideals of such men as Berzelius and Gilbert. At the same time, they were eager to perceive an underlying unity in nature that could be expressed in mathematical form. This was probably the principal attraction of chemical atomism, for it seemed to provide a major step toward the dream, cherished since Newton's day, of a thoroughgoing mathematization of chemistry.

1. Karl Hufbauer, "Social Support for Chemistry in Germany during the Eighteenth Century: How and Why Did it Change?" *Hist. Stud. Phys. Sci.* 3 (1971): 205-31 (210-13); Karl Hufbauer, *The Formation of the German Chemical Community* (Berkeley: University of California Press, 1982). See also R. S. Turner, "University Reformers and Professorial Scholarship in Germany, 1760-1806," in L. Stone, ed., *The University in Society* (Princeton: Princeton University Press, 1974), 2:495-531; and C. E. McClelland, *State, Society and University in Germany, 1700-1914* (Cambridge: Cambridge University Press, 1980).

2. J. Liebig, "Der Zustand der Chemie in Preussen," *Ann. Chem. Pharm.* 34 (1840): 97-136 (100).

3. E.g., R. Stauffer, "Speculation and Experiment in the Background of Oersted's Discovery of Electromagnetism," *Isis* 48 (1957): 33-50; L. P. Williams, *Michael Faraday* (London: Chapman & Hall, 1965), passim; H. A. M. Snelders, "Romanticism and *Naturphilosophie* and the Inorganic Natural Sciences: An Introductory Survey," *Stud. Romant.* 9 (1970): 193-215; B. Gower, "Speculation in Physics: The History and Practice of *Naturphilosophie*," *Stud. Hist. Phil. Sci.* 3 (1973): 301-56; S. G. Brush, *The Temperature of History: Phases of Science and Culture in the Nineteenth Century* (New York: Franklin, 1978); and R. Löw, "The Progress of Organic Chemistry During the Period of German Romantic Naturphilosophie," *Ambix* 27 (1980): 1-10.

4. J. J. Winterl, *Prolusiones ad chemiam saeculi decimi noni* (Budapest, 1800); L. W. Gilbert, "Einige kritische Aufsätze über die in München wieder erneuerten Versuche mit Schwefelkies-Pendeln, Wünschelruthen, u.d.m.," *Ann. Phys.* 26 (1807): 370-449; R. Chenevix, "Kritische Bemerkungen, Gegenstände der Natur betreffend, geschrieben während seines Aufenthalts in Deutschland," *Ann. Phys.* 20 (1805): 417-96; A. F. Fourcroy, L. B. Guyton de Morveau, C. L. Berthollet, and L. M. Vauquelin, "Rapport sur une prétendue

découverte de M. Vinterl, professeur de chimie à Pest," *Ann. chim.* 71 (1809): 225-53. See Snelders, "The Influence of the Dualistic System of Jakob Joseph Winterl (1732-1809) on the German Romantic Era," *Isis* 61 (1970): 231-40.

5. K. Caneva ("From Galvanism to Electrodynamics: The Transformation of German Physics and Its Social Context," *Hist. Stud. Phys. Sci.* 9 [1978]: 63-159) and A. Hermann ("Dynamismus und Atomismus—Die beiden Systeme der Physik in der 1. Hälfte des 19. Jahrhunderts," *Erkenntnis* 10 [1976]: 311-22) have argued for sharp generational and philosophical divisions between "concretizing" and "abstracting" physicists and between atomists and dynamists in early nineteenth-century German physics. I do not see these categories as applying so unambiguously to German chemists in the same period.

6. R. Bošković, *Theoria philosophiae naturalis* (Vienna, 1758).

7. J. Berzelius, "Versuch, die bestimmten und einfachen Verhältnisse aufzufinden, nach welchen die Bestandtheile der unorganischen Natur mit einander verbunden sind," *Ann. Phys.* 37 (1811): 249-334, 415-72; "Ueber die bestimmten und einfachen Verhältnisse, nach welchen die bestandtheile der unorganischen Natur verbunden sind," *J. Chem. Phys.* 2 (1811): 297-326.

8. J. G. F. Schrader, *Grundriss zur Experimentalnaturlehre nach den neuesten Entdeckungen*, 2d ed. "verbessert, ergänzt und grossen Theils umgearbeitet" by Gilbert (Hamburg, 1804), pp. 14-15, cited in Hans Schimank, "Ludwig Wilhelm Gilbert und die Anfänge der 'Annalen der Physik'," *Sudhoffs Archiv* 47 (1963): 360-72.

9. Chenevix, "Kritische Bemerkungen," ed. n. pp. 418-19.

10. Gilbert, "Kritische Aufsätze," pp. 397, 411.

11. Gilbert, *Dissertatio historico-critica de mistionum chemicarum simplicibus et perpetuis rationibus earumque legibus nuper detectis* (inaugural diss. at Leipzig, 24-25 September 1811); trans. by Gilbert as "Historisch-kritische Untersuchung über die festen Mischungs-Verhältnisse in den chemischen Verbindungen, und über die Gesetze, welche man in ihnen in den neuesten Zeiten entdeckt hat," *Ann. Phys.* 39 (1811): 361-428.

12. Ibid., "Untersuchung," p. 362.

13. Ibid., *Dissertatio*, p. 39; "Untersuchung," pp. 427-28.

14. Besides Berzelius' passing mention in the paper cited in n. 7 above, there were references to Dalton's theory in Gay-Lussac's 1809 paper on gaseous combining proportions, trans. in *Ann. Phys.* 36 (1810): 6-36, and in two papers by Davy (*Phil. Trans. Roy.Soc.* 100 [1810]: 16-74 and 231-57) trans. in *Ann. Phys.* 37 (1811): 34-63, 155-207, and 39 (1811): 1-42.

15. Gilbert, *Ann. Phys.* 39 (1811): 24n. Gilbert referred the reader to his September 1811 dissertation, a work, however, that provided no information on the calculation of atomic weights.

16. Schweigger, fn. to Davy, "Ueber einige Verbindungen des Phosphors und Schwefels . . . ," *J. Chem. Phys.* 7 (1813): 494-517 (505n).

17. H. G. Söderbaum, ed., *Jac. Berzelius Bref*, 6 vols. (Stockholm/Uppsala: Almqvist & Wiksell, 1912-32), 3, i, 11-13.

18. T. Thomson, *System of Chemistry*, 3d ed. (Edinburgh, 1807), 3: 424-29; *Système de chimie*, trans. J. Riffault (Paris, 1809); *System der Chemie*, trans. F. B. Wolff, 5, *Zusätze und Erweiterungen der Wissenschaft seit*

1805 (Berlin, 1811). Gilbert cited none of these in his well-documented dissertation; in the same month (September 1811) Schweigger mentioned only the French edition (*J. Chem. Phys.*, 3:112-13n.).

19. J. Dalton, *Ein neues System des chemischen Theiles der Naturwissenschaft*, trans. F. B. Wolff (Berlin, 1812-13). In addition to the Thomson and Dalton editions, Wolff (1766-1845) also translated Davy's *Elements of Chemical Philosophy* (in 1814), as well as works by L. N. Vauquelin and J. A. Chaptal. He wrote his own *Lehrbuch der Chemie* (3 vols., Berlin, 1818-21), modeled closely on recent French and English treatises, and was coeditor, with Klaproth, of a popular *Chemisches Wörterbuch*, 5 vols. (Berlin, 1807-10); *Supplemente* in 4 vols. (1815-19).

20. Schweigger, "Ueber Daltons Messkunst der chemischen Elemente," *J. Chem. Phys.* 10 (1814): 355-81 (355-36). This is the earliest German reference to this volume I have found; distribution may have been delayed by the conditions in Germany at that time.

21. Schweigger, "Ueber einige noch unerklärte chemische Erscheinungen," *J. Chem. Phys.* 5 (1812): 49-74.

22. For Schweigger's theory of crystal electricity, see Snelders, "J. S. C. Schweigger: His Romanticism and His Crystal Electrical Theory of Matter," *Isis* 62 (1971): 328-38. For Berzelius' reaction to Schweigger's theory, see *J. Chem. Phys.* 9 (1813): 296-97n. (excerpt of Berzelius' 4 December 1812 letter to Schweigger). Berzelius' and Schweigger's theories were independent and approximately simultaneous.

23. Schweigger, "Erscheinungen," pp. 63n., 65-66, and the editorial fnn. to Davy's translated article, "Ueber einige Verbindungen des Phosphors," pp. 500-505n., 509-11n., 516n.

24. Schweigger, "Daltons Messkunst," pp. 378-79; *J. Chem. Phys.* 11 (1814): 423n.

25. Schweigger, "Daltons Messkunst," p. 359. Cf. Schweigger, "Ueber die festen chemischen Mischungsverhältnisse nebst stöchiometrische Tafeln," *J. Chem. Phys.* 14 (1815): 497-516 (498).

26. Schweigger, "Daltons Messkunst," pp. 361-62, 370. He wrote: "An hypothesis designed to facilitate mathematical calculations is quite distinct from a physical hypothesis" (p. 358). He regarded chemical atomism as analogous to calculus, which utilizes infinitesimals to calculate macroscopic quantities, whence the term *Körperdifferential* (p. 377).

27. Ibid., pp. 360-62, 371-72.

28. Ibid., pp. 355-56, 379-80; "Ueber stöchiometrische Tafeln," *J. Chem. Phys.* 11 (1814): 449-64 (453-54); "Ueber die Verfertigung und Benutzung der logarithmischen Rechenstäbe . . . ," *J. Chem. Phys.* 14 (1815): 115-29 (125-26); "Mischungsverhältnisse," pp. 498, 501; "Ueber stöchiometrische Scale überhaupt und insbesondere über Döbereiners und Meineckes neuere, auf diesen Gegenstand sich beziehende Schriften," *J. Chem. Phys.* 15 (1815): 495-500; ed. fnn., *J. Chem. Phys.* 14 (1815): 448n. and 462n. In 1818 Schweigger wrote: "When Richter arrived home from foreign soil in the atomistic clothing that Dalton provided him, like Ulysses he was no longer recognized in the fatherland . . . " (*J. Chem. Phys.* 24 [1818]: 356).

29. W. H. Wollaston, trans. *J. Chem. Phys.* 12 (1814): 85-105. By this time

the "Continental System" was rapidly eroding, and communications between Britain and Germany were beginning to normalize.

30. Schweigger, "Tafeln," p. 454n.; "Mischungsverhältnisse," pp. 498-501.

31. Schweigger, "Scale," p. 496; "Mischungsverhältnisse," pp. 498-500; ed. fnn., *J. Chem. Phys.* 14 (1815): 462n. and 472n.; "Ueber allgemeine Körperanziehung, mit Hinsicht auf die Theorie der Krystallelektricität," *J. Chem. Phys.* 39 (1823):231-50.

32. C. G. Bischof, *Lehrbuch der Stöchiometrie* (Erlangen, 1819), pp. 154-63; P. T. Meissner, *Chemische Aequivalenten- oder Atomenlehre* (Vienna, 1834), 1: 57-58, 106.

33. F. C. Vogel, "Beiträge zu der Lehre von den bestimmten chemischen Mischungs-Verhältnissen," *J. Chem. Phys.* 7 (1813): 1-42, 175-250; A. F. Gehlen, "Ueber das electro-chemische System und den Grund der bestimmten Verhältnissmengen," *J. Chem. Phys.* 12 (1814): 403-11; G. H. Schubert, "Einige Beiträge zu den stöchiometrischen Berechnungen des Mischungsverhältnisses der Fossilien," *J. Chem. Phys.* 15 (1815): 200-231; J. N. Fuchs, "Ueber den Gehlenit, ein neues Mineral aus Tyrol," *J. Chem. Phys.* 15 (1815): 377-86.

34. For (meager) biographical information, see J. C. Poggendorff, *Biographisch-literarisches Handwörterbuch* 1 (Leipzig, 1863), p. 1,231; *Allgemeine deutsche Biographie* 15 (Leipzig, 1882), p. 439; and R. Löw, *Pflanzenchemie zwischen Lavoisier und Liebig* (Munich: Donau-Verlag, 1977), pp. 297-98, 329-32, 337.

35. See, e.g., his unkind comments regarding Kastner transcribed by his biographer Jakob Volhard: *Justus von Liebig* (Leipzig: Barth, 1909), 1:19-20.

36. G. F. Hildebrandt, *Anfangsgründe der allgemeinen dynamischen Naturlehre*, 2 vols., (Erlangen, 1807). On Hildebrandt, see H. Kopp, *Geschichte der Chemie* (Brunswick, 1844), 2:324-27; Snelders' biography in *Dictionary of Scientific Biography* 6 (New York: Scribner's 1972); and J. R. Partington, *A History of Chemistry* (London: Macmillan, 1962), 3:638-39.

37. On Döbereiner, see Partington, *History* (1964), 4:178-80 and references on 178n.; Eduard Farber's biography in *DSB* 4 (1971); H. Döbling, *Die Chemie in Jena zur Goethezeit* (Jena: Fischer, 1928), most of which concerns Döbereiner and is not mentioned in the above sources; and Löw, *Pflanzenchemie*, passim.

38. F. Chemnitius, *Die Chemie in Jena von Rolfinck bis Knorr* (Jena: Fromann, 1929), pp. 28-31; W. Prandtl, *Deutsche Chemiker in der ersten Hälfte des neunzehnten Jahrhunderts* (Weinheim: Verlag Chemie, 1956), pp. 37-72.

39. Prandtl, *Deutsche Chemiker*, p. 39. Schweigger was then (1809) professor of mathematics and physics at the Bayreuth *Gymnasium*. Cf. Schweigger's letter to Döbereiner, printed as "Vermischte Bemerkungen," *J. Chem. Phys.* 17 (1816): 326-39, where Schweigger uses the familiar "Du."

40. J. W. Döbereiner, "Einige stöchiometrische Untersuchungen," *J. Chem. Phys.* 14 (1815): 206-23: also ed. fn., *J. Chem. Phys.* 16 (1816): 16n.

41. For the sake of comparison, Wollaston's weights, recalculated on the basis of H = 1, are as follows: O = 7.58, S = 15.2, N = 13.3, C = 5.71, and Fe = 26.1. Döbereiner even printed a logarithmic scale similar to Wollaston's

(*J. Chem. Phys.* 15 [1815]:500, reported by Schweigger), and dedicated his *Neueste stöchiometrische Untersuchungen* (Jena, 1816) to the English scientist. In 1816 he tripled his "proportion" for nitrogen, thus deciding on a set of weights virtually identical to Wollaston's ("Stöchiometrische Untersuchungen," *J. Chem. Phys.* 17 [1816]: 241-57).

42. All were published at Jena in 1816. They were printed in small editions and generally were worn out through heavy student use, so are now very rare; I have been unable to locate any copies in the United States. Prandtl, *Deutsche Chemiker*, pp. 51, 55-56; J. Schiff, ed., *Briefwechsel zwischen Goethe und Johann Wolfgang Döbereiner (1810-1830)* (Weimar: Böhlaus, 1914), pp. xiii, 28, 39, 41.

43. Döbereiner, "Ueber die thierische Kohle," *J. Chem. Phys.* 16 (1816):86-91; "Ueber die Pflanzenkohle und die metallische Grundlage derselben," *J. Chem. Phys.* 16 (1816): 92-104; "Das Daseyn einer Zusammensetzung aus Kohlensäure und Kohlenoxyd bewiesen," *J. Chem. Phys.* 16 (1816):105-10; "Ueber die Zusammensetzung des Zuckers und des Alkohols," *J. Chem. Phys.* 17 (1816): 188-89; "Ueber die Anwendung des Kupferoxyds zur Zerlegung organischer Substanzen und über die Zusammensetzung und Sättigungs-Capacität der Weinsäure," *J. Chem. Phys.* 17 (1816): 369-75; *Neueste stöchiometrische Untersuchungen* (1816), cited in Löw, *Pflanzenchemie*, pp. 300-302; "Vermischte chemische Bemerkungun," *J. Chem. Phys.* 23 (1818): 67-97 (67-80).

44. "Auszug eines Briefes von Hofrath [Ferdinand] Wurzer, Prof. der Chemie zu Marburg," *Ann. Phys.* 56 (1817): 331-34 (letter of 16 July 1817, describing Döbereiner's recent investigations); Döbereiner, "Versuch zu einer Gruppirung der elementaren Stoffe nach ihrer Analogie," *Ann. Phys.* 91 (1829): 301-7.

45. "Aus einem Schreiben des Herrn Prof. und Bergrath Döbereiner an den Prof. Gilbert," *Ann. Phys.* 57 (1817): 435-38; "Chemische Bemerkungen und Versuche," *Ann. Phys.* 59 (1818): 318-27; "Vermischte Bemerkungen," pp. 80-93.

46. Döbereiner, "Schreiben," pp. 436-38 (H = 1, C = 5.7, O = 7.5, Ca = 20). Döbereiner's formula for acetic acid is correct as the anhydride. He regarded the "base of the acid," C_3H_3O, as composed of an oxidized polymer of olefiant gas, CH (modern C_2H_4). The formulas reproduced here were described only verbally, not symbolically. I have transcribed them into modern chemical notation, retaining the atomic weights used.

47. See Döbereiner's letter to Goethe, 7 December 1812, printed in Döbling, *Chemie in Jena*, pp. 160-67. For Döbereiner as a chauvinist, see ibid., pp. 162-63, and "Bemerkungen und Versuche," pp. 321-22.

48. Döbereiner, *Grundriss der allgemeinen Chemie*, 3d ed. (Jena, 1826), p. 29.

49. Döbereiner, "Nachtrag zu den vermischten chemischen Bemerkungen," *J. Chem. Phys.* 23 (1818): 219-33 (225).

50. Löw, *Pflanzenchemie*, pp. 332-33, 362-66, 377-78.

51. Berzelius, *Bref*, 3, ii, 58; Thomson was the other (letter of 23 July 1821).

52. Poggendorff, 2:103; Bischof, *Lehrbuch*, pp. 173-84; Meissner, *Atomenlehre* 1:112-26; [L. Dobbin and J. Kendall, eds.,] *Prout's Hypothesis*,

Alembic Club Rpt. no. 20 (Edinburgh: Alembic Club, 1932), pp. 7-13; Partington, *History* 4:224-25; and Löw, *Pflanzenchemie*, passim.

53. In January of that year, he and Döbereiner became temporary coeditors of Schweigger's journal while the latter traveled.

54. Published at Halle in 1815, *Chemische Messkunst* was dedicated to Kastner, Schweigger, and Karl Mollweide.

55. Ibid., pp. 3-4, 5n.

56. Bischof, *Lehrbuch*, pp. 173-84.

57. Meinecke, "Ueber die Entwicklung der Salze aus den gediegenen Verbindungen," *J. Chem. Phys.* 25 (1819): 269-89 (270).

58. Meinecke, "Anleitung zur chemischen Messkunst," *J. Pharm.* 25:ii (1816), 56-214. Four continuations and a concluding summary of this article were published in the same journal: [2] 1:i (1817), 72-130; [2] 1:ii (1817), 3-98; [2] 2:ii (1818), 3-41; [2] 3:i (1819), 21-59; and [2] 3:ii (1819), 475-534. Unless otherwise noted, all subsequent references to the "Anleitung" refer to the initial installment. "Ueber die Producte der Weingährung," *J. Chem. Phys.* 17 (1816): 177-87. "Das specifische Gewicht der elastischen Flüssigkeiten nach stöchiometrischen Berechnungen," *Ann. Phys.* 54 (1816): 159-75.

59. Meinecke, *Erläuterung zur chemischen Messkunst* (Halle, 1817), issued as part 2 of *Chemische Messkunst* but textually quite distinct.

60. Meinecke, "Anleitung," pp. 61-64; *Erläuterungen*, pp. 2-5.

61. Meinecke cited only the first four chemists; his failure to mention Prout is discussed below.

62. E.g., *Erläuterungen*, pp. 37-43 and 187-204; "[Auszug aus einem Briefe] vom Herrn Prof. Meinecke in Halle," *J. Pharm.* 25:ii (1916), 241-43; "Anleitung," [2] 3: ii (1819), 505-11; "Ueber die Dichtigkeit der elastischflüssigen Körper im Verhältniss zu ihren stöchiometrischen Werthen," *J. Chem. Phys.* 22 (1818): 137-59 (146-47); "Ueber den stöchiometrischen Werth der Körper, als ein Element ihrer chemischen Anziehung," *J. Chem. Phys.* 27 (1819): 39-47.

63. Meinecke, *Erläuterungen*, p. 43.

64. Löw, *Pflanzenchemie*, p. 299, citing *Erläuterungen*, p. 195. Meinecke merely repeated, without citation, a statement given earlier in somewhat different form by Dalton, Gay-Lussac, and Avogadro, and applying solely to elementary permanent gases. Meinecke's atomic weights as given in a table of elementary solids (*Erläuterungen*, p. 198) reveal no inverse proportionality between "stoichiometric values" and specific heats.

65. The values listed here are meant merely to indicate approximate and relative weights; Wollaston, e.g., determined the "equivalent" of hydrogen as 1.32 on the scale of oxygen equals ten.

66. Meinecke used Berzelius' weights in the "Anleitung" and in "Weingährung"; Wollaston's weights in "Gewicht," *Erläuterungen*, "Dichtigkeit," and in "Ueber die Bestandtheile und die Dichtigkeit der vorzüglichsten Gase und Dunste," *J. Pharm.* [2] 2:i (1818), 325-32; he returned to Berzelius' weights in "Anleitung," [2] 2:ii (1818), 3-41, and returned back again to Wollaston's weights in "Anleitung," [2] 3:ii (1819), 475-534.

67. Meinecke, "Anleitung," pp. 71-72; discussed by Dobbin and Kendall, *Prout's Hypothesis*, pp. 8-9.

68. Meinecke, "Gewicht," pp. 161-63; *Erläuterungen*, pp. 19-23; "Anleitung," [2] 3:ii (1819), 477-78.

69. See O. T. Benfey, "Prout's Hypothesis," *J. Chem. Educ.* 29 (1952):78-81; and R. Siegfried, "The Chemical Basis for Prout's Hypothesis," *J. Chem. Educ.* 33 (1956): 263-66.

70. In his *Chemische Messkunst* (pp. 2-3), Meinecke lamented that there were no unifying laws for elemental weights similar to those revealed so strikingly in compounds. The integral multiples hypothesis supplied that desideratum.

71. [Prout], *Ann. chim.* [2] 1 (1816):411-16.

72. J. B. Biot and F. Arago, "Mémoire sur les affinités des corps pour la lumière . . . ," *Mém. Inst. France* 7 (1806): 301-87 (320).

73. In his "Gewicht," published in October 1816, Meinecke mentioned that his *Erläuterungen* was in press.

74. Meinecke, *Erläuterungen*, pp. 23-26; "Gewicht," pp. 164-75.

75. Thomson, *Ann. Phil.* 12 (1818): 9-11; Bischof, *Lehrbuch*, pp. 164-84, esp. pp. 164, 173, 183.

76. Schweigger, "Ueber die eben erschienene Schrift 'Lehrbuch der Stöchiometrie' . . . ," *J. Chem. Phys.* 24 (1818): 338-66 (342-55).

77. Meinecke, "Ueber den stöchiometrischen Werth," pp. 46-47; "Entwicklung der Salze," pp. 269-70.

78. His mental health had been weak since 1814. See obituary in F. A. Schmidt, ed., *Neuer Nekrolog der Deutschen* 1, ii (Ilmenau, 1823), pp. 860-63.

79. W. Higgins, *Phil. Mag.* 51 (1818): 161-72; Meinecke, "Beilage," *J. Chem. Phys.* 26 (1819): 296. Meinecke gave 1799 as the publication date for Higgins' *Comparative View of the Phlogistic and Antiphlogistic Theories*. In the English article, which Meinecke cited, it is printed correctly as 1789, i.e., earlier than any of Richter's stoichiometric writings.

80. E.g., Wolff's article "Verwandtschaft," in *Supplemente zu dem chemischen Wörterbuch* 4 (Berlin, 1819), pp. 118-202; Bischof, *Lehrbuch*, pp. 26, 77-80.

81. E.g., H. Buff, *Versuch eines Lehrbuchs der Stöchiometrie* (Nuremberg, 1829), pp. 15-30, and 2d ed. (*Lehrbuch* . . .), 1842, pp. 18-28; Meissner, *Atomenlehre*, 1:46-54. Kopp's treatment (*Geschichte*, 2:353-98) is very good.

82. E.g., Schweigger's electrochemical theory, the "triad" concept, Döbereiner's and Meinecke's hypotheses regarding the "constitutions" of organic compounds and Meinecke's suggestion of what came to be known as isomerism (*Erläuterungen*, pp. 142-45); these last ideas will be discussed in later chapters.

83. Berzelius, *Reiseerinnerungen aus Deutschland*, ed. G. Klingemann (Weinheim: Verlag Chemie, 1948), pp. 3-4.

84. Liebig, "Zustand," pp. 100, 118-21, 134n.

85. David Knight has written: "To [*Naturphilosophen*], any kind of atomic theory was anathema"; "German Science in the Romantic Peirod," in *The Emergence of Science in Western Europe*, ed. M. P. Crosland (New York: Science History, 1976), pp. 161–78 (165).

86. "*Naturphilosophie* is thus neither dynamical in the previous meaning of the word, nor atomistic, but rather [advocates] *dynamical atomism*," *Sämmtliche Werke*, pt. 2, vol. 3 (Stuttgart, 1858), pp. 22–23. For a discussion of Schelling's conception of dynamical atomism, see Gower, "Speculation in Physics," pp. 322–24.

6

The Consolidation of Berzelian Atomism

In the first two decades of the nineteenth century, many of the brightest minds in the younger generation of European chemists could be counted among the elaborators of chemical atomism. All, however, had to come to terms with an inevitable lacuna in the conceptual structure of the new doctrine. Stoichiometry, the empirical basis of chemical atomism, was a straightforward product of gravimetric analysis. But stoichiometry has nothing direct to say about the twin problems of formula assignment and selection of the proper multiples of chemical equivalents to represent correct atomic weights. Early nineteenth-century chemists were forced either to assume atomic weights in order to determine formulas, or, more usually, to assume formulas in order to calculate atomic weights from the stoichiometry of the relevant reactions.[1] Formulas were chosen on the basis of simplicity assumptions (Dalton), conventions (Wollaston), semi-empirical chemical rules (Berzelius), or a combination of these.

Such procedures were nothing more than a *pis aller* in the absence of physical methods of inferring molecular magnitudes. To be sure, one physical theory had had some application—the gas theory of Avogadro and Ampère—but apparent contradictions, exceptions, and inconsistencies hindered acceptance of either major tenet of this theory, the equal volumes-equal numbers and the submolecularity hypothesis. This situation changed abruptly in 1819 when two physical approaches to the formula problem emerged. The reader is familiar with the law of atomic heats; the second discovery will now be discussed.

MITSCHERLICH AND ISOMORPHISM

Eilhard Mitscherlich (1794–1863) initially was an orientalist but switched to chemistry when his career goals were frustrated. After a second Göttingen Ph.D. (with Stromeyer), he habilitated at Berlin and worked in H. F. Link's botanical laboratory. There, in the last week of the year 1818, he discovered to his surprise that certain phosphates and arsenates had precisely identical crystalline forms.[2]

The phenomenon seemed surprising because it contradicted a mineralogical principle routinely assumed at this time, especially by the Abbé R. J. Haüy: the unique correlation between crystal form and chemical (or mineralogical) identity. Ironically, it was only Mitscherlich's neophyte standing that explains his ignorance of the fact that exceptions to this principle had been discovered by a number of mineralogists during the past generation; examples of both isomorphism (different substances crystallizing in the same form) and polymorphism (the same substance crystallizing in more than one form) were becoming increasingly familiar.[3] These anomalies were assimilated to the assumption of the composition-form correlation by asserting that accidental immixture of a slight impurity in a crystal could dramatically influence its form or that minute differences in angles were sufficient to differentiate apparently isomorphic crystals.

Melhado has rightly stressed that Mitscherlich's "achievement was not to have discovered the existence of isomorphs, but to have accorded them a theoretical position they previously lacked." The theoretical key was Berzelian atomism.[4] Upon making his initial discovery, Mitscherlich immediately sought and found an explanation for the isomorphism in the identical degree of oxidation of the bases combining with the same acid, or of acids combining with the same base. In the original German paper, he gave preference to the conventional and ontologically cautious terminology of "proportions" and "volumes," but in the French version he also used the word "atom."[5]

Berzelius was in Paris when Dulong and Petit discovered the

law of atomic heats. On his return trip to Stockholm, Berzelius visited several German cities, including Berlin. Having just heard from Gustav Rose that a young Privatdocent had made a significant mineralogical discovery, Berzelius sought out and befriended Mitscherlich. By this time (August 1819) Mitscherlich had verified but not publicly announced a number of cases of isomorphism, and Berzelius immediately perceived the theoretical significance of the new phenomenon. Greatly impressed with Mitscherlich, Berzelius urged that he be installed in the Berlin professorship vacant since Klaproth's death. The Prussian authorities offered instead to send the novice chemist to spend a year in Berzelius' laboratory in Stockholm, a welcome suggestion to both Mitscherlich and Berzelius. The single year stretched into nearly two; after Mitscherlich's return to Berlin in 1821 he was made in rapid succession *ausserordentlicher* and *ordentlicher* professor, holding the latter post until his death in 1863.[6]

In the summer of 1821, shortly before his departure from Stockholm, Mitscherlich summarized his research in a detailed memoir. As was the case with the Dulong-Petit law, Mitscherlich found that the law of isomorphism was not quite exact. Yet he was able to provide compelling evidence against Haüy's doctrine of accidental immixture, and he forthrightly concluded that compounds possessing analogous formulas in general crystallize in similar forms. At Berzelius' suggestion he here introduced the term "isomorphism" to denote this phenomenon.[7]

Mitscherlich and his mentor Berzelius did not neglect the important consequences of isomorphism for atomic theory. For one thing, it seemed to prove that "crystallization depends solely on the numbers of the atoms [in a molecule], and not at all on their nature."[8] A second conclusion implicit in the memoir is that the isomorphism of the salts studied tended to corroborate the analogous formulas previously assigned them by Berzelius (e.g., bases of the form MO_2, phosphoric acid PO_5, and arsenic acid AsO_5). Isomorphic substances must have analogous formulas. For this reason the new law could also be used heuristically, to infer the formula of a substance that is isomorphic to another with a known formula. That Mitscherlich immediately perceived this is clear from his first memoir.[9] Finally, a similar sort of atomistic reasoning was applied to explain dimorphism: the atoms of a given substance can combine in two different spatial arrangements.[10]

BERZELIUS' ELABORATION OF CHEMICAL ATOMISM

Earlier we mentioned a revealing published exchange of thoughts between Dalton and Berzelius that appeared in Thomson's *Annals of Philosophy* in 1814-15.[11] Berzelius there indicated his commitment to physical as well as chemical atomism, and both chemists stressed a number of points of agreement. But both also recognized their dissimilar scientific methodologies, particularly regarding the knotty question of the interrelation of theory and experiment. Berzelius wrote: "Mr. Dalton has chosen the method of an inventor, by setting out from a first principle, from which he endeavors to deduce the experimental results. For my own part, I have been obliged to take the road of an ordinary man, collecting together a number of experiments, from which I have endeavored to draw conclusions more and more general. I have endeavored to mount from experiment towards the first principle; while Mr. Dalton descends from that principle to experiment."[12] This depiction of Dalton's method as classically hypothetico-deductive and his own as empirically inductive contains an element of truth, but it cannot stand without qualification. It tends to mask Berzelius' passionate devotion to chemical theory, as well as his inveterate hypothesizing. His good friend Marcet complained about this characteristic of Berzelius' work: "How you love chemical theory! Shall I tell you the fault that I find in your writings? They cover too much at one time. . . . Your writings thus have a richness that dazzles ordinary philosophers. . . . In a word, you write for the Davys and Wollastons, but keep in mind, dear friend, that there are many more Marcets than those kind."[13]

Metaphysically as well as methodologically there were important differences between Dalton and Berzelius. Both proceeded from Lavoisierian principles, but Dalton also participated in the British tradition of popular Newtonianism with its predilection for straightforward mechanical models. Dalton thus complacently accepted the ontological status of operationally-defined elements and did not share Lavoisier's own qualms about the possible complexity of the "simple substances." Only in a few cases where the evidence seemed very strong indeed— chlorine and certain earths—did he suggest that hitherto undecomposed substances were probably compounds. This casual nonreductionism was undoubtedly heuristically important for Dalton's conceptual journey to his atomic theory.

Berzelius, by contrast, was much closer to Davy's and Prout's conception of the chemical elements, which was influenced by reductionist and mechanist principles. Such thinkers frowned on the conceptual inelegance of a long list of Daltonian (ontological) elements. In the period 1809-12, Berzelius published speculations on the possible compound nature of hydrogen, phosphorus, nitrogen, and chlorine. Although after 1812 he relinquished the idea that hydrogen contains oxygen, he strongly defended the complexity at least of nitrogen and chlorine for several more years; in this period he maintained that the evidence he adduced amounted to virtual proof of his claim, though conceding that it remained largely circumstantial.

But such reductionist programs were irrelevant for chemical atomism. For instance, Dalton pointed out on the last page of the 1808 volume of his *New System of Chemical Philosophy* that Davy's discovery of the compound nature of soda and potash could be incorporated into his system with no essential change. Berzelius' "nitric suboxide" (NtO) had similar stoichiometric and eudiometric relations to everyone else's "nitrogen." And Davy put the matter in a nutshell, providing literally the final word in his and Berzelius' troubled personal correspondence regarding the elementarity of chlorine: "All your proportions coincide with mine, and it is merely a question about to which body the *same definite* proportions belong."[14]

Berzelius was an extraordinarily skillful laboratory chemist and had uncommon patience. These virtues enabled him gradually to recognize and verify a number of empirical laws even before he was fully acquainted with Daltonian atomism: the stoichiometric relations themselves, the constancy of basic oxygen, and the oxide rule, to name the most important. The last in particular he defended tenaciously, since he regarded it as the critical conceptual connection between oxides and salts. The oxide rule stated that the oxygen content of the least oxidized component of the substance must be regarded as a unity that enters only in integral multiples into all other oxide components. He then extended the rule by analogy to the first-order level, by denying the existence of any compound of the form A_2B_3. This "single-atom axiom" he regarded as both empirically verified and a sine qua non for the very existence of chemical atomism. In this he was justifiably taken to task by Dalton and others.[15]

Notwithstanding these disagreements over details, however, Berzelius and Dalton had much more in common than most historians have recognized. To be sure, the oxide rule was unmotivated by Dalton's theory, but Berzelius himself regarded this as in no way inconsistent with chemical atomism, but simply as a lacuna for atomists to study and eventually fill. Similarly, when Berzelius advocated the "volume theory" of Gay-Lussac for its empirical basis, he simultaneously and repeatedly stressed that it was "absolutely the same thing" as the corpuscular theory of Daltonian atomism; in other words, the two theories were entirely equivalent means of viewing the same set of phenomena.[16] This implies, and even requires, full acceptance of the equal volumes-equal numbers hypothesis for elements.

It was Berzelius' faith in the oxide rule that led him to assert that nitrogen gas was the suboxide of the "nitric" radical (N = NtO). Well-known analogies between nitrogen and phosphorus led him in 1816 to carry out a systematic examination of the oxides of phosphorus. Dulong and Thomson independently performed overlapping and complementary experiments about the same time.[17] Dulong's work resulted in the discovery of two new acids (hypophosphoric and hypophosphorous), and Berzelius clarified the stoichiometry of a variety of neutral, acid, and basic phosphates and phosphites. The analogies between nitrogen and phosphorus compounds were dramatically extended. Ironically, however, Berzelius' faith in his oxide rule and single-atom axiom was at the same time shaken rather than strengthened.

In 1813 Berzelius had not been able to decide on a definitive formula for the highest oxide, phosphoric acid, and ended up wavering between PO_2 and PO_3. The crucial datum was the ratio of oxygen in phosphoric to that in phosphorous acid relative to a given weight of phosphorus; the ratio seemed to range above 1.5 but below 2.0, troublesomely nonintegral. Now Berzelius was able to measure the ratio more precisely, and it came to exactly 1.67, or $5/3$. The only option consistent with the single-atom axiom was to double the presumed weight of phosphorus and write the two acids PO_5 and PO_3, and this Berzelius did. However, this created anomalies for his oxide rule. Some acid phosphates gave canonical oxide-rule multiples of 5, but neutral and basic phosphates had such nonintegral multiples as $5/2$ and $5/3$.[18] Faced with the prospect of losing his cherished oxide rule, Berzelius turned to the same strategy that had saved the orthodoxy of the nitrogen oxides: suggesting the possibility of the presence of undetected oxygen within "elemental" phosphorus. If a phosphorus "molecule" contains one oxygen atom, then the oxide-rule multiples become 6, $6/2 = 3$, and $6/3 = 2$, all beautifully integral.

It was, however, troubling that this oxygen remained completely undetectable. No experimental stratagem seemed capable of flushing it from its hiding place, and Berzelius finally admitted defeat. His empiricist commitment forced him to conclude that phosphorus was probably elemental, so that the oxide rule was not in fact general—though he added that exceptions to it seemed to be limited to compounds of phosphorus, and, by analogy, just possibly nitrogen as well. Given Berzelius' strong correlation between first- and higher-order compounds, this failure of the oxide rule provided the first weakening of its first-order analogue, the single-atom axiom. Here we see an uncharacteristically hesitant Berzelius: phosphorus is *probably* elemental, nitrogen and chlorine are (only) *probably* compound, the oxide rule is not completely general, though the single-atom axiom can *probably* still hold.[19] In an article written within a few weeks or months after his phosphate paper was published, Berzelius stated these conclusions more forthrightly: the oxide rule had a few undoubted exceptions, but the single-atom axiom still appeared general. Curiously, he now seemed to recognize the provisionality and conventionality of the latter rule: "So far as actual experience allows us to decide, it is reasonable to suppose that no [complex] combinations occur."[20]

Less than a year after composing his phosphate paper, Berzelius discovered a similar pattern in the arsenates. In 1813 he thought he recognized a suboxide containing one-sixth the oxygen of arsenic acid and determined the oxidation of arsenic versus arsenous acid as in the ratio of three to two. This pattern was the same as that found for the oxides of sulfur and led him to formulate the three compounds either as AsO, AsO_4, and AsO_6, or as As_2O, AsO_2, and AsO_3. Now he was no longer able to form any suboxide at all—it does not in fact exist—and a more precise determination of the oxidation ratio of the acids of arsenic revealed the same pattern as for the acids of phosphorus, five to three. Once again he noted that the resulting exceptions to the oxide rule could be removed if it were supposed that oxygen existed undetected in the arsenic "radical"; he by no means dismissed the possibility that this oxygen might someday be discovered, not just in arsenic, but also in phosphorus and nitrogen. However, he noted that he had already demonstrated the improbability that phosphorus contains any oxygen, and for arsenic an oxygen content was even less likely considering its pronounced metallic character.[21]

The year 1818-19 was an important watershed for Berzelius in a number of respects. He was ill with gout and various nervous and circulatory disorders throughout the early months of 1818, brought on most likely by a combination of overwork and selenium poisoning. In July he left Stockholm for an extended visit to Paris, financed by the Swedish government. In Paris he found rest, recuperation, fruitful collaboration with Dulong and others, and the leisure to produce a French version of the important third volume of his *Lärbok i kemien*, which had been published in Stockholm shortly before his departure.[22] During his absence he was elected secretary of the Swedish Royal Academy of Sciences, both an honorific and a remunerative post. His new position of power allowed him to cease publishing the *Afhandlingar i fysik, kemi och mineralogi*—he and Hisinger had started the journal in 1806 to bypass the stodgy and dilatory Academy *Handlingar*—but it also added a new literary duty. Beginning in 1821 he composed every March a detailed *Årsberättelse om framstegen i fysik och kemi* ("annual report on the progress of physics and chemistry"), which discussed virtually all newsworthy events of the preceding calendar year. These book-length reports continued until Berzelius' death in 1848; the German translations in particular (*Jahresberichte*) exerted enormous influence on European chemistry in the 1820s and 1830s.[23]

Berzelius was impressed by the vitality of the Parisian chemical community, but appalled by its lack of attention to his laws of proportions. For both of these reasons he thought it important to edit a French version of his ideas on chemical atomism and electrochemical dualism. The work appeared in June 1819, shortly before his departure from Paris, as *Essai sur la théorie des proportions chimiques, et sur l'influence chimique de l'electricité*. This small volume provides a compact summary of the theoretical ideas and experimental data that Berzelius had been assiduously developing for the preceding twelve years.

For instance here, as earlier, Berzelius stressed the identity of the volume theory and the corpuscular theory, and for the first time noted that this "would seem to prove that every gas of a simple body contains an equal number of atoms in the same volume, measured at the same temperature and pressure," though this relation does not hold for gases of compound substances. Berzelius pointed out that those who believe water is HO (hence O = 8) are implicitly asserting that the particle density of oxygen is twice that of hydrogen. "This being merely a gratuitous assumption whose truth is not even susceptible to examination, it seems to me simpler and more probable to assume the same weight ratio between the volume and the atom in combustible bodies as in oxygen, since nothing leads us to suspect that there is a difference between them."[24] This is empty rhetoric. The proponents of the view that water is HO (such as Wollaston, Thomson, and L. Gmelin) could (and did) reply with equal justice that it was Berzelius who was making the gratuitous assumption—that of equal volumes-equal numbers (EVEN) for elemental gases.

Berzelius also discussed his oxide rule and single-atom axiom; for both he admitted some lack of generality. The single-atom axiom seemed initially to hold so often, he wrote, "that I first attributed the [exceptions] to our uncertainty concerning the relative weights of the atoms; but greater experience—

admittedly still insufficient to decide the issue—seems nonetheless to indicate to me that elementary atoms in inorganic nature can combine in other ratios, although these occur but seldom."[25] He did not further elaborate, but one senses a certain fluidity of opinion here. This feeling is reinforced by a passage in another context where Berzelius mentions his "secret fear" that several of his atomic weights are double what they should be, an anticipation of future events.[26] A reasonable conjecture is that Berzelius was influenced in this direction by the law of Dulong and Petit, which was announced to the Paris Academy of Sciences while Berzelius was editing the *Essai*.[27]

A similar hesitancy characterized his treatment of the oxide rule. Berzelius admitted that the rule was indeed violated for the acidic oxides of phosphorus and arsenic and also for the oxides of nitrogen gas (if it be considered as a simple substance). If these three substances contain undetected oxygen, then the anomalies would vanish, but—Berzelius averred—he wanted to treat only known facts. Nonetheless, the known fact that nitrogen gas had never been decomposed failed to prevent him from retaining his hypothesis of the "nitric" radical throughout the *Essai*. As in his 1813 publication, this meant that nitrogen gas had to be taken at two volumes rather than one (NtO = N = 28). Presumably this was not anomalous for Berzelius since he denied the EVEN hypothesis for "compounds" such as "nitric suboxide."

The doctrinal fluidity noticeable in the *Essai* originated in his own work on the phosphates and arsenates, and probably also in the Dulong-Petit law. This indecision presaged a slow transformation in many of Berzelius' atomistic assumptions over the next several years. In an 1820 letter to Gay-Lussac, he admitted that the reactions of the novel substance we now call potassium thiocyanate (KNCS) "can only be explained according to a theory analogous to that which you have adopted for the muriates."[28] Berzelius was unable to detect any oxygen in thiocyanic acid, providing a corroborating parallel to other hydracids such as hydrochloric acid and therefore strengthening the view of the elementarity of chlorine. Gmelin's 1822 study of potassium ferricyanide (modern $K_3Fe[CN]_6$) he regarded as definitively deciding the issue.[29] Accepting the elementarity of chlorine resulted in further exceptions to the oxide rule. Considering the undisputed exceptions represented by the phosphates and arsenates, Berzelius' resolve to retain the "nitric" radical began to crumble, since its principal justification had been to maintain the generality of the oxide rule. By 1823 he fully accepted the elementarity of nitrogen and had essentially relinquished his oxide rule in its general form.[30]

The overthrow of the oxide rule had major consequences for Berzelius' system of chemical atomism. For one thing, the elementarity of nitrogen now made it susceptible to the EVEN rule, which meant that Berzelius was forced to halve its atomic weight (from $NtO = 28$ to $N = 14$) and double the number of nitrogen atoms in all relevant formulas. This he did immediately, but without even a word of justification or explanation.[31] The formulas for (anhydrous) nitrous and nitric acids now had to be written N_2O_3 and N_2O_5; chemical analogy implied that a similar modification was necessary for the weights and formulas for oxides of phosphorus, arsenic and antimony. Though this meant the fall of the single-atom axiom, a certain consistency with the formulas for the halogen oxides now was evident. Berzelius' longstanding opposition to the elementarity of chlorine had been based on his objection to formulas like that for chloric acid, Cl_2O_5, which violated his unitary laws. By 1823, however, there was clear evidence for the existence of a general class of oxides R_2O_3 and R_2O_5, a class that now included compounds of nitrogen, phosphorus, antimony, chlorine, and iodine. All of these consequences and more flowed from Berzelius' conversion to Davy's chlorine theory in 1820.

The imminent publication of a revised German translation of the third volume of his textbook led Berzelius to undertake the first major revision of his ideas since 1812. He had had trouble getting an acceptable German edition of his *Lärbok* into print; no less than three imperfect and partial translations appeared in the years 1816–24, and none of these continued past the second Swedish volume (first edition, 1812; second edition, 1822). Finally Berzelius found an ideal German editor, Friedrich Wöhler, a promising young Heidelberg medical doctor who spent the year 1823–24 in his Stockholm laboratory and learned fluent Swedish. Back in Germany, Wöhler shortly took over the translation of the *Jahresberichte* from C. G. Gmelin and began editing a definitive German translation of the *Lärbok*. For this work Berzelius meticulously revised the latest edition of each volume and sent the Swedish copy piecemeal to Wöhler, who translated it and saw it through the press. The first two German volumes, in four parts and nearly two thousand pages, appeared in 1825-26.[32]

It was at this juncture that Berzelius was faced with the unavoidable task of revising the third volume, which contained the important general theoretical overview represented in the 1819 *Essai*. Major alterations had, we have seen, been foreshadowed in the French work and indicated in a desultory fashion in the first few *Jahresberichte*. In fact it took Berzelius only about four weeks of work in the summer of 1826 to produce a strikingly altered system of chemical atomism. The changes, as Berzelius wrote to Wöhler on 11 July 1826, constituted "ganz und gar eine Revolution" and resulted in "überall hübschere Formeln." He asked Wöhler to translate and transmit the new section to Poggendorff for insertion in the *Annalen der Physik*. Two weeks later Wöhler replied: the "divine" piece was already translated and would appear in three monthly installments of the *Annalen* beginning with the next month's issue (August 1826). "Poggendorff can congratulate himself—and does in fact do so—to have gotten it for his journal."[33]

The goal of the article was to review and judge the various means of determining formulas and finally to decide on the single most probable set of atomic weights. Berzelius outlined five experimental approaches to formula determinations. The first of these, the comparison of vapor densities, is the only one that "gives results not subject to any doubt." This evidence, incorporating the EVEN assumption for elementary gases, provides the means to determine unambiguously the atomic weights of the crucial elements hydrogen, oxygen, nitrogen, and chlorine, and the formulas of all compounds formed solely from these substances. By contrast, Dalton's simplicity assumption (which leads to the HO water formula and $O = 8$) was characterized as arbitrary, unnatural, and inconsistent with everything else we know about common weight and volume relations.[34]

Berzelius' second and third approaches were also ones he had been using regularly and fruitfully for sixteen years: comparison of the degrees of oxidation of a given element, and the (modified) oxide rule. Particularly if the deductions from these two methods agree, one can be "fairly certain" to have inferred the correct number of atoms in the molecule. For instance, since the degrees of oxidation of sulfuric and sulfurous acids are as three to two and the oxide-rule multiples for the sulfates and sulfites are generally three and two respectively, it is a safe conclusion that the correct acid formulas contain three and two atoms of oxygen respectively. Chemical analogies to water (known from EVEN to be H_2O) leads to an H_2S formula for hydrogen sulfide and to the conclusion that molecules of each acid contain but one atom of sulfur. In the case of such nonintegral oxide rule multiples as $3/2$ and $5/2$, 3 and 5 are the probable numbers of oxygen atoms in the acids.[35]

It was Berzelius' fourth and fifth approaches that were revolutionary: isomorphism and atomic heats. Immediately upon meeting Mitscherlich, Berzelius had understood the significance of isomorphism for the atomic theory. To Marcet he wrote in the fall of 1819: "These observations are extremely valuable for the corpuscular theory, since they will provide a method to verify and correct speculations on the internal composition of bodies; I consider this one of the most beautiful discoveries ever made in chemistry." And to Dulong he wrote: "You see that this sort of research will be conjoined with yours to verify our speculative ideas regarding the number of elementary atoms contained in compound bodies and their mode of combination."[36] In his first *Jahresbericht*, written when Mitscherlich was still working in his laboratory, Berzelius reported at length regarding the new discovery, concluding: "I have made such demands on the Academy's patience . . . because I regard it as the most important [discovery] since the theory of chemical proportions was proposed, a theory which formed the basis for its development."[37]

In his 1826 article Berzelius pointed out that isomorphism of two compounds, one of whose formula is unknown, proves the identity of the two formulas, since "isomorphism is a mechanical consequence of the identity of atomic construction." For instance, since alumina is the only oxide of aluminum, its formula only became satisfactorily established when Mitscherlich demonstrated its isomorphism with oxides of iron and manganese known from other evidence to be trioxides.[38]

Berzelius also had quickly understood the phenomenon of atomic heats as a means of determining formulas: this is shown by his 1819 letters to Dulong and Marcet and hints in the *Essai* cited above.[39] Of thirteen atomic heats measured by Dulong and Petit, two led to atomic weights consistent with Berzelius' 1819 table and so resulted in no alterations. Berzelius rejected three as anomalous.[40] The other eight determinations argued for a halving of his 1819 metallic atomic weights,[41] and for these cases he accepted the Dulong-Petit verdict. What this meant was that Berzelius was forced to abandon his general MO_2 formula for basic metal oxides and adopt monoxide formulas. Chemical analogy dictated halving the atomic weights of *all* the metals, and Berzelius was never one to neglect analogy. All were halved.

But analogical reasoning was only one of the methodological principles utilized so effectively by Berzelius; another was the construction of a consistent general pattern by applying widely disparate and imaginative approaches. Berzelius did not adopt the results of atomic heats or isomorphism on their own merits alone, but rather because these phenomena provided the encouragement he needed to continue along a path he had already been treading. The problem can be simply stated. Berzelius had hitherto avoided monoxide formulas for the basic oxides because many metals also exhibit higher sesquioxides which would then have to be formulated M_2O_3, contrary to the single-atom axiom. But we have seen that his acceptance of the elementarity of nitrogen and chlorine resulted in such oxide formulas as N_2O_3, N_2O_5, and Cl_2O_5, and by analogy, P_2O_5 and As_2O_5. The existence of "radicals" consisting of two atoms rather than just one appeared inescapable. Therefore, accepting metal sesquioxides would constitute extension of a general class rather than a multiplication of anomalies.

Once he was forced down this path, he could think of all sorts of reasons to oppose the simple MO, MO_2, MO_3 . . . pattern. The oxygen atoms rapidly increase to unwieldy numbers. The number of missing or undiscovered stages of oxidation was appalling. Above all, it was inconsistent with the new physical methods of isomorphism and atomic heats. Perhaps Berzelius' clearest example of the use of disparate methods for formula assignment and atomic weight determination was the case of chromium. To wit: (a) The two stages of oxidation were as 1:2. (b) The higher oxide forms salts with an oxide rule multiple of three, hence probably has three oxygen atoms. (c) The lower oxide is isomorphic with oxides of iron and manganese known to be trioxides, hence is also a trioxide. Therefore, the two chromium oxides must be Cr_2O_3 and CrO_3 — rather than his earlier assignments CrO_3 and CrO_6 — and his earlier atomic weight for chromium had to be halved. Q.E.D.[42]

Berzelius had never been faced systematically with double atoms of inorganic radicals, and, true to his style, he immediately coined a notational symbol to represent them.

> In all cases where 2 atoms of the radical combine with 1, 3, or 5 atoms of oxygen, as for example sulfur, the clarity of the formulas will be greatly increased if we have a special symbol for the double atom. The most appropriate symbol would be to double the initial letters but in such a way that they remain together and constitute one rather than two symbols. However, in order to indicate these formulas in print I have found it far easier to accomplish and just as clear to draw a straight line through the lower third of the initial letters to indicate two atoms; so for example P represents a single and P̶ a double atom of phosphorus, As a single and A̶s a double atom of arsenic.[43]

Combined with Berzelius' "dot" notation for oxides, this produced such formulas as ĊuS̈ for copper sulfate and

$$\dot{K}\ddot{S}^3 + \ddot{A}l\ddot{S}^3 + 24\overline{H}$$

for alum. I will henceforth routinely and silently translate such formulas into, e.g., $KO·3SO_3 + Al_2O_3·3SO_3 + 24H_2O$.

THE EMERGENCE OF ISOMERISM

Ever since chemists accepted the law of definite proportions in the first years of the century it was assumed that the chemical and physical properties of a substance are always directly correlated with its composition.[44] This belief was a consequence of the error of assuming that the truth of a statement implies the truth of its converse: all sugar samples have the empirical formula CH_2O, but not all substances having the composition CH_2O exhibit the properties of sugar. Carbohydrates are examples of a phenomenon that from 1830 was designated "isomerism." In fact, the correlation of composition with properties was further weakened by four other phenomena encountered in the first third of the nineteenth century: polymerism, allotropy, isomorphism, and polymorphism. Historically as well as conceptually, these five phenomena were connected; since ultimately all were explained by reference to atomic and molecular arrangements, their history is an important theme in the development of atomic theory.

The idea that corpuscular arrangements could affect or even determine the properties of matter extends back to the ancient atomists and was well represented in the seventeenth century. The resurgence of mechanist thought at the beginning of the nineteenth century led, not surprisingly, to a resurgence of speculations of this sort. In 1808, for instance, Wollaston speculated on the appearance of three-dimensional atomic arrangements.[45] In 1812 Davy thought that "Matter may ultimately be found to be the same in essence, differing only in the arrangement of its particles."[46]

In addition to these reductionist speculations, differing atomic arrangements were also invoked quite early as explanations for specific anomalies to the property-composition correlation. Biot and Thenard suggested this structural explanation to account for the dimorphism of calcium carbonate (in hexagonal calcite and rhombic aragonite).[47] Davy's combustion of the diamond in 1814 thoroughly convinced him that the difference between the allotropic modifications of carbon were simply the result of the differing arrangements of their ultimate particles.[48]

The most fruitful field for isomerism was, of course, organic chemistry. Only a handful of organic compounds had been crudely analyzed by 1811 when Gay-Lussac and Thenard developed a successful general method and published nineteen accurate organic analyses. These included an instance of three very different substances with the same composition: sugar, starch, and gum arabic.[49] A baffled Berzelius wrote Gay-Lussac: "I am struck by the identity of the constituent proportions of gum, sugar and starch. It would be difficult to make two experiments on the same substance which agree better than these analyses of different substances. What produces the differences in their chemical characters? The small amount of fixed matter which they leave behind [i.e. as ash residue in combustion]?"[50] Three years later Gay-Lussac offered a different explanation. He had encountered a similar instance in the apparent identity of elemental composition of acetic acid and cellulose ("matière ligneuse"); reminding his readers of sugar, starch, and gum, he wrote: "This is a new proof that the arrangement of the molecules in a compound has the greatest influence" on its properties.[51]

Nor was Gay-Lussac the only chemist to explain isomerism on the basis of differing arrangements of constituent particles. Davy independently opined that the identity of proportions of gum and sugar must depend on "the different arrangement, or degree of condensation of their elements," and affirmed that, in general, acid and alkaline properties "must depend upon a peculiar corpuscular arrangement." J. E. Bérard thought that the identity in composition of wax and spermaceti, and of fish oil and copal, could only be accounted for by a "different arrangement of their molecules." In 1818 M. E. Chevreul defined unique substances as having both identical compositions and identical arrangements and in 1823 averred that known cases of isomerism must be explained by "different arrangements, either in their elementary atoms or in their compounds atoms or particles."[52]

The German romantic "stoichiometers" also seized on this structural explanation, as it seemed consonant with the idealist monistic metaphysics by which they were strongly influenced. Apparently following Gay-Lussac,[53] Meinecke wrote that in sugar, starch, and gum "the components are in different states, i.e. they are bound and ordered among each other in different ways." Asserting that only three possibilities exist, he proceeded to assign the following "chemical constitutions": starch = $2C \cdot 2H_2O$; gum = $CO \cdot CH_2 \cdot H_2O$; sugar = $CO_2 \cdot CH_4$.[54] Although the basis for these assignments is unclear and unstated, Kastner and Döbereiner immediately accepted them.[55]

With the assistance of the *Naturphilosoph* botanist C. G. Nees von Esenbeck and the mathematician H. A. Rothe, C. G. Bischof carried out a detailed—though completely speculative—investigation of organic isomerism and polymerism. Bischof began by suggesting that all organic compounds are formed by combinations of five simple substances: water, the two oxides, and the two known hydrides of carbon. Assuming no more than three of any one component in a molecule, Rothe calculated that there are no less than 961 possible different nonreducible permutations of these components. Over two-thirds represented (hypothetical) cases of isomerism or polymerism. The largest number of permutations for a single empirical formula was 21, for the carbohydrate formula $C_nH_nO_n$ (C = 6, O = 8); Bischof suggested this must be the "Grundverhältniss" for all sugars. In any case it was now clear how the great variety of organic compounds could be produced from combinations of relatively few simple components.[56]

A number of chemists, then, had proposed structural interpretations of isomerism by 1819. But Berzelius, for one, was hesitant. Since 1811 he had been urging an electrochemical theory of affinity that traced the chemical differences between compounds to the electrical characteristics of their constituent atoms. This fundamentally materialistic conception seemed distant from—and perhaps inconsistent with—the mechanist-reductionist structural approach, which stressed the arrangements rather than the natures of the ultimate particles. In his 1811 letter to Gay-Lussac, for example, Berzelius implied that it was to electrochemistry that one must look to provide an explanation for the sugar-starch-gum isomerism.[57] Similarly, when he and Gay-Lussac discovered two different tin oxides that seemed identical in composition, Berzelius worried aloud in two letters to Marcet: "What I can't account for is that these two oxides have such different properties.... This circumstance appears inexplicable for the moment and incompatible with our chemical experience in general.... It is quite clear that there must [also] be a difference of composition between these different species [of sugar investigated by Gay-Lussac]."[58]

On the other hand, Berzelius' strong commitment to atomic theory and to certain aspects of the reductionist program made him sympathetic to the atomistic proclivities of the structuralists. In his first major paper on organic chemistry (1814), he wrote: "We may then form the idea that the organic atoms [molecules] have a certain mechanical structure. . . . It is only by such a structure that we can explain the different products . . . composed of the same elements, and in proportions (stated in per cents.) but little different from each other. I am persuaded that an attempt to study the probabilities of the construction of organic atoms . . . would be of the greatest importance, and might be even capable of correcting analysis."[59] In a similar spirit, he warmly welcomed the discovery of isomorphism, even though this phenomenon seemed to indicate that the arrangements of atoms were at least occasionally more important than their natures, electrochemical or otherwise.[60]

Clearly isomerism had a lively prehistory before it entered the mainstream of chemical thought. That transformation was brought about in the course of the 1820s particularly by Berzelius and two young German chemists, Wöhler and Liebig. As students these two happened to select research topics in a currently "hot" field — cyanogen derivatives — that brought them into a collision course.

Liebig's attention was drawn to the metal salts of fulminic acid (*Knallsäure*, HONC) by watching a peddler in Darmstadt entertain a crowd. As a student of Kastner in Bonn and subsequently (1821) Erlangen, Liebig's first two publications concerned silver and mercury fulminate, but he was justifiably suspicious of his own analyses. On a fellowship in Paris, he happened to come to the attention of Gay-Lussac, who took him into his laboratory and collaborated with him on the first accurate analysis of the fulminates. Liebig and Gay-Lussac concluded that "fulminic acid" was nothing other than oxygenated cyanogen, or what they proposed to call "cyanic acid"; silver fulminate was thus $C_2NO \cdot AgO$.[61]

The best way to see that this formula is the same as modern AgOCN is to note that Gay-Lussac (following Wollaston) used doubled weights for N and Ag compared to C and O, so $C_2NO \cdot AgO$ is equivalent in modern weights to $C_2N_2O \cdot Ag_2O$, reducible to AgOCN. In general, use of doubled atomic weights for metals forming salts with monobasic organic acids required a doubling of the presumed formula for the acid. Furthermore, writing the metal oxide as a theoretically separable component of the salt resulted in an odd number of oxygen atoms in the anhydrous acid, which tended to confirm the nonreducibility of the double formula. For instance, to the end of his life Berzelius wrote acetic acid [CH_3CO_2H] as $C_4H_6O_3$, and lead acetate as $PbO \cdot C_4H_6O_3$ [= $Pb(CH_3CO_2)_2$]. In this way four-volume weights and formulas became standard in organic chemistry between about 1815 and 1850.

Wöhler, then a student of Leopold Gmelin in Heidelberg, was investigating the product of alkaline hydrolysis of cyanogen just at the time Liebig was first investigating the fulminates.[62] Wöhler's third paper on this subject was written in Stockholm, reporting the synthesis of the silver salt of what he called "cyanic acid," a salt that had very different properties from Liebig's silver fulminate. Analysis gave the same result and led to the same formula, $2C_2N_2O \cdot AgO_2$ (in Berzelius' weights of the early 1820s).[63] Coincidentally, this paper was published in Poggendorff's *Annalen der Physik* for May 1824 directly following a German translation of the Liebig-Gay-Lussac paper. When Wöhler's paper was reprinted in French in the *Annales de chimie*, Gay-Lussac added an editorial comment regarding the two compounds: "One must, to explain their difference, assume a different mode of combination between their elements."[64] Although vaguely expressed, Gay-Lussac was doubtless making a reference to the "arrangement" hypothesis he had proposed ten years earlier.

But Berzelius, Liebig and Wöhler were all more troubled than Gay-Lussac. Berzelius suggested that the Liebig-Gay-Lussac analysis was in error; perhaps the sample was wet and fulminic acid is actually cyanous (*cyanichte*) acid, C_4N_4O. Liebig retaliated precisely the same way: a new analysis purported to show that Wöhler's data were faulty, fulminic is the true cyanic acid, and Wöhler's "cyanic" is actually cyanous acid, $3C_2N \cdot O_2$ in conventional equivalents (= C_3N_3O in Berzelius' weights). It was then Wöhler's turn for rebuttal: he was able to show a probable source of error in Liebig's analysis of silver cyanate, and a new series of analyses by an improved method verified his own earlier work. Privately to Berzelius he expressed doubt about the published analysis of silver fulminate: "Just consider how hard it is to manipulate and dry these devilish things." In the spring of 1826 Berzelius summarized the dispute for the Swedish Academy, tactfully but firmly supporting the work of his former student.[65]

The impasse was broken following the first personal meeting between Wöhler and Liebig in Frankfurt in March 1826. Liebig again repeated careful analyses of both fulminic and cyanic acids, the latter using Wöhler's improved method. He now found that they were unquestionably identical in composition, a conclusion subsequently accepted by Berzelius, Wöhler, and others.[66] Another pair of substances had to be added to the list of violations to the composition-properties correlation. This one drew the attention of chemists as never before, since it had provoked a public squabble between Liebig and Wöhler and implicitly between their mentors, Berzelius and Gay-Lussac, who were then the most prominent chemists in Europe.[67]

Three more cases of isomerism were discovered by 1830. One of these, ortho-, meta-, and pyrophosphoric acid, was later found to apply only to the schematic anhydrides. Wöhler encountered the second when he found that artificially prepared ammonium cyanate is spontaneously transformed into urea. Brooke has rightly argued that the discovery became renowned not because it provided clear evidence against vitalism, but principally because it was another indisputable instance of isomerism, all the more striking since in this case the two substances seemed interconvertible. Finally, Berzelius confirmed Gay-Lussac's 1826 result that the novel racemic acid was identical in composition and saturation capacity to the well-known tartaric acid. The discovery of all three new cases of isomerism were the occasions of structural explanations.[68]

Berzelius summarized the situation in 1830. He recalled the various cases of identical composition and different properties known to that time and argued that the phenomenon appeared to be a rather general one. He coined the word "isomer" and adduced the structural explanation.[69] Two years later he distinguished isomerism in the strict sense from compounds whose formulas form simple multiples of each other (for which he coined the term "polymer") and from compounds whose apparent identity in composition could be explained on the basis of different division into electrochemical components (which he proposed to call "metamers").[70] To further clarify his ideas, he subsequently distinguished what he called "empirical" from "rational" (i.e., theoretical) formulas, the latter reproducing the chemist's conception regarding the grouping of the atoms within the molecule. Alcohol, for example, might be represented by the empirical formula C_2H_6O, or by one of the rational formulas $C_2H_4 \cdot H_2O$ or $C_2H_6 \cdot O$. Although attaining specific ideas of "the relative positions of the elementary atoms" is fraught with uncertainty, it is by no means a hopeless goal, even now (in 1833).[71]

THE OLDER RADICAL THEORY AND THE BERZELIAN HEGEMONY

It was no coincidence that Berzelius introduced the concept of rational formulas just at the time when chemists were beginning to form systematic theories regarding the groupings of atoms within organic molecules. The earliest of these was the "etherin" theory of Dumas and Polydore Boullay. The central tenets of this theory were that (1) alcohol is a double and ether a single hydrate of etherin ("hydrogène bi-carboné", i.e., ethylene); (2) esters are formed from ether and anhydrous acids; and (3) etherin is similar to ammonia in that both form analogously constituted compounds with acids.[72]

In his *Jahresbericht* Berzelius treated the theory respectfully, but rightly noted that the theory as a whole, and especially the third tenet, resulted from schematic manipulation of formulas in a manner that was often only distantly related to the empirical data. He thought that such an approach might be heuristically useful, but was unlikely to lead to an adequate general theory. It was generally recognized at this time that organic chemistry was in its infancy, and in 1828 Berzelius for one was not particularly sanguine about any early success in the theoretical realm.[73]

More to Berzelius' liking was the benzoyl theory. In 1832 Liebig and Wöhler found oil of bitter almonds to be the hydride of a radical of the composition $C_{14}H_{10}O_2$, which they christened benzoyl; they were able to make no less than nine derivatives of this compound and to show the mutual relationships among all ten substances. Berzelius greeted the discovery with great joy, though remarking that it would be better to consider the true radical as the electropositive portion $C_{14}H_{10}$, benzoyl being the oxide of this hydrocarbon.[74]

With the advent of ethyl in 1833–34 the theory of organic radicals emerged from adolescence.[75] Although not the first proponent of the ethyl theory, Liebig gave it its most vigorous development.[76] Like Dumas he viewed alcohol as the hydrate of ether, but denied that water formed any part of the latter's constitution, instead asserting that it is the oxide of ethyl, C_4H_{10}:

Dumas	Liebig
$C_4H_8 \cdot 2H_2O \rightarrow C_4H_8 \cdot H_2O$	$C_4H_{10}O \cdot H_2O \rightarrow C_4H_{10}O$
alcohol ether	alcohol ether

Berzelius' viewpoint was different from both of these: setting ethyl = Ae = C_2H_5, ether is $Ae_2 + O$, "wood spirit" is $Ae + O$, and alcohol is the oxide of a unique radical, $C_2H_6 + O$.[77] He thus accepted the ethyl radical, but broke with both Liebig and Dumas by regarding ether and alcohol as being unrelated by simple hydration or dehydration. His reasoning here was complex.[78]

Ethyl proved to be a beautifully orthodox organic radical, orthodox in the Lavoisierian-Berzelian sense. Electropositive like most inorganic "radicals" (i.e., elements), it could be distinguished unaltered in a whole series of compounds: ethyl oxide (ether), ethyl chloride, ethyl cyanide, ethyl esters, and so on. Ethyl was at once more empirical and more easily assimilable to dualist ideas than Dumas' etherin, facts that Liebig and Berzelius emphasized. New radicals appeared in rapid succession: methyl, acetyl, cacodyl, and so on, and for a brief but heady period in the 1830s it looked as if the radical theory would soon unlock the dark secrets of organic chemistry.[79] When Dumas finally came over to the ethyl formulations in 1837, Liebig joined with him in a common proclamation: "In mineral chemistry the radicals are simple, while in organic chemistry the radicals are compound; that is the only difference."[80]

Berzelius' reputation and influence were then at their height, especially in Germany. German chemistry had experienced a remarkable resurgence during the 1820s, in part due to Liebig's efforts at Giessen from 1824 and in part due to Berzelius' German students. Four others in addition to Wöhler and Mitscherlich spent a year at Berzelius' Stockholm laboratory; C. G. Gmelin (1814-15), the brothers Heinrich and Gustav Rose (1820-21), and Gustav Magnus (1827-28). Each then returned to spread the master's doctrines. Oddly, most congregated in one city—Berlin. For a few months in 1831 there were no fewer than five Berzelius pupils there: one full professor (Mitscherlich), two *ausserordentliche* professors (the Roses), a *Privatdocent* (Magnus), and a professor at the Gewerbeschule (Wöhler, who left late that year for a university professorship at Kassel).

Liebig, of course, had been a pupil of the eminent Gay-Lussac, Berzelius' chief rival. In the 1820s Liebig utilized not Berzelian atomic weights, but rather Wollastonian conventional equivalents, which Gay-Lussac himself had begun to use after Berthollet's death, and which were most popular in Germany ever since the early atomistic contributions of Schweigger and Döbereiner. Berzelius' early impressions of Liebig's work were not entirely favorable, but after their first personal meeting in Hamburg in September 1830, the relations between the two chemists became extremely cordial.[81] In the issue of Poggendorff's *Annalen der Physik* for the same month appeared a paper by Liebig that signalled his acceptance of Berzelian atomic weights. That autumn Liebig invented a much improved method for organic analysis, incorporating the ingenious five-bulbed *Kaliapparat*.[82] From that time onward, an astonishing number of precise organic analyses emanated from Giessen, particularly after Liebig took over the editorship of Geiger's *Annalen der Pharmacie* at the beginning of 1832.

In the early 1830s, Liebig was rightly viewed as an advocate of Berzelius' chemical theories. With his enormous experimental and theoretical productivity, Wöhler's tireless translations of Berzelius' *Jahresberichte* and *Lehrbuch der Chemie*, and the work of the Berlin Berzelians, the Swedish chemist exercised a hegemony in German and even European chemistry that has probably never been matched. But it was of short duration. One of the developments that led to a sharp decline in his influence was the theory of substitution, which will be discussed in chapter 7; another was a resurgence of German empiricism, centering on the unlikely figure of Wöhler's first chemistry professor, the Heidelberg chemist Leopold Gmelin (1788–1853).

GMELIN'S INSURGENCY

The first edition of Gmelin's famous *Handbuch der theoretischen Chemie* appeared in 1817–19, and for the next three decades it constituted the chief competitor to Berzelius' *Lehrbuch*. Like the German stoichiometers, Gmelin was attracted to the apparent empirical certification of Wollaston's conventional equivalents, which he called *Mischungsgewichte* (mixing weights); he admitted nonetheless that the determination of these weights "is subject to some arbitrariness, since it cannot be determined with certainty whether A is united with 1, 1½, 2, etc., of B." In the third edition he asserted that *Mischungsgewichte* were "more probable" than atomic weights, although the water formula HO versus H_2O "cannot be decided with certainty." He placed some distance between his position and that of the atomists, without denigrating the latters' views.[83]

Wollastonian conventional equivalents were enormously popular and used nearly exclusively in Great Britain throughout the first half of the nineteenth century. In France the situation was complex, due to the introduction by Gay-Lussac of yet a third system, incorporating the assumed formula of carbonic acid as CO.[84] After a period of vogue of "Aequivalenten" and "Mischungsgewichte" most German chemists by the early 1830s converted to Berzelian atomic weights, led by Liebig and the Berzelius-*Schüler*. In general they followed the Berzelian notational conventions as well, e.g., barred letters for double atoms, dots for oxygen atoms, and superscripts.

The first sign of insurgency against the Berzelian orthodoxy was Liebig's decision to reject many of these notational conventions: for Berzelius' $\text{Ꮳ}^2\text{H}^4$ Liebig now wrote C_4H_8, $\overline{Ag}$ became AgO, and so on. He had various reasons for this reform: bars were typographically inconvenient, superscripts might be mistaken by mathematicians for exponents, dots could get lost, and organic and inorganic formulas would now appear unified.[85] From 1837 Liebig began to alter the papers of Berzelius and Wöhler published in his *Annalen der Pharmacie* — against the authors' wishes — to conform with the new style.

This might seem to be a rather minor rebellion, but it presaged a much deeper apostasy. The scene of the drama was the idyllic Black Forest town of Freiburg im Breisgau, where the Gesellschaft deutscher Naturforscher und Aerzte held their annual *Versammlung* in September 1838. Wöhler and Heinrich Rose picked up Liebig in Giessen and Gmelin in Heidelberg on their way south and west. In Freiburg they invited Magnus into their rented digs, even though this made the accommodations crowded. Wöhler wrote to Berzelius that he and his four companions behaved like "most merry students" rather than sober university professors. One has the impression of rowdy good spirits; others at the meeting misinterpreted their behavior as arrogance and christened them the "chemical aristocracy." They avoided the boring sessions, left before the end of the meeting, and for eight days they traveled, all five in Wöhler's carriage, through Alsace the long way home. The weather was magnificent, the scenery breathtaking, and all were in the best of moods. The oldest of the group and the only one who could not be regarded as a disciple of Berzelius, Leopold Gmelin, "entered completely into our cheerful, frivolous high spirits, but which did not prevent us from speaking often and seriously of chemistry, from which, I think, we have mutually profited more than from all the learned papers in Freiburg."[96]

A year later Liebig revealed to Berzelius one of the subjects of their discussions.

> I have long been occupied by a matter which is of the greatest importance for science. It came up during a journey to Freiburg, which I made in the company of L. Gmelin, Wöhler, Magnus, and H. Rose, and we all found ourselves in agreement on it. Namely, we should all decide to use equivalents in our formulas instead of the atoms of the volume theory. Instead of H_2 or $\underline{H}$ simply use H. In fact, if you would decide to join and lead us, you would be doing an extraordinary service for science; we agreed to ask you this, since it must happen sooner or later anyway. Why do we carry a theoretical viewpoint in the signs with which we express the composition, the constitution of a body. Theoretical views will change, but our nota-

tions ought not. What does theory, philosophy of chemistry, have to do with notation. We will never agree on the weight of an atom, a law of nature knows no exceptions . . . The distinction between atom and equivalent is especially disadvantageous for students. I beg you to consider this subject most carefully; you will celebrate a great triumph if the suggestion comes from you; all chemists, all the ones with whom I have spoken about this wish it.[87]

Berzelius replied:

> You seek to persuade me to join and lead those who reject the double hydrogen atom, because you think this must eventually happen, and would happen sooner if I should become the mouthpiece of this movement. Once more I must disagree with you. He who wishes to discard atoms and only speak of equivalents, can certainly do so. But then he should never speak of atoms. The concept of equivalents is relative to a specific series of compounds, and would be quite sufficient if bodies only combined in a single proportion. But since this is not the case, it is no longer positive, but merely conventional.[88]

This was really the nub of the whole matter, but unfortunately it was lost on Liebig. Berzelius was one of the few at this time who understood that what were known as equivalents were really nothing more than a rival set of atomic weights.

In June 1839 Wöhler remonstrated with Liebig concerning the latter's unauthorized alterations of his and Berzelius' notational styles.[89] The next paper by Berzelius to appear in the *Annalen* was accompanied by the following editorial "Remarks":

> I must first of all apologize that the unfortunate lack of barred letters has prevented us from reproducing the formulas in the manner in which they were written in Berzelius' manuscript. At this opportunity I cannot but cite the main reason why I consider the Liebig-Poggendorff form of expressing the composition of bodies as a better one; I do not deny that Berzelius' formulas for the composition of organic compounds are shorter and more convenient, but to introduce them into organic chemistry would be to give form and color to a large number of hypothetical assumptions. . . . Formulas, if they are not to be changed every moment, really ought not express theoretical views, and I myself think that our notation has won little, but lost in simplicity and conciseness, when we include in our notation of compositions theoretical conclusions derived from volume theory. Equivalents will never change; and I

greatly doubt if complete agreement over the expression of weight relations by atomic weights will ever be obtained. The study of chemistry would be infinitely facilitated if all chemists agreed to return to equivalents. If only one or two do this by themselves, it would cause more confusion than anything else. It is to be hoped that the time is no longer distant when all chemists will do this.[90]

Four years later the issue arose again, and this time Berzelius and Wöhler were able to persuade Liebig to instruct his printer to procure the necessary barred and dotted symbols.[91]

However, in 1844 all of these efforts at reconciliation unraveled. Liebig published a long uncomplimentary review article entitled "Berzelius und die Probabilitätstheorien." He wrote: "Surely no one could charge me with a crime . . . when, following my convictions, I declared it a mistake to make our notation an expression of changing theoretical conceptions (for example the volume theory)."[92] And now he carried through on his resolve made six years earlier and returned to the Wollaston-Gmelin formulations in all his published work.

Led now by Liebig as well as Gmelin, there was a general movement to conventional equivalents in the mid-1840s. The English — and, following their example, the Americans — had always preferred equivalents. New French and German textbooks now adopted them in increasing numbers.[93] An international crop of bright young chemists — Hermann Kolbe, Edward Frankland, Marcellin Berthelot — joined the trend. By 1850 they were clearly the most popular basis of formula notation.

It is therefore highly ironic that in the preface to the monumental fourth edition of his *Handbuch* (1843) Gmelin announced his "decisive conversion" to the atomic theory, exchanging *Mischungsgewichte* for *Atomen*.[94] He meant this quite literally. In previous editions he had refused to make any sharp distinctions between mixtures and compounds, in a sense reverting to Berthollet's discredited ideas. Now his distinction was clear, and he wholeheartedly accepted stoichiometry and chemical atomism. Despite his conversion, however, he retained conventional equivalents. He pointed out that Berzelius himself introduced barred atoms because certain elements (such as hydrogen) *always* enter into combination in pairs.[95] Is it not then more consistent to assume that such a presumed double atom is in fact a single atom of twice the weight? Thus the confusing necessity to distinguish between equivalent and atom would vanish.[96] This is evidently the sort of argument Gmelin had used to convince Liebig, Wöhler, Magnus, and Rose of the desirability of a return to conventional equivalents.

One might well inquire as to why this small revolution was successful precisely at this time, the late 1830s and the 1840s. One possible factor was Faraday's discovery in 1834 of the laws of electrolysis: the weight of matter electrodeposited is proportional to the quantity of electricity passed and electrochemical equivalents coincide with chemical equivalents. Faraday regarded this as strong evidence for the reality of (conventional) equivalents and for the idea that all electrolytes contain exactly one (conventional) equivalent of each ion.[97] But Berzelius and others disputed Faraday's interpretations,[98] and Stanley Guralnick has convincingly argued that most contemporary chemists did not view Faraday's laws in this light.[99]

A more compelling reason for the success of conventional equivalents was a resurgence in positivism among European scientists and intellectuals, datable quite precisely to the late 1830s. In 1837, for instance, were published William Whewell's *History of the Inductive Sciences* and Dumas' *Leçons de philosophie chimique*, both of which immediately acquired reputations for skeptical attitudes toward atoms (although I have argued above that in both cases this skepticism did not apply to *chemical* atoms).[100] At the British Association meeting in Liverpool the same year a concerted attack on atoms seems to have taken place.[101] In the following year the third volume of Auguste Comte's *Cours de philosophie positive* appeared, much of which dealt with chemistry. Comte argued that Wollaston's "transformation . . . of the atomic theory per se into that of *chemical equivalents*, which offers a much more positive statement . . . would definitely constitute a major improvement, if it is not reduced to a simple artifice of language, the actual concept itself remaining essentially unchanged."[102]

Comte's work was immediately popular and quite widely read. It cannot be coincidence that Berzelius began using the term "Positivismus" just at this time.[103] Liebig too began to turn away from theorizing, or at least he claimed to do so. He declared to Berzelius in July 1839 that he suffered from "a real fear of theoretical discussions" because of the intrinsic uncertainties; after reading Dumas' and Persoz' speculations on substitution the following spring he wrote: "I have become totally sober—colder and more rational than you can imagine . . . I was cured, it was an emetic, everything is disgorged and emptied, I am resolved never to mention it in my journal." He now turned resolutely from theory to practice. This was an indication of his, and others', drift toward empiricism around 1840.[104]

1. For a recent discussion of this problem, see N. W. Fisher, "Avogadro, the Chemists, and the Historians of Chemistry," *History of Science* 20 (1982): 77–102, 212–31 (79–81).

2. E. M. Melhado, "Mitscherlich's Discovery of Isomorphism," *Hist. Stud. Phys. Sci.* 11 (1980): 87–123; J. R Partington, *History of Chemistry*, 4 vols. (London: Macmillan & Co., 1961–70), 4:205–11, and works cited therein. Probably the most insightful biography is W. Prandtl, *Deutsche Chemiker in der ersten Hälfte des neunzehnten Jahrhunderts* (Weinheim: Verlag Chemie, 1956), pp. 242–85 (for the date of discovery, see Mitscherlich's letter reproduced on p. 247).

3. H. W. Schütt, "Zum Prioritätsproblem der Entdeckung des chemischen Isomorphismus," *Physis* 16 (1974): 5–22; Melhado, "Isomorphism," pp. 88–111.

4. Melhado, "Isomorphism," pp. 89, 111. What Melhado calls stoichiometry is in many cases what I refer to as chemical atomism.

5. Mitscherlich, "Ueber die Krystallization der Salze, in denen das Metall der Basis mit zwei Proportionen Sauerstoff verbunden ist," *Abh. Akad. Wiss. Berlin, 1818–19*, 427–37 (read 9 December 1819); "Sur la rélation qui existe entre la forme crystalline et les proportions chimiques," *Ann. chim. phys.* [2] 14 (1820): 172–90.

6. A. Mitscherlich, ed., *Gesammelte Schriften von Eilhard Mitscherlich* (Berlin, 1896), pp. 2–4; E. Mitscherlich, *Ueber das Verhältniss zwischen der chemischen Zusammensetzung und der Krystallform arseniksaurer und phosphorsaurer Salze*, Ostwalds Klassiker nr. 94 (Leipzig, 1898), ed. nn. by P. Groth, pp. 55–57.

7. Mitscherlich, "Om förhållandet emellan kemiska sammansättningen och krystallformen hos arseniksyrade och phosphorsyrade salter," *Kungl. Svenska Vetenskapsakademiens Handl., 1822*, 4–79; *Gesammelte Schriften*, pp. 133–73, and *Verhältniss*, pp. 3–54, esp. p. 3 and 58n.

8. Mitscherlich, *Verhältniss*, pp. 3, 54.

9. Mitscherlich, "Sur la rélation," p. 174.

10. Mitscherlich, *Verhältniss*, p. 52; what differs in the dimorphic sub-

stances is "die Lagen der Atome zu einander," or "die verschiedene Stellung, welche die Atome zu einander haben."

11. Dalton, "Remarks on the Essay of Dr. Berzelius on the Cause of Chemical Proportions," *Ann. Phil.* 3 (1814): 174-80; Berzelius, "An Address to those Chemists who wish to examine the Laws of Chemical Proportions, and the Theory of Chemistry in general," *Ann. Phil.* 5 (1815):122-31; see above, p. 78.

12. Berzelius, "An Address to those Chemists. . . . ," p. 122.

13. H. G. Söderbaum, ed., *Jac Berzelius Bref*, 6 vols. (Uppsala, 1912-61), 1:iii, 43 (Marcet to Berzelius, 5 May 1813).

14. Ibid., 1:ii, 64 (Davy to Berzelius, 19 October 1813). Davy's defense of the elementarity of chlorine must not be construed as a retreat from reductionism; he treated chlorine only as provisionally elemental, as "undecompounded" rather than necessarily as truly simple.

15. Dalton, "Remarks," pp. 176-77; T. Thomson, *Ann. Phil.* 9 (1817): 38. The single-atom axiom hinges only on the choice of atomic weights: AB_3 is the same as A_2B_3 with a doubled weight for element A. It is, thus, an axiom or convention rather than an empirically verifiable law.

16. See above, pp. 73-78.

17. Berzelius, "Sur la composition des acides phosphorique et phosphoreux, et sur leurs combinaisons avec les bases salifiables," *Ann. chim. phys.* [2] 2 (1816): 151-76, 217-40, 329-39; Berzelius, "Untersuchungen über die Zusammensetzung der Phosphorsäure, der phosphorige Säure, und ihrer Salze," *Ann. Phys.* 53 (1816): 393-446, 54 (1816), 31-55; Dulong, "Extrait d'un mémoire sur les combinaisons du phosphore avec l'oxigène," *Ann. chim. phys.* [2] 2 (1816): 141-50; Thomson, *Ann. Phil.* 9 (1817): 34-40; *Berzelius Bref*, 1:iii, 137-40, and 3:i, 43-46.

18. PO_5 and PO_3 are correct anhydride formulas with a doubled atomic weight for phosphorus: $2H_3PO_4 - 3H_2O = P_2O_5$, and $2H_3PO_3 - 3H_2O = P_2O_3$. The cited oxide-rule multiples emerge from the following (the last column gives Berzelius' formulations and oxide-rule multiples):

$2NaH_2PO_4 - 2H_2O$ = $Na_2O \cdot P_2O_5$ = $NaO_2 \cdot PO_5$ (5)
$2Na_2HPO_4 - H_2O$ = $2N_2O \cdot P_2O_5$ = $NaO_2 \cdot PO_5$ ($^5/_2$)
$2Na_3PO_4$ = $3Na_2O \cdot P_2O_5$ = $3NaO_2 \cdot 2PO_5$ ($^5/_3$)

19. Berzelius, "Untersuchungen," pp. 40-42, 51-52; *Berzelius Bref*, 1:iii, 139-40 (Berzelius to Marcet, 9 May and 26 June 1816).

20. Berzelius, "Proportions, determinate," in D. Brewster, ed., *Edinburgh Encyclopaedia* (1830), 17:175-90 (184-85), dated and discussed by E. Melhado, *Jacob Berzelius: The Emergence of His Chemical System* (Madison: University of Wisconsin Press, 1981), pp. 271-73. Melhado's consideration of Berzelius' "unitary laws" (pp. 270-78) is excellent; I have adopted his assertion of the conceptual connection between the single-atom axiom and the oxide rule.

21. Berzelius, "Nouvelles recherches sur les proportions chimiques," *Ann. chim. phys.* [2] 5 (1817): 174-81 (179-80); "Försök att närmare bestämma åtskilliga oorganiske kroppars sammansättning, till vinnande af en närmare

utveckling af läran om de kemiske proportionerna," *Afh. fys. kemi* 5 (1818): 379-520 (455-75); "Versuche die Zusammensetzung verschiedener unorganischer Körper näher zu bestimmen. . . . " *J. Chem. Phys.* 23 (1818): 160, 170-86; *Berzelius Bref*, 1:iii, 151 (Berzelius to Marcet, 14 April 1817). See Melhado, *Berzelius*, pp. 248-61.

22. *Berzelius Bref*, 1:iii, 173-75 (Berzelius to Marcet, 22 April and 17 July 1818); J. E. Jorpes, *Jac. Berzelius* (Stockholm: Almqvist & Wiksell, 1966), pp. 45-47, 81-84, 97; Berzelius, *Autobiographical Notes* (Baltimore: Williams & Wilkins, 1934), pp. 96-107. Also see C. A. Russel's introduction to the 1972 Johnson rpt. of Berzelius' *Essai sur la théorie des proportions chimiques* (Paris, 1819).

23. *Jahresberichte über die Fortschritte der physischen Wissenschaften*; the first three vols. (1822-24) were translated by C. G. Gmelin, and the rest by Wöhler. Each report was published in German the year following its composition in Swedish and therefore treats works published two years earlier than the imprint. To avoid ambiguity I will hereafter cite this series as, e.g., *Jahresbericht* for 1820, 1 (1822).

24. Berzelius, *Essai*, pp. 48-49.

25. Ibid., p. 29.

26. Ibid., p. xv

27. *Berzelius Bref*, 1:iii, 193-94. In this letter, dated 27 April 1819, Berzelius told Marcet that he was intentionally avoiding "une connaissance plus particulière" of the law of atomic heats, in order not to be accused of using Dulong and Petit's work without citation. But, he added, if there are discrepancies between the atomic weights deduced, "Dulong's must be preferred to mine." The passage cited from p. 29 of the *Essai* also appears in the 1818 *Lärbok* (3:25), but mention of his "crainte secrète" is new to the *Essai*. It would seem that Dulong and Petit's work encouraged him to continue his drift toward acknowledging exceptions to the single-atom axiom.

28. *Berzelius Bref*, 3:ii, 123 (May 1820); also 1:iii, 207 (Berzelius to Marcet, 12 July 1820).

29. Berzelius, "Om sammansättningen af svafvelhaltige blåsyrade salter . . . ," *Kungl. Svenska Vetenskapsakademiens Handl.* 1820, pp. 82-99, and *Jahresbericht* for 1820, 1 (1822): 45-49. Berzelius formulated potassium thiocyanate in his 1818-20 atomic weights as $K + 2NC^2S^2$. L. Gmelin, "Ueber ein besonderes Cyaneisenkalium," *J. Chem. Phys.* 34 (1822): 325-46.

30. For details regarding Berzelius' gradual shift in the years 1816-23 to the acceptance of the elementarity of chlorine and nitrogen, see Partington, *History*, 4:167-68, and H. G. Söderbaum, *Berzelius' Werden und Wachsen* (Leipzig, 1899), pp. 116-27. For Berzelius' own summary of this shift, see *Jahresbericht* for 1821, 2 (1823): 53-61.

31. Berzelius, *Jahresbericht* for 1822, 3 (1824):80; ibid., 4 (1825):92, 113n.

32. Berzelius, *Lehrbuch der Chemie*, trans. F. Wöhler, 4 vols. in 8 (Dresden, 1825-31). This represents a translation of the Swedish second edition of vols. 1 and 2 (1817, 1822), and a revised translation of the first edition of vols. 3-6 (1818, 1827, 1828, 1830). It is properly referred to as the German second edition, though this does not appear on the title page. For details on this and other editions, see A. Holmberg, *Bibliografi över J. J. Berzelius* (Stockholm,

1933), pp. 1-3; Jorpes, *Berzelius*, pp. 94-96; and O. Wallach, ed., *Briefwechsel zwischen J. Berzelius und F. Wöhler*, 2 vols. (Leipzig, 1901), passim (hereafter cited simply "Wallach").

33. Berzelius, "Ueber die Bestimmung der relativen Anzahl von einfachen Atomen in chemischen Verbindungen," *Ann. Phys.* [2] 7 (1826): 397-416, 8 (1826):1-24, 177-90; *Lehrbuch*, 3:i, 87-131 (1827). See correspondence in Wallach, 1:121-31 (127, 130), and *Berzelius Bref*, 2:i, 69-70 (Berzelius to Dulong, 9 Sept. 1826).

34. Berzelius, "Bestimmung," pp. 397-401.

35. Ibid., pp. 401-2.

36. *Berzelius Bref*, 1:iii, 198-99, 207-9 (Berzelius to Marcet, 11 October 1819 and 12 July 1820); ibid., 2:i, 10-11 (Berzelius to Dulong, 5 November 1819).

37. *Jahresbericht* for 1820, 1 (1822):67-74 (74); ibid., 2 (1823):41-43.

38. Berzelius, "Bestimmung," p. 402.

39. See above, n. 27. Cf. also *Berzelius Bref*, 2:1, 69-70 (Berzelius to Dulong, 9 September 1826).

40. Silver, tellurium, and cobalt, of which the latter two truly are anomalous. Silver caused a problem for Berzelius because it was the only monovalent element among the thirteen, and his 1819 atomic weight had been four times rather than merely twice too high (by assuming AgO_2 rather than modern Ag_2O).

41. Except bismuth, which was 50% too high; the seven halved weights were those for iron, lead, copper, nickel, zinc, tin and gold; "Bestimmung," pp. 413-15.

42. Ibid., pp. 411-13, 415-16, 22-23.

43. Ibid., pp. 8-9.

44. Partington, *History*, 4:203, 256, 751; J. H. Brooke, "Wöhler's Urea, and its Vital Force? — A Verdict from the Chemists," *Ambix* 15 (1968): p. 110.

45. See above, p. 64-65.

46. Davy, *Elements of Chemical Philosophy*, pp. 181-82; cf. similar statement on pp. 488-89.

47. J. B. Biot and L. J. Thenard, "Sur l'analyse comparée de l'arragonite et du carbonate de chaux rhomboïdal," *Bull. Soc. Philomatique* 1 (1807): 32-35.

48. Davy, "Some Experiments on the Combustion of the Diamond . . . ," *Phil. Trans. Roy. Soc.* 104 (1814): 557-70; R. Siegfried, "Sir Humphry Davy on the Nature of the Diamond," *Isis* 57 (1966): 325-35.

49. Gay-Lussac and Thenard, *Recherches physico-chimiques*, 2 vols. (Paris, 1811), 2:288-92. Gum arabic is a water-soluble polysaccharide resin derived from the acacia tree.

50. *Berzelius Bref*, 3:ii, 115-16 (Berzelius to Gay-Lussac, 25 September 1811).

51. Gay-Lussac, "Mémoire sur l'iode," *Ann. chim.* 91 (1814): 5-160 (148-49); see M. P. Crosland, *Gay-Lussac* (Cambridge University Press, 1978), pp. 134-35 for a discussion of Gay-Lussac and isomerism.

52. Davy, *Elements of Agricultural Chemistry* (London, 1813), p. 113, and

"On the Analogies between the undecompounded Substances," *J. Sci. Arts* 1 (1816): 283-88 (285); J. E. Bérard, "Essai sur l'analyse des substances animales," *Ann. chim. phys.* [2] 5 (1817): 290-98 (296-97); for Chevreul, see Partington, *History* 4:256-57.

53. Meinecke failed to cite Gay-Lussac, but it is improbable that he had not read the passage just cited.

54. Meinecke, *Erläuterungen zur chemischen Messkunst* (Halle, 1817), pp. 142-45; "Fortsetzung der Anleitung zur chemischen Messkunst," *J. Pharm.* [2] 2:ii (1818), 3-41 (3-15); 3:i (1819), 21-59 (39). I have transcribed Meinecke's unique formula notation into a more familiar style, making no essential alterations.

55. Kastner, "Bemerkungen," *J. Chem. Phys.* 26 (1819): 253-61; Döbereiner, *Anfangsgründe der Chemie und Stöchiometrie*, 2d ed. (Jena, 1819), pp. 4-5.

56. C. G. Nees von Esenbeck, C. G. Bischof, and H. A. Rothe, *Die Entwickelung der Pflanzensubstanz physiologisch, chemisch und mathematisch dargestellt, mit combinatorischen Tafeln der möglichen Pflanzenstoffe und den Gesetzen ihrer stöchiometrischen Zusammensetzung* (Erlangen, 1819), esp. pp. 35, 47, 67, 75, 108-9, 133-63, 178.

57. After the passage cited above (*Berzelius Bref*, 3:ii, 116), Berzelius added: "How could we ever come to clear ideas on this subject without electrochemical science?"

58. *Berzelius Bref*, 1:iii, 151-52, 155-57 (14 April and 23 September 1817); Berzelius, "Lettre . . . sur la combinaison de l'oxigène avec le fer, le manganèse et l'étain," *Ann. chim. phys.* [2] 5 (1817): 149-60 (156).

59. Berzelius, "Experiments to determine the Definite Proportions in which the Elements of Organic Nature are combined," *Ann. Phil.* 4 (1914): 323-31, 401-9; 5 (1815): 93-101, 174-84, 260-75 (274). This work occupied Berzelius the entire fall and winter 1813-14 and was sent to Thomson in April 1814 (*Berzelius Bref* 1: iii, 88-89, 94). The context makes clear that Berzelius is *not* offering here the structural interpretation of isomerism.

60. *Berzelius Bref*, 1:i, 73; 1:iii, 198-99, 207-9; 2:i, 10-11, 69-70; 3:ii, 124; *Jahresbericht* for 1820, 1 (1821): 67-74.

61. Liebig and Gay-Lussac, "Analyse du fulminate d'argent," *Ann. chim. phys.* [2] 25 (1824): 285-311; the authors expressed this as a verbal equivalent formula.

62. Wöhler, "Ueber die eigenthümliche Säure, welche entsteht, wenn Cyan (Blaustoff) von Alkalien aufgenommen wird," *Ann. Phys.* 71 (1822): 95-103; "Bildung der Cyansäure auf neuem Wege, und fernere Untersuchungen über die Cyansäure und deren Salze," *Ann. Phys.* 73 (1823):157-72; "Zerlegung der Cyansäure," *Ann. Phys.* [2] 1 (1824): 117-24.

63. I.e., $C = 12$, $O = 16$, $N = 14$, $Ag = 430$; *Jahresbericht* for 1822, 3 (1824), 75-80; ibid., 4 (1825), 69-71, 91-92, 110-17.

64. Wöhler, "Recherches analytiques sur l'acide cyanique," *Ann. chim. phys.* [2] 27 (1824): 196-200; Gay-Lussac, n. on pp. 199-200.

65. Berzelius, *Jahresbericht* for 1824, 5 (1826): 62-65, 85-94; ibid., 6 (1827): 104-8; Liebig, "Neue Analyse von Wöhler's Cyansäure," *Archiv für die gesamte Naturlehre* 6 (1825): 145-53; Wöhler, "Ueber die Zusammenset-

zung der Cyansäure," *Ann. Phys.* [2] 5 (1825): 385-88; Wallach, 1:100-102 (Wöhler to Berzelius, 11 December 1825).

66. Liebig, "Ueber Cyan- und Knallsäure," *J. Chem. Phys.* 48 (1826): 376-81; Berzelius, *Jahresbericht* for 1826, 7 (1828): 120-21; J. Volhard, *Justus von Liebig*, 2 vols. (Leipzig: Barth, 1909), 1:403.

67. I disagree with Brooke's conclusion that the cyanate-fulminate isomerism was "inconclusive" ("Verdict," p. 110n).

68. Wallach, 1:207, 305-6; Partington, *History* 4:258-60, 272, 751; Brooke, "Verdict," 108-13.

69. Berzelius, "Ueber die Zusammensetzung der Weinsäure und Traubensäure . . . nebst allgemeinen Bemerkungen über solche Körper, die gleiche Zusammensetzung, aber ungleiche Eigenschaften haben," *Ann. Phys.* [2] 19 (1830): 305-35, esp. 319, 326-35; *Jahresbericht* for 1830, 11 (1832): 44-48.

70. *Jahresbericht* for 1831, 12 (1833): 63-64. Berzelius' example of metamerism was stannous sulfate and the yet unknown stannic sulfite, $SnO \cdot SO_3$ and $SnO_2 \cdot SO_2$.

71. Ibid., 13 (1834), 186-88.

72. Dumas and Boullay, "Mémoire sur la formation de l'éther sulfurique," *Ann. chim. phys.* [2] 36 (1827): 294-310; Dumas and Boullay, "Mémoire sur les éthers composés," *Ann. chim. phys.* 37 (1828): 15-53. The name "etherin" was coined by Berzelius.

73. *Jahresbericht* for 1827, 8 (1829): 286-97; *Lehrbuch der Chemie*, 3: i (1827), 147-48; Wallach, 1:457, 604 (Berzelius to Liebig and Wöhler, 2 September 1832; Wöhler to Berzelius, 28 January 1835).

74. Wallach, 1:455-59; *Jahresbericht* for 1832, 13 (1834): 190, 197-208.

75. Partington, *History* 4:347-52.

76. Liebig, "Ueber die Constitution des Aethers und seiner Verbindungen," *Annalen der Pharmacie* (cited hereafter as *Ann.*) 9 (1834): 1-18, 31-39. J. Carrière, ed., *Berzelius und Liebig: Ihre Briefe von 1831-1845* (Munich, 1898; hereafter cited as Carrière), pp. 61, 74-75 (Liebig to Berzelius, 30 May and 26 November 1833).

77. Berzelius, "Ueber die Constitution organischer Zusammensetzungen," *Ann.* 6 (1833): 173-76; *Jahresbericht* for 1832, 13 (1834): 191-97; Carrière, pp. 55-56 (Berzelius to Liebig, 21 May 1833). "Wood spirit" was impure acetone.

78. Wallach, 1:514; Volhard, *Liebig* 1:267.

79. Partington, *History* 4:252-53, 283-86, 347-58, 367-75.

80. Dumas and Liebig, "Note sur l'état actuel de la chimie organique," *Comptes rendus* 5 (1837): 567-72 (569).

81. Wallach, 1:304; Carrière, pp. 1-2.

82. Liebig, "Ueber die Zusammensetzung der Camphersäure und des Camphers," *Ann. Phys.* [2] 20 (1830): 41-47 (44); Liebig, "Ueber einen neuen Apparat zur Analyse organischer Körper . . . ," *Ann. Phys.* [2] 21 (1831): 1-43; Carrière, pp. 3-5 (Liebig to Berzelius, 8 January 1831).

83. L. Gmelin, *Handbuch der theoretischen Chemie*, 3 vols. (Frankfurt, 1817-19), 1:28-29, 286; 3d ed. (1827-30), 1:ix, 31-35.

84. All three systems were in use for many years: J. Persoz, *Introduction à*

l'étude de la chimie moléculaire (Paris, 1839) used Berzelian atomic weights, Gay-Lussac used conventional equivalents, and Dumas and most other French organic chemists used the "French system" of H = 1, C = 6, O = 16, N = 14, Na = 46, etc.

85. Liebig, "Constitution des Aethers," p. 3n.; Liebig and J. C. Poggendorff, *Handwörterbuch der reinen und angewandten Chemie*, 9 vols. (Brunswick, 1842-62), 1: ix (fascicle published 1836).

86. Wallach, 2:59-62 (Wöhler to Berzelius, 14 October 1838).

87. Carrière, pp. 201-2 (Liebig to Berzelius, 5 September 1839); see the very similar letter, Liebig to J. Pelouze, 14 October 1838, printed by R. Fox, *The Caloric Theory of Gases* (Oxford: Clarendon, 1971), pp. 319-20.

88. Ibid., pp. 205-6 (letter of 27 September 1839).

89. Ibid., pp. 189-90 (Wöhler to Liebig, 27 June 1839). This letter does not appear in the expurgated Liebig-Wöhler correspondence published by Hofmann.

90. Liebig, *Ann.* 31 (1839): 35-36n.

91. Carrière, pp. 247-52 (October-November 1843).

92. Liebig, *Ann.* 49 (1844): 295-335 (300).

93. E.g., J. Pelouze and E. Fremy, *Cours de chimie générale*, 3 vols. (Paris, 1848-50); H. Buff, *Lehrbuch der Stöchiometrie* (Nuremberg, 1842); F. F. Runge, *Grundriss der Chemie* (Munich, 1848).

94. Gmelin, *Handbuch der Chemie*, 4th ed., 12 vols. (Heidelberg, 1843-70), 1:iv-vi, 40-65. Five volumes were published by Gmelin's death in 1853; the rest were edited by colleagues and students.

95. For instance, Berzelius wrote hydrogen chloride as $\text{H}\hspace{-0.5em}=\hspace{-0.5em}\text{Cl}$, a four-volume formula. This was a result of his doubled atomic weights for monovalent metals ($H_2Cl_2 + KO = H_2O + KCl_2$), in just the same way that four-volume formulas for most organic compounds derived from this assumption.

96. Ibid., p. 48.

97. For a convenient summary of Faraday's research, see L. P. Williams, *Michael Faraday* (London: Chapman & Hall, 1965), ch. 6, or Partington, *History*, 4:112-28, 173-74.

98. *Jahresbericht* for 1834, 15 (1836): 30-38; ibid., 16 (1837): 29-36; Wallach, 1:591, 607-8.

99. S. M. Guralnick, "The Contexts of Faraday's Electrochemical Laws," *Isis* 70 (1979): 59-75, esp. 61-66.

100. See above, pp. 86 and 115-18.

101. For a discussion, see Fox, *Caloric Theory*, p. 291, as well as his able treatment of Dumas' atomic skepticism, pp. 286-91. Liebig attended the Liverpool meeting but did not participate in the discussions, as his English was too poor: Volhard, *Liebig*, 1:143-46.

102. A. Comte, *Cours de philosophie positive*, 1st ed. (Paris, 1838), 3:149-50. Comte, too, was not as skeptical toward atoms as has been assumed; he seems to have regarded the atomic theory as a useful convention, and so he can be regarded as a chemical atomist. See L. Laudan, "Towards a Reassessment of Comte's 'Methode Positive,'" *Phil. Sci.* 38 (1971):35-53.

103. Wallach, 2:34 (Berzelius to Wöhler, 28 June 1838); Carrière, p. 206 (Berzelius to Liebig, 27 September 1839).

104. Specifically, Liebig began to occupy himself with agricultural and physiological chemistry from 1840 on: Carrière, pp. 191, 210-11 (Liebig to Berzelius, 26 April 1840); cf. also ibid., p. 94 (22 July 1834), where Liebig complains that "The loveliest theories are overthrown by these damned experiments; it's no fun at all being a chemist anymore." Liebig's turn to empiricism is also documented by comparing the first and second editions of his *Familiar Letters on Chemistry*. In the first edition (New York, 1843, p. 18), he speaks highly of atoms and the atomic theory, but in the second edition (*Chemische Briefe*, Heidelberg, 1845, pp. 62-65), he praises equivalents and in general takes a considerably more positivistic approach.

7

Organic Chemistry in France

As the fourth decade of the nineteenth century began, chemical atomists found themselves divided into several distinct theoretical schools, loosely segregated along national boundaries. The predominantly English system of conventional equivalents was beginning to spread to the Continent but was being resisted by the Berzelian atomists; the hybrid system developed by Gay-Lussac and Dumas was largely confined to France. It required another generation before a majority of European chemists were able to agree on a single set of atomic weights. The scientific basis for that agreement was provided largely from within the rapidly maturing field of organic chemistry. This and the following two chapters describe these developments.

HYDRACIDS, SUBSTITUTION, AND TYPES

During a soirée in the Tuileries Palace, probably in 1827, the noble guests of King Charles X were annoyed by some mysterious vapors emitted by the chandelier candles. Entrusted with investigating the incident, the unofficial royal chemist Alexandre Brongniart asked his son-in-law Jean-Baptiste Dumas to analyze the candles. Dumas found that some of the chlorine used to bleach the wax had chemically combined with it; the irritating vapor was hydrogen chloride, released during combustion. Dumas, to whom we indirectly owe the account, said that this incident provided the inspiration for his extensive experiments on the chlorination of organic compounds, work that led in the course of the 1830s and 1840s to a decisive change of direction in chemical theory.[1]

Even a third of a century after the death of Lavoisier, chemists remained faithful, by and large, to the basic theoretical assumptions of the father of the chemical revolution. Lavoisier had made oxygen the centerpiece of his system as the principle both of combustion and of acidity. Bases were regarded as lower oxides of metals, acids as higher oxides of nonmetals. Organic acids consisted of hydrocarbon "radicals" united with oxygen. Salts, as the union of acids and bases, were therefore thought to consist of two differently oxidized substances combined together.[2] The existence of a few exceptions to the generalization that all acids could be shown to contain oxygen — such as prussic and muriatic acids — were not unreasonably thought merely to reflect the limitations of chemical analysis of the day.[3]

The first decade of the nineteenth century saw the development of chemical atomism by Dalton and of electrochemistry by Davy; the second decade witnessed Berzelius' synthesis of electrochemical atomism into the oxygen-centered dualistic system of Lavoisier. Berzelius instinctively applied these dualist precepts to the emerging field of organic chemistry. He routinely subtracted a molecule of water from the formulas of hydrated organic acids to obtain the formulas of the hypothetical anhydrides, regarded as the true acids; these relatively electronegative compounds could combine additively to the relatively electropositive bases to form salts. Analysis of the lead salt — his usual procedure — gave the anhydride formula directly. This had the effect of doubling the formula of the acid, relative to modern ideas.[4]

Ironically, the Lavoisierian scheme was weakened when Davy found that the alkalies and alkaline earths *were* oxides of hitherto unknown metals; sodium and potassium were thus shown to be converted into the strongest bases known by a dose of the "acidifying" oxygen principle. Furthermore, increasingly sophisticated analytical techniques failed to find any oxygen in prussic or muriatic acids. In 1810 Davy — perhaps inspired by Lavoisier's operationalism — declared oxymuriatic gas a provisional element or "undecompounded substance," and muriatic acid a hydride of "chlorine."[5]

Davy soon expanded the new class of acids represented by hydrogen chloride to include chloric and iodic acids. In his painstaking investigation of iodine and its compounds, Gay-Lussac agreed with Davy on the nature of chlorine and iodine; he discovered and investigated hydriodic acid, whose name he introduced, along with the name of the general class of "hydracids." Dulong even suggested shortly thereafter that oxalic acid was a hydracid.[6] Now acids without oxygen were anomalous for the Lavoisierian theory of acidity, but they could be assimilated into electrochemical dualism simply by asserting, as Davy and Gay-Lussac did, that chlorine was another "negative principle" *like* oxygen.[7]

However, other orthodox Lavoisierian concepts were challenged by the new type of acid. Instead of combining with bases by simple addition, as oxyacids were thought to do, hydracids had to undergo substitution of their hydrogen by an electropositive substance, such as sodium. What could prove an even more critical anomaly arose when Gay-Lussac published his classic memoir on prussic (hydrocyanic) acid: one of the substances that could replace the hydrogen of the acid was nothing other than chlorine itself! The fact that one of the most electronegative elements could substitute indifferently for one of the most electropositive would ultimately lead to the toppling of dualistic organic chemistry. But Gay-Lussac did not draw any special theoretical attention to the phenomenon in 1815, merely noting that "it is quite remarkable that two bodies with such different properties play the same role in combining with cyanogen."[8]

Berzelius initially tried to save orthodoxy by retaining the muriatic radical in chloric and hydrochloric acids. However, the discovery around 1820 of salts of the novel ferro-, ferri-, and thiocyanic acids convinced him of the existence of at least a limited class of hydracids. Others sought to follow Dulong's lead in expanding the hydracid concept to oxyacids, but the idea was not picked up by the leading theorists.[9]

In the summer of 1820, Davy's protégé at the Royal Institution, Michael Faraday, found that under the influence of sunlight chlorine could replace the hydrogen of the "Dutch oil" (ethylene chloride) to form "perchloride of carbon" (perchloroethane) and that the reaction proceeded such that "for every volume of chlorine that combines, an equal volume of hydrogen is separated."[10] In 1828 Gay-Lussac noted a similar phenomenon with respect to the chlorination of oils and waxes: chlorine "takes the place of the liberated hydrogen," and iodine and bromine react similarly. For this reason, he wrote, it may be necessary to forego bleaching wax with chlorine.[11]

We see that Dumas' anecdote about the "case of the corrosive candles" did not represent a novel observation; in fact, he admitted that he was *consciously* following Gay-Lussac's lead.[12] In 1834 he published a major paper on chlorine substitution, wherein a new "theory or law of substitutions or metalepsy" was announced.[13] The principal phenomenon exemplifying this "theory or law" was the conversion of alcohol into chloral by the action of chlorine. According to Dumas' etherin theory, alcohol consisted of a molecule of olefiant gas united with water, i.e., $C_8H_8 \cdot H_4O_2$; two-thirds of the hydrogen was thus represented as united with carbon, and the remaining one-third as combined with oxygen, forming water. Dumas maintained that the chloral reaction supported this conception. The first action of the chlorine, he suggested, was to remove the hydrogen of the water moiety, without replacement; three-fourths of the remaining hydrogen bonded to carbon was then substituted by chlorine atoms. Hence:

(1) $C_8H_8 \cdot H_4O_2 + Cl_4 = C_8H_8O_2 + Cl_4H_4$
(2) $C_8H_8O_2 + Cl_{12} = C_8H_2Cl_6O_2 + Cl_6H_6$

Dumas generalized this (assumed) phenomenon by asserting that whenever a substance containing water is subjected to chlorine, it first loses the hydrogen belonging to the water without replacement, after which chlorine substitutes volume for volume for hydrogen.[14]

Dumas made it clear here that he was not proposing an alternative to electrochemical dualism; indeed, he thought that these experiments "confirmed" his dualistic etherin theory. But few non-French chemists agreed. In particular, Justus Liebig continued to subject etherin to searching criticisms. On his return trip from Great Britain in October 1837, Liebig spent seventeen days in Paris; he met Dumas for the first time and managed to persuade him — only temporarily as it would turn out — to reject etherin and embrace ethyl.[15] Suffused with amity, the two promised to sheath their swords and henceforth work in unison. The détente lasted less than five months, but it formed a critical stage in the theoretical thinking of both chemists.

This *rapprochement* was more than symbolic. Liebig, too, had been approaching the phenomenon of substitution, but by another route: an expanded hydracid theory. In the benzoyl paper of 1832, Liebig and Wöhler had managed to replace the nonbenzoyl hydrogen in oil of bitter almonds by oxygen, cyanogen, halogen, and so on. They argued that this showed that the hydrogen "is combined with the other elements in a unique way," and used the radical concept to indicate that fact — the radical had to be $C_{14}H_{10}O_2$.[16] For all his praise of the benzoyl paper, Berzelius could not accept Liebig's interpretation, arguing that the benzoyl radical must be the hydrocarbon portion alone. As Liebig and Wöhler wrote the series of formulas, the implication was that substitution rather than electrochemical addition was taking place. Even worse, the force binding the radical and replaceable adduct could not be electrochemical, since positive hydrogen and negative chlorine, for example, seemed to be bound in the same manner and with about the same force.

The affinity between the benzoyl radical theory and the expanded hydracid theory, and of both to substitution theory, have not been much noted by historians, but the connection was evident to both Liebig and Berzelius. On 23 November 1837 Liebig wrote to Wöhler: "I have for years been dominated by the absurd idea that all organic acids are hydracids. You remember such expressions from our benzoyl paper."[17] To Wöhler Berzelius expressed doubts regarding the correctness of viewing oil of bitter almonds as a hydrogen compound analogous to hydracids.[18] To Liebig he wrote: "You further speak ... of hydrogen inside or outside of the radical. Here we fall into Laurent's thick fog-chemistry. I know quite well that you base this idea on your lovely experiments on benzoyl hydride."[19]

Liebig's nascent hydracid theory was expanded and developed a few years after the benzoyl radical theory. When T. J. Pelouze visited Liebig for several weeks in the late summer of 1836, one of the subjects of their collaboration was mellitic acid (modern $C_6(CO_2H)_6$). Citing the precedent of Dulong's view of oxalic acid, they formulated the substance as a dibasic hydracid, $C_4O_4H_2$.[20] Late that year Liebig outlined for Berzelius his emerging theory. In a second letter of 5 January 1837, he declared that cyanuric and citric acids "belong to the phosphoric acid series", which had been declared tribasic by Graham in 1833; furthermore, their hydrogen must be "outside the radical." Indeed, if *all* acids, organic and inorganic, were considered to be hydracids, "chemistry would acquire an admirable simplicity." He noted several facts, some familiar and some new, which were anomalous for dualistic acids but would not be for hydracids.[21]

Hydracid theory must have been one of the topics of conversation between Liebig and Dumas in October; after his return to Giessen, Liebig quickly performed a number of experiments, expanding the theory to include meconic acid. Again Liebig wrote Berzelius: "I am dominated by the idea that all organic acids are in a certain sense hydracids; you will say this is an old story, but listen. . . . " On 18 December Dumas read a co-authored paper on hydracids to the Paris Academy.[22]

Berzelius finally gave Liebig his opinion of the new theory; not surprisingly, he disagreed with Liebig, claiming that all of the new facts could easily be subsumed under the old theory. The formulas of all acids would have to be transformed if Liebig's theory were accepted, and salt formation would have to be viewed as substitution rather than addition. Berzelius could not see how Liebig's research necessitated such heroic alterations in the time-tested orthodoxy.[23]

Berzelius' critique cut little ice for Liebig, and he proceeded in the following spring to publish a major paper on the subject.[24] At this point Berzelius lost patience with his younger, less experienced friend. The theory was "strange" and even "nonsensical"; "to me it is impossible," he complained. "You have built a ramshackle structure that will topple in the first scientific wind to come along. And that will serve as a well-deserved punishment for having so capriciously abandoned the beautiful, simple, certain theory which earlier you proclaimed as a creed. If you only hold fast to it, you cannot help but produce beautiful things."[25]

After this Liebig, too, had no more patience left for his beloved mentor. Exasperated, he wrote to Wöhler: "Berzelius has been napping while we've been working. The reins have fallen from his hands, and that has awakened him; the old dull-toothed lion utters . . . a roar that could not frighten a mouse. Don't write any more about Berzelius' letter—he can only refer to things about which he is absolutely and totally wrong; he is dominated by the fixed idea that all my experiments concern the abstraction of water from certain salts with the help of heat; I have twice attempted to disabuse him of this error, but in vain."[26] Berzelius' "roar" was in the form of a letter to Pelouze, published in the *Comptes rendus*, that attacked the Dumas-Liebig hydracid paper.[27] He purported to be attacking Dumas alone, who he professed to believe was the real sole author of the piece. Liebig quite understandably felt personally rebuked, and he wrote Berzelius both privately and publicly to verify that he in fact was the principal author, and that he intended fully to defend the theory.[28]

It is yet another indication of the close historical and conceptual connection between hydracid theory and chlorine substitution that in the same open letter Berzelius also vituperated against the recent work in the latter field by Dumas, Laurent, and Malaguti. Like Liebig, Dumas rose to the defense publicly, claiming that he had been misrepresented. Repeatedly he protested that Berzelius was putting the words of his student Laurent in his (Dumas') mouth: "M. Berzelius attributes to me an opinion on this subject precisely contrary to that which I have always expressed, namely that in these circumstances chlorine takes the place of hydrogen without changing the nature of the substance."[29] The law of substitution, Dumas stressed, was a mere "empirical rule" that summarized "universal experience," and he was not responsible for those (i.e., Laurent) who asserted the *chemical* equivalence of chlorine and hydrogen in an organic compound. Furthermore, he reneged on his assurances to Liebig by reasserting the validity of the etherin theory.[30]

Dumas was soon to regret his words. Three months later he announced the discovery of a chlorinated acetic acid,[31] and the following year he sketched his theory of types.[32] He was astonished that the chlorinated acid was so little altered. "The chlorinated vinegar is still an acid, like ordinary vinegar; its acid power has not changed. It saturates the same quantity of base as before . . . So here is a new organic acid in which a very considerable quantity of chlorine has entered, and which exhibits none of the reactions of chlorine, in which the hydrogen has disappeared, replaced by chlorine, but which experiences by this remarkable substitution only a gradual change in its physical properties." He concluded: *"In organic chemistry there exist certain types which are conserved even when one has introduced in the place of the hydrogen which they contain equal volumes of chlorine, bromine, or iodine."*[33]

Berzelius replied to this assault.[34] The properties of acetic and chloracetic acids were quite different, in Berzelius' eyes, and he provided a comparative table that emphasized their divergent properties. By comparing chloracetic acid to inorganic chlorine compounds, he attempted to show that Dumas' new substance was easily assimilable into the dualistic scheme. He also chided Dumas for abandoning electrochemical organic chemistry.

Dumas' rebuttal conceded that the two acids, written $C_8H_8O_4$ and

$$C_8H_2{{O_4}\atop{Cl_6}},$$

had somewhat different properties, but their "fundamental" properties were the same, and so they belonged to the same type, $A_8B_8C_4$.[35] This was to become the key to the theory of types, as well as its chief weakness, for the assignment of which properties were "fundamental" opened the door to subjectivity and even capriciousness.[36]

The mature theory of types was developed in three papers published in the first two months of 1840.[37] Dumas now defined types as "bodies which are shown to be formed from the same number of chemical equivalents united in the same manner."[38] These types were conserved in substitution reactions — for substitution reactions follow the laws of chemical equivalents — and so the "fundamental chemical properties" were retained in all compounds of a given type. But Dumas was forced to modify this simple version for it was clear that many groups of compounds existed that satisfied this definition but which had widely differing properties, for example, formic acid $C_4H_2O_3$ and methyl ether C_4H_6O.[39] Dumas asserted that such substances were not to be thought of as belonging to the same "chemical" type, but did indeed belong to the same "mechanical" or "molecular" type.[40] Now one could imagine three modes of chemical reactions: (1) substitution which causes no alteration in the fundamental chemical properties, thus conserving the chemical type; (2) substitution which does result in such an alteration, producing a new chemical type but retaining the mechanical type; and (3) addition or elimination of atoms, which produces a new mechanical type. When the second sort of reaction takes place, "the molecule remains unchanged, in that it forms a group, a system, wherein an element has pure and simply taken the place of another. According to this entirely mechanical viewpoint, which M. Regnault has developed, all bodies formed by substitution would possess the same grouping and would belong to the same mechanical type. In my eyes they constitute a *natural family*."[41] Dumas asserted that *all* the atoms of a compound could be successively replaced without destroying the mechanical type, "as long as the molecule remains unchanged."[42]

Dumas concluded this important paper with a disclaimer as to the originality of his enunciation of the law of substitution,[43] and emphasized that the theory of types was not in irreconcilable conflict with the older orthodoxy.

> We have two systems before us: the one ascribes a principal role to the nature of the elements, the other to their number and arrangement. Pushed to extremes, each of these systems, in my opinion, can lead to absurdity. . . . The nature of the elements, their weight, form and arrangement, all must exert a real influence on the properties of the body. The influence of the nature of molecules was well defined by Lavoisier, that of their weight was characterized by Berzelius' immortal work. Mitscherlich's discoveries relate to the influence of their form, and only the future knows whether the present work of the French chemists is destined to yield the key to the role which their arrangement plays.[44]

As prophetic as these words turned out to be and as fruitful as the type theory in its general outlines became, the specific system that Dumas developed to classify organic compounds was not widely adopted.

Ultimately the introduction of the concept of mechanical types was the theory's turning point. Previous to this event, the theory had excellent predictive capacity ("substitution does not alter the fundamental chemical properties"), but poor accord with experience; afterwards, the theory's empirical fit was adequate, but it became useless as a predictive tool ("substitution *may* alter fundamental chemical properties"). Dumas pointed out that it was easy to assign bodies to the same "chemical system" — they must simply possess very similar chemical properties. But the assignment of mechanical types, he admitted, was more problematical: "I leave to competent authority [à qui de droit] the difficult decision of *which* experimental series should demonstrate that two bodies belong to the same mechanical system, be their apparent chemical properties ever so different."[45]

So much for mechanical types in this paper. But Dumas apparently soon decided that he was the "competent authority," for in the next two papers he considerably elaborated on his definition, giving a number of examples.[46] The chief difficulty of assigning compounds to particular mechanical types was the fact that such assignments called for judgments on similarity of what we would today call molecular structures. Although Dumas was willing to make such judgments, he recognized that the chemistry of his day was strained to its limits, and perhaps beyond, to do so.

LAURENT AND GERHARDT

Dumas based his theory of types not just on earlier experimental work on chlorine substitution, but also on the ideas of Auguste Laurent (1808–53). Laurent was Dumas' assistant from 1830 to 1832 when he studied the reactions of naphthalene. This research led him to what he called the theory of fundamental and derived radicals,[47] later christened the nucleus theory.[48] Naphthalene, for instance, was regarded as a fundamental radical that could chemically be altered to form various derived radicals, either by substitution (which took place according to

Dumas' recent law) or by addition. Certain of these alterations take place inside the radical, others outside, and such differences could be determined from the reactions of the substance.[49]

The most important factor in this scheme is not the identity of an atom but its position. Hydrogen or chlorine could play the same role inside a radical, but conversely a chlorine atom inside or outside of the radical would exhibit different chemical properties. Similarly, oxygen could replace hydrogen inside the radical with no great alteration of properties; but oxygen introduced *outside* the radical would make the compound acidic. This explained how, for instance, neutral carbohydrates could have so much oxygen (over 50%) whereas fatty acids could have so little (as low as 10%).[50]

Laurent and Dumas influenced each other in significant ways. Two of the roots of Laurent's theory were Dumas' work on etherin and on chlorine substitution; conversely, soon after roundly censuring Laurent's view that chlorine "plays the role" of hydrogen, Dumas proclaimed this idea as his guiding principle. Dumas' types, especially his mechanical types, were closely related to Laurent's radicals. However, Laurent's classificatory principle was not conservation of the total number of equivalents, as for Dumas, but rather conservation of the nuclear radical, a corollary of which was the conservation of the number of carbon atoms.[51] This criterion would prove much more fruitful than Dumas'.

Laurent's ideas may also have influenced Liebig in his modification of the orthodox radical theory. Liebig's radicals around 1838 were nonelectrochemical, and they could be indifferently substituted by chlorine, by metals, or by oxygen without acidification. He distinguished between atoms inside or outside of his radicals. All of these aspects were also characteristic of Laurent's radicals. In fact, Liebig and Laurent—and by 1839 Dumas as well—were tacitly proposing to reject the accepted electrochemical-dualist definition of a radical as the electropositive or hydrocarbon portion of a molecule, and to replace it by a more operational definition: the portion of a molecule that is conserved in a series of chemical reactions. It is ironic that none of these chemists could find much to agree upon.

Laurent became bitter about the lack of recognition for his very real contributions. He pointed out that chemists were eager to lay the failures of the theory at his door, but they ascribed its successes to Laurent's borrowings from Dumas. Certainly, Laurent admitted, he had been influenced by Dumas' work, in the same way that Dumas had built on his predecessors' accomplishments; but such an argument leads to infinite regress, all the way back to Moses, Hermes, and Tubal-Cain![52]

By the early 1840s, Laurent had a compatriot in the battle against dualism: Charles Gerhardt (1816-56). The striking similarities in the careers of Laurent and Gerhardt have often been noted. Zealous in their desire to reform organic chemistry, forceful and quite unawed by authority in their critique of accepted dogma, imaginative and daring in their theoretical conceptions, both were ostracized by the Parisian chemical establishment, both spent a few years teaching at provincial universities before finding private employment in Paris, and neither lived to the age of fifty. After their first meeting in the autumn of 1843 the two became close friends; in 1851 they founded a private school of chemistry in Paris. This collaboration was important for the development of both chemists, but perhaps especially so for Gerhardt. He wrote to his friend Cahours: "You must really get to know Laurent; you have to know him to appreciate him. . . . I assure you, he's the best fellow in the world, a bit touchy when it comes to questions of priority, but think how much this man has been wronged! . . . I am so proud that he has been willing to come down to my level; I have never met a man so filled with ideas; his letters are very precious to me, I study them as gospel."[53]

Gerhardt's first book, *Précis de chimie organique* (2 vols., Paris, 1844-46), exhibits an important theme of his career, namely his rejection of dualistic rational formulas and reliance on purely empirical formulas. Gerhardt denied the existence of water in hydrated acids and metal oxides in salts and pointed out that most of the organic radicals portrayed in dualistic formulas were completely hypothetical; he was skeptical about the possibility of determining atomic arrangements within molecules. In addition, the idea of homologous series developed by Gerhardt in the early 1840s naturally led to a purely numerical treatment of molecular formulas, analogous to a mathematical series, hence to the use of empirical formulas.

Gerhardt's structural agnosticism was shared by A. E. Baudrimont.[54] In the introductory section of his *Précis*, Gerhardt gave a sympathetic account of Baudrimont's ideas, and added: "Up to a certain point we share the opinion of M. Baudrimont, but without entirely rejecting the idea of molecular predisposition, as he does."[55] Nevertheless, *rational* formulas were entirely inadmissable for Gerhardt as for Baudrimont. His position at this time was that chemical molecules do have determinate arrangements of atoms, but that arrangement cannot be discovered by chemical reactions, and so one must rely on empirical formulations. Baudrimont, and M. A. Gaudin as well, shared this conviction, but maintained that structures could be determined by *physical* methods.

In an unpublished review of the *Précis*, Laurent praised the straightforward character of Gerhardt's taxonomy. "Nevertheless, we fear that the advantages of such a classification approach too nearly those of a simple dictionary." He noted that according to Gerhardt's system, methyl acetate, for instance, would not belong to the acetic acid family, but rather to the family of next-higher carbon number. Laurent expressed his disagreement with Gerhardt's total rejection of rational formulas; he asserted that certain groupings do subsist throughout a series of reactions, for instance ammonium and phenyl. Furthermore, Gerhardt did not remain consistent with his agnostic proclamations, assuming in certain cases the presence of water, as well as the nitro, sulfuric acid, hydrochloric acid, and other groups.[56]

In their correspondence Laurent castigated Gerhardt for the sterility of his positivist outlook.

> Don't reject every attempt to determine arrangement, for in spite of yourself it dominates you in a multitude of cases. . . . Your classification is bad. . . . Without a dominating idea, it is impossible to do anything. Will you ever get anything from your classification? No, nothing, absolutely nothing, because there is no idea there. A classification must show a series of relationships. And I am persuaded that, whatever may be the point of departure, one will always be able to come to interesting relationships. But this point of departure must be an idea. . . . Now, there is not a single relationship to

comprehend in your classification. It is well and good to keep repeating that we need neutral ground on which the whole world can meet. Well! Good grief! How about alphabetical order!⁵⁷

In Laurent's view there were two basic requirements for a useful chemical system: an unambiguous classificatory principle and a constitutional theory. The most promising candidate for the first of these was the number of carbon atoms in a compound, and Laurent and Gerhardt had each developed this idea in his own fashion, since his first theoretical paper. Laurent objected to Gerhardt's system primarily because there was no theory behind it, corresponding to his own theory of fundamental and derived radicals. Like Liebig, Laurent viewed theories relativistically and heuristically; even false theories were bound to lead the scientist to fruitful discoveries, whereas strict positivism leads nowhere. He wrote in 1854: "I do not reject researches into causes, though they may perhaps be nothing but a perpetual mirage destined to lead us incessantly to the exploration of new countries."⁵⁸

It is probable that the appearance of Gerhardt's *Précis* induced Laurent to publish his own views on chemical classification.⁵⁹ He emphasized the need for classificatory principles that are both rational and mnemonic, principles that would constitute "if not a method, at least an artificial system . . . without falling into the inconveniences of the dictionary or of arbitrariness."⁶⁰ The traditional practice of classifying according to chemical function was highly unsatisfactory, Laurent emphasized, because it was both ambiguous and nonheuristic. "What do we learn from such a system? that an acid is an acid, a base is basic, a dye dyes — and nothing more."⁶¹ Botanists, Laurent pointed out, do not classify all roots together, all stems, leaves, flowers in separate classes, but rather deal with the entire plants themselves; a root generates a stem which generates a leaf, and all belong to a single entity. This method is applicable to organic chemistry: Laurent's system arranges in one class seemingly unlike members that are mutually interconvertible, classes that form "veritable chemical trees," the key being "the principle of mutual generation."⁶² For example, acetic acid and chloracetic acid are mutually interconvertible, so they belong in the same class; but benzoic acid loses carbon dioxide irreversibly to form benzene, so these two compounds must be arranged in different classes.

THE REFORM OF EQUIVALENTS

These taxonomic efforts led Gerhardt and Laurent to a reform of the atomic weight-molecular formula problem. The first step in this direction was contained in a paper by Gerhardt read to the Paris Academy on 5 September 1842. Gerhardt noted that many organic reactions release or absorb "the elements of C_2O_4, H_4O_2, N_2H_6, or a multiple thereof" and never a *sub*multiple.[63] But chemists were accustomed to writing the formulas of these compounds as CO_2, H_2O, and N_2H_6, so for consistency the formulas of the first two really ought to be doubled.[64] On the other hand, if one reduces the atomic coefficients to least factors, one would have to write CO_2, H_2O, and NH_3. These would then be internally self-consistent, but to maintain consistent molecular magnitudes in organic reactions it would become necessary to halve the formulas of virtually all organic compounds—in other words to go from four–volume to two–volume organic formulas.[65]

This is a reasonably direct paraphrase of Gerhardt's reasoning, but that reasoning was by no means clearly expressed, nor was the paper semantically consistent. Here, again in paraphrase, is an example of a semantically confused argument. According to the atomic theory (volume theory), water consists of two volumes (atoms) of hydrogen and one volume (atom) of oxygen; according to the theory of equivalents, on the other hand, water contains equal equivalents of hydrogen and oxygen—that is, one equivalent of oxygen equals one volume (atom) of oxygen, but one equivalent of hydrogen equals two volumes (atoms) of hydrogen. (So far so good . . .) Now the water molecule that takes part in organic reactions is written H_4O_2, but in this expression H_4 equals two equivalents of hydrogen and O_2 equals one equivalent of oxygen(?); but it is more consistent with inorganic chemistry to say that water contains two equivalents (volumes, atoms) of hydrogen and one equivalent (volume, atom) of oxygen, i.e., that water should always be expressed as H_2O. But this also implies halving all organic formulas. In such a case atoms, volumes, and equivalents would all become synonymous expressions.[66]

This makes sense neither from a modern nor from an 1840s perspective. But perhaps more serious still was the *ex cathedra* style of the paper. Gerhardt proposed to double the atomic weights of all the metals and assume M_2O oxide formulas, but he did not bother to defend the proposal in any but the briefest way, nor to refute Berzelius' thoughtful and detailed rationale for such formulas as KO, NaO, and AgO. Understandably, the paper made a uniformly bad impression on his colleagues.[67] A second more detailed paper published early in 1843 was clearer and less confused, but it also made little headway in the chemical world.[68]

Gerhardt did of course make one convert: Laurent. Both in a private communication and in a published paper in 1844 Laurent declared his adherence to Gerhardt's new "equivalents."[69] In the autumn of 1846, Laurent published an article that fully clarified Gerhardt's proposals and contained many new and valuable thoughts.[70] He asserted that when formulas are written using Gerhardt's weights, the total number of hydrogen plus nitrogen atoms must be divisible by two (Gerhardt had stated a similar rule for the case where the number of nitrogen atoms is zero). Expanding this rule, he divided all atoms into two classes: "monasides" (later called "monads") such as carbon, oxygen, sulfur, silicon, selenium, and tellurium, and "dyodides" (later called "dyads") such as hydrogen, nitrogen, the halogens, phosphorus, arsenic, antimony, sodium, potassium, and silver. His "even-number rule" in its most general form asserted: "The sum of all dyads in a molecule must be an even number."[71]

Laurent used the even-number rule to deduce that radicals such as ethyl and methyl could never exist as stable compounds, and that the received formulas for cyanogen (CN) and amidogen (NH_2) must be doubled; further, water must be H_2O and not HO, metal oxides must be written Ag_2O, K_2O, and Na_2O, and the correct formulas for ethyl and methyl ethers are $C_4H_{10}O$ and C_2H_6O. He pointed out that his rule consistently dictated formulas taken at the same number of volumes, a procedure that is justified in its own right, since all formulas, organic and inorganic, were thereby related to a single standard. He explained the rule by supposing that dyadic elements exist in the form of "binary" molecules, as each atom must always be joined to a complementary atom ("moitié complémentaire"); monads, by contrast, can exist either singly or in pairs. Thus dyads always appear in twos, whereas monads may appear in any integral number.[72]

Laurent now distinguished clearly between atoms and molecules: "M. Gerhardt's atom represents the smallest quantity of a simple body which can *exist in a compound*. My molecule represents the smallest quantity of a simple body which must be employed to *produce a combination*; this quantity splits in two during the act of combination."[73] For example, a molecule of oxygen splits into two atoms in forming two molecules of water; a molecule of chlorine splits into two atoms to form a chlorine-substituted organic molecule and a molecule of hydrogen chloride. The atom Cl may enter into a combination, but the smallest quantity of chlorine that can participate in a reaction is Cl_2; this was an empirical law with no known exceptions. As for the rare instances when vapor density measurements lead to two-volume formulas with fractional atoms, as in the case of ammonium chloride, Laurent suggested that the molecule splits in two in the vapor state (this was sixteen years before Leopold von Pebal established this phenomenon experimentally). Ladenburg put the matter well when he said that here Laurent "first clearly stated what Gerhardt wished to advance."[74] Laurent cited Ampère (but not Avogadro), whom he was in some respects following.

Equivalents, Laurent emphasized, refer merely to combining weights and often say nothing about molecular magnitudes.[75] He pointed out that $H_2N_2O_6$ is "equivalent" to H_2SO_4 since they combine with equal weights of soda; but this does not mean that these are proper molecular formulas. What is required, he maintained, is a strict distinction between "monobasic" and bibasic" acids. The latter—including water and hydrogen sulfide as well as sulfuric, carbonic, oxalic, and tartaric acids—form acid and double salts, acid and double esters, acid and double amides, and anhydrides; the former—including the hydrogen halides as well as nitric, formic and acetic acids—form none of these substances, since they possess but a single replaceable hydrogen atom.[76]

Laurent recognized that the most fundamental task facing chemists was the assignment of correct molecular formulas from the empirical data. Although his even-number rule was his most trustworthy guideline (he repeatedly emphasized that there was not a single authenticated counterexample to the rule), he also relied on many other types of data: volume considerations, isomorphism, specific heats and volumes, and the concept of series (à la Gerhardt); and it was important to him that all these diverse methods agree with each other as much as possible. He stressed that *all* relevant facts must be used to arrive at consistent formulas, that there is no absolute method. His fundamental axioms were that compounds must be given the simplest possible formulas consistent with these data (rejecting, of course, any formulas containing fractional atoms) and that analogous substances must be given analogous formulas.[77]

The Avogadro-Ampère gas theory appears *not* to have been one of these fundamental axioms, but rather more of a consequence of them. Laurent's conviction that all formulas should be related to the same number of volumes was not predicated on the equal volumes-equal numbers hypothesis, but rather on the observation that chemical and physical anomalies, above all a lack of analogy between organic and inorganic formulas, disappear when one does so. The decision had little to do with belief in the truth of a physical axiom, but rather with a desire to relate all chemical substances to a uniform standard, at the same time satisfying *chemical* precepts—at the head of which stood the even-number rule. Laurent rarely mentioned "Ampère's idea"; I am not aware that Gerhardt ever mentioned it. It was a pencil-and-paper, a mathematical, a taxonomic sort of chemistry that led Gerhardt and Laurent to their reforms of atomic weights and molecular formulas; perhaps even more significant, it was an almost esthetic desire for consistency, simplicity, and beauty in the formulas used by chemists.

By 1846, then, Laurent had clarified the concepts of atoms, molecules, and equivalents. But he never distinguished between true and merely conventional equivalents, and his words fell on deaf ears — the general confusion merely increased in volume. By this time most chemists were dead set against the revolutionary ideas of "les deux," and did not invest the effort to understand what was not clearly expressed.

The irony of this story is rarely noted. But Ladenburg's insight was clear:

> It must appear striking and peculiar to any unprejudiced person, that the numbers which Gerhardt proposed as the "equivalents" of the elementary substances (with the exception of the numbers for the metals), agree almost completely with the atomic weights of Berzelius, of the year 1826. It is also noteworthy that Gerhardt does not mention Berzelius, and is obviously quite unaware that he, to a large extent, adopted his numbers. On the other hand, the Swedish chemist did not appear to have noticed this agreement, since he violently attacked Gerhardt's paper. What I consider as most remarkable, however, is the fact that, at the time when Gerhardt made his proposal, very eminent chemists (I only mention Liebig and his pupils) were actually employing atomic weights for the most important elements, such as carbon, hydrogen, chlorine, etc., with the ratios which Gerhardt recommended as new; but that, a few years afterward, the equivalents of Gmelin, against which Gerhardt's paper was directed, were almost universally adopted.[78]

Ladenburg was wrong in one respect: Gerhardt's paper was directed not against Gmelin's conventional equivalents — Ladenburg himself pointed out that such an attack was entirely unnecessary at that time, since these equivalents were not in fact in general use — but rather against the inconsistent application of Berzelius' atomic weight values, i.e., against the arbitrary doubling and halving of molecular formulas. Gerhardt's (and Laurent's) view was that they should all be based on a single standard. Of course, Berzelius saw nothing arbitrary about his procedure, so he felt he had good reasons to attack the Gerhardt-Laurent reform.[79]

Berzelius failed to see how much he had in common with the French reformers: the latter's work with regard to molecular magnitudes forms an analogy to the work of Berzelius with regard to atomic magnitudes. The fertility of their comprehensive approaches seeking to utilize all possible physical and chemical evidence, as opposed to the sterility of more positivistically inclined scientists who severely and intentionally limited their own horizons, became evident by the 1860s. It is tragic that Berzelius and the French reformers, whose methodologies were so similar, ended their lives in intellectual exile both from the chemical world as a whole and from each other.

1. Dumas told A. W. Hofmann, who related the anecdote in the obituary "Zur Erinnerung an Jean Baptiste André Dumas," *Berichte* 17R (1884): 630-760. The date is uncertain, but must have been after February 1826 (Dumas' marriage) and before July 1830 (abdication of Charles X); Hofmann commented that the incident occurred while Dumas was engaged in his work on the ethers, which makes 1827 the most likely year (p. 667). See L. J. Klosterman, "Studies in the Life and Work of Jean Baptiste André Dumas (1800-1884): The Period up to 1850" (Ph.D. diss., University of Kent at Canterbury, 1976, p. 299).

2. A. L. Lavoisier, *Traité élémentaire de chimie* (Paris, 1789); *Elements of Chemistry*, trans. R. Kerr (Edinburgh, 1790; rpt. New York: Dover, 1965), pp. 80, 159-64, 176-93. Lavoisier used the word "radical" to designate the nonoxygen moiety of any substance; his belief that the alkalies were metallic oxides was frankly labeled speculation.

3. See H. E. Le Grand, "Lavoisier's Oxygen Theory of Acidity," *Ann. Sci.* 29 (1972): 1-18, and esp. M. P. Crosland, "Lavoisier's Theory of Acidity," *Isis* 64 (1973): 306-25.

4. J. Berzelius, "Experiments to determine the Definite Proportions in which the Elements of Organic Nature are combined," *Ann. Phil.* 4 (1814): 323-31, 401-9; 5 (1815): 93-101, 174-84, 260-75. For instance, he wrote acetic acid as "6H + 4C + 3O" (pp. 174-76) from the analysis of lead acetate; our $Pb(C_2H_3O_2)_2$ therefore was $2C_4H_6O_3 \cdot PbO_2$ to Berzelius, or $C_4H_6O_3 \cdot PbO$ after he halved his atomic weight of lead in 1826.

5. H. Davy, "Researches on the oxymuriatic Acid . . . ," *Phil. Trans. Roy. Soc.* 100 (1810): 231-57 (243); "On some of the Combinations of Oxymuriatic Gas and Oxygene . . . ," *Phil. Trans. Roy. Soc.* 101 (1811): 1-35 (32).

6. Davy, "Some experiments on a solid Compound of Iodine and Oxygene, and on its chemical Agencies," *Phil. Trans. Roy. Soc.* 105 (1815): 203-13; "On the action of Acids on the Salts usually called Hyperoxymuriates, and on the Gases produced from them," *Phil. Trans. Roy. Soc.* 105 (1815): 214-19; J. L. Gay-Lussac, "Mémoire sur l'iode," *Ann. chim.* 91 (1814):5-160 (117, 130-52); P. L. Dulong, secondhand extract in *Mém. Inst.* (1815, publ. 1818), pp. cxcviii-cc. J. H. Brooke has recently demonstrated that Davy never developed a hydracid theory ("Davy's Chemical Outlook: The Acid Test," in Sophie Forgan, ed., *Science and the Sons of Genius: Studies on Humphry Davy* [London: Science Reviews, 1980], pp. 121-75).

7. Davy to Berzelius, 4 August 1813, in H. G. Söderbaum, ed., *Jac. Berzelius Bref*, 6 vols. (Uppsala, 1912-61), 1:ii, 60; Gay-Lussac, "Mémoire sur l'iode," p. 148; M. P. Crosland, *Gay-Lussac* (Cambridge University Press, 1978), pp. 78-88, 129-34.

8. Gay-Lussac, "Recherches sur l'acide prussique," *Ann. chim.* 95 (1815): 136-231 (155, 210).

9. J. R. Partington, *History of Chemistry* (London: Macmillan & Co., 1964), 4:275-77, cites Murray, Chevreul, L. Gmelin, Clark, Baudrimont, and Griffin as advocating an expanded hydracid theory between 1818 and 1834.

10. M. Faraday, "On two new compounds of Chlorine and Carbon . . . ," *Phil. Trans. Roy. Soc.* 111 (1821): 47-74 (63).

11. Gay-Lussac, *Cours de chimie* (Paris, 1828), Leçon 28, pp. 11-12, 22.

12. Hofmann, "Erinnerung," p. 669; Dumas, "Recherches de chimie organique," *Ann. chim. phys.* [2] 56 (1834): 113-50 (140); Dumas, *Comptes rendus* 6 (1838): 695.

13. Dumas, "Recherches"; "Considérations générales sur la composition théorique des matières organiques," *J. pharm.* 20 (1834): 261-94. On Dumas' substitution theory, see S. C. Kapoor, "Dumas and Organic Classification," *Ambix* 16 (1969): 1-65; N. W. Fisher, "Organic Classification Before Kekulé," *Ambix* 20 (1973): 106-31, 209-33; and Klosterman, "Dumas," passim.

14. Dumas, "Recherches," pp. 140-43; "Considérations," pp. 285-86.

15. Klosterman, "Dumas," p. 265n.

16. F. Wöhler and J. Liebig, "Untersuchungen über das Radikal der Benzoësäure," *Ann.* 3 (1832): 249-87 (280).

17. *Aus Justus Liebig's und Friedrich Wöhler's Briefwechsel*, ed. A. W. Hofmann, 2 vols. (Brunswick, 1888), 1:114.

18. *Briefwechsel zwischen Jakob Berzelius und Friedrich Wöhler*, ed. O. Wallach, 2 vols. (Leipzig, 1901), 1:672 (letter of 20 December 1836).

19. *Berzelius und Liebig: Ihre Briefe von 1831-1845*, ed. J. Carrière (Munich, 1898), p. 161 (letter of 4 May 1838); hereafter cited as Carrière.

20. Liebig and Pelouze, "Vermischte Notizen," *Ann.* 19 (1836): 241-90 (252); Wallach, *Briefwechsel*, 1:663

21. Carrière, pp. 121-23. The earlier letter mentioned is missing, but its contents can be inferred from the 5 January 1837 letter.

22. Liebig to Berzelius, 26 November 1837, Carrière, p. 134; Dumas and Liebig, "Note sur la constitution de quelques acides," *Comptes rendus* 5 (1837): 863-66.

23. Berzelius to Liebig, 19 December 1837 and 20 February 1838, Carrière, pp. 142-45.

24. Liebig, "Ueber die Constitution der organischen Säuren," *Ann.* 26 (1838): 113-89.

25. Berzelius to Liebig, 4 May 1838, Carrière, pp. 158-59.

26. Liebig to Wöhler, 18 May 1838, Carrière, p. 165 (this letter does not appear in the expurgated Hofmann collection).

27. Berzelius, "Lettre . . . à M. Pelouze," *Comptes rendus* 6 (1838): 629-44.

28. Liebig to Berzelius, 27 May 1838, Carrière, pp. 166-68; Liebig, "Lettre . . . à M. le Président," *Comptes rendus* 6 (1838): 823-29.

29. Dumas, "Réplique," *Comptes rendus* 64 (1838): 646-48 (647); Dumas, "Réponse . . . à la lettre de M. Berzelius," ibid., 689-702 (695-96, 699).

30. Ibid., p. 698.

31. Dumas, "Acide produit par l'action du chlore sur l'acide acétique," *Comptes rendus* 7 (1838): 474.

32. Dumas, "Mémoire sur la constitution de quelques corps organiques et sur la théorie des substitutions," *Comptes rendus* 8 (1839): 609-22.

33. Ibid., pp. 621-22.

34. Berzelius, "Lettre . . . à M. Pelouze," *Ann. chim. phys.* [2] 71 (1839): 137-46; *Jahresbericht* for 1838, 19 (1840): 361-77.

35. Dumas, "Note sur la constitution de l'acide acétique et de l'acide chloracétique," *Comptes rendus* 9 (1839): 813-15.

36. For a discussion of this point, see Fisher, "Classification," p. 129.

37. Dumas, "Premier mémoire sur les types chimiques," *Ann. chim. phys.* [2] 73 (1840): 73-100; Dumas and Stas, "Second mémoire . . . ," *Ann. chim. phys.*, 113-66; Dumas, "Mémoire sur la loi des substitutions et la théorie des types," *Comptes rendus* 10 (1840): 149-78.

38. Dumas, "Premier mémoire," pp. 74-75, 91-92, 99.

39. In Dumas' weights. Formic acid is represented as the hypothetical anhydride, and methyl ether as a two-volume formula.

40. For example, acetic and chloracetic acids belong to the same chemical type; acetic acid and alcohol belong to the same mechanical type.

41. Dumas, "Mémoire sur la loi . . . ," pp. 163-64.

42. Ibid., p. 164.

43. He did not refuse full credit for the theory of *types*, as some historians have claimed.

44. Ibid., pp. 177-78.

45. Dumas, "Premier mémoire," p. 74.

46. Dumas, "Second mémoire"; "Mémoire sur la loi."

47. See esp. his paper "Sur la nitronaphtalase, la nitronaphtalèse et la naphtalase," *Ann. chim. phys.* [2] 59 (1835): 376-97, where he first used the phrase "radical fondamental" and "dérivé" (389).

48. Laurent, "Théorie des combinaisons organiques," *Ann. chim. phys.* [2] 61 (1836), 125-46, summarized in *Comptes rendus* 2 (1836): 130-32; "Recherches diverses de chimie organique," *Ann. chim. phys.* [2] 66 (1837): 136-213, 314-36; *Thèse de chimie et de physique* (Paris, 1837); previously unpublished portion reprinted by J. Jacques, "La thèse de doctorat d'Auguste Laurent," *Bull. Soc. Chim.* [5] 1 (1954): D31-39; summarized by Laurent in his *Méthode de chimie* (Paris, 1854), pp. 237-39; *Chemical Method*, trans. Wm. Odling (London, 1855), pp. 194-96.

49. For recent discussions see Fisher, "Classification," and S. C. Kapoor, "The Origins of Laurent's Organic Classification," *Isis* 60 (1969): 477-527.

50. Laurent, "Théorie," pp. 127-28. This may sound mysterious to a modern chemist. However, if the "radical" is defined as the carbon skeleton together with all atoms bonded completely and directly to it, then the criterion works: replacement of methylene hydrogen by carbonyl oxygen does not acidify, but introduction of hydroxyl oxygen often does. In 1859 August Kekulé defined his radicals in precisely this way.

51. Laurent, "Essai sur l'action du chlore sur la liqueur des Hollandais et

sur quelques éthers," *Ann. chim. phys.* [2] 63 (1836): 377-89. For a strong defense of Dumas, see Klosterman, pp. 323-24.

52. Laurent, "Recherches," pp. 326-27.

53. *Correspondance de Charles Gerhardt*, M. Tiffeneau, ed., 2 vols. (Paris, 1918-25), 2:38 (letter of 29 May 1845). Cf. an earlier letter (Gerhardt to Cahours, 15 August 1841) wherein Gerhardt accuses Laurent of doing "steam chemistry" ("chimie à la vapeur") (ibid., p. 11).

54. For which see Fisher, "Classification," pp. 124-26, and Kapoor, "Laurent's Classification," pp. 492-94.

55. *Précis de chimie organique*, 2 vols. (Paris, 1844-46), 1:10-12.

56. Tiffeneau, *Correspondance*, 1: 270-88. For an enlightening discussion, see J. H. Brooke, "Laurent, Gerhardt, and the Philosophy of Chemistry," *Hist. Stud. Phys. Sci.* 6 (1975): 405-29.

57. Ibid., pp. 5, 19-20 (letters of 4 July 1844, 12 and 24 February 1845). Fisher, "Classification," pp. 215-17, provides a perceptive analysis of the relations between the two French chemists.

58. Laurent, *Méthode*, p. xiii; *Method*, p. xvi; see Brooke, "Philosophy of Chemistry."

59. Laurent, "Classification chimique," *Comptes rendus* 19 (1844): 1,089-1,100.

60. Ibid., p. 1,090.

61. Ibid., p. 1,091.

62. Ibid., p. 1,092.

63. Gerhardt, "Recherches sur la classification chimique des substances organiques," *Comptes rendus* 15 (1842): 498-500; detailed paper published simultaneously in *Rev. sci.* 10 (1842): 145-218. Gerhardt actually wrote "C^4O^4, H^4O^2, Az^2H^6," using Dumas' atomic weights and symbols.

64. The N_2H_6 formula originated indirectly in the (incorrect) formula for the caustic alkalies: $KO + H_2Cl_2 = KCl_2 + H_2O$; by analogy, then, $N_2H_6 + H_2Cl_2 = N_2H_8Cl_2$.

65. Gerhardt, "Recherches," pp. 498, 151-52.

66. Ibid., pp. 160-61.

67. Thenard's almost violent reaction was vividly described by Gerhardt in a letter to Cahours: E. Grimaux and C. Gerhardt, Jr., *Charles Gerhardt: sa vie, son oeuvre, sa correspondance* (Paris, 1900), p. 65.

68. Gerhardt, "Considérations sur les équivalents de quelques corps simples et composés," *Ann. chim. phys.* [3] 7 (1843): 129-43; 8:238-45.

69. Laurent, "Classification," p. 1,100; Laurent to Gerhardt, July 1844, in Grimaux, *Gerhardt*, p. 468.

70. Laurent, "Recherches sur les combinaisons azotées," *Ann. chim. phys.* [3] 18 (1846): 266-98.

71. Ibid., pp. 267-68, 294; *Méthode*, pp. 57-58, 77; *Method*, pp.46-48, 69. The rule was stated earlier in a restricted form by Gerhardt: "Recherches," p. 163; "Considérations," pp. 142-43; and by Laurent: "Sur les combinaisons organiques azotées," *Comptes rendus* 20 (1845): 850-55. In modern terminol-

ogy, the rule states that, while the number of atoms of even valence (monads) may be even or odd, the number of atoms of odd valence (dyads) must always be even. The reason why this rule subsists is simple: since each chemical bond engages two valences, the total number of valences of all the atoms in a stable molecule must be an even number, hence atoms with odd valences must be present in even numbers. As we shall see, Laurent placed much more emphasis on the even-number rule than on the Avogadro-Ampère gas theory as a means of correcting molecular formulas. It is puzzling why this rule, which constitutes an important step toward the theory of valence, has been largely ignored by historians of chemistry.

72. Laurent, "Recherches sur les combinaisons azotées," pp. 294-95; "Sur les combinaisons organiques azotées," p. 854.

73. Laurent, "Recherches", p. 296.

74. A. Ladenburg, *Lectures on the History of the Development of Chemistry* (Edinburgh: Alembic Club, 1905), p. 187.

75. Laurent, "Recherches," pp. 284-85.

76. Ibid., pp. 289-91; *Méthode*, pp. 59-76; *Method*, pp. 46-70.

77. Laurent, "Recherches," pp. 289, 292; *Méthode*, pp. 4, 16; *Method*, pp. 3, 12. The latter axiom was especially important for Laurent; in a letter to Gerhardt he referred to it as his "battle horse," and wrote: "analogous formulas [must] correspond to analogous bodies, *even if one must thereby assume hypothetical bodies* (Tiffeneau, *Correspondance*, 1:41-42, letter of 11 May 1845).

78. Ladenburg, *Lectures*, pp. 187-88.

79. Berzelius, *Jahresbericht* for 1842, 23 (1844): 319-20.

8

Newer Types and Radicals

WILLIAMSON'S WATER TYPE

Alexander Williamson is known best as the discoverer of a useful method of synthesizing ethers from alcohols and as the developer of the empirical basis of the "water type," which was incorporated into Charles Gerhardt's comprehensive theory of types in the 1850s. However, many of Williamson's contemporaries regarded his work as far more fundamental and influential than these assessments would imply.[1] Recent historical work has drawn attention to Williamson's important dynamical speculations of 1851, and to his participation in the "atomic debates" of the 1860s; it is now recognized how vital a role he played, both as scientist and as propagandist, in the history of nineteenth-century chemical atomism.[2] Far from being a mere empirical prelude to Gerhardt's theory, Williamson's research exerted a profound theoretical influence on such chemists as Gerhardt himself, William Odling, Adolphe Wurtz, and August Kekulé.

Williamson (1824–1904) was born in London to Scottish parents. He studied chemistry with Leopold Gmelin in Heidelberg (1841–44) and Liebig in Giessen (1844–46), receiving his doctoral degree in 1845. He then moved to Paris, set up a private laboratory, and became acquainted with Dumas, Laurent, Gerhardt, and Thomas Graham. Despite physical handicaps (paralysis in one arm, blindness in one eye, and myopia in the other), Williamson soon established a modest reputation in inorganic chemistry; in 1849, on Graham's endorsement, he was appointed Professor of Analytical Chemistry at University College, London. He remained there until his retirement thirty-eight years later.

About the time of his appointment, Williamson became interested in the theory of etherification. The reader will recall that Dumas and Boullay regarded ether and alcohol as composed of etherin (ethylene) plus one and two molecules of water respectively. The process of etherification consisted in abstraction of one water molecule from alcohol through the dehydrating action of sulfuric acid. Liebig, although firmly denying any validity to the etherin theory, decided upon a similar scheme: alcohol was the hydrated oxide of the ethyl radical, $C_4H_{10}O \cdot H_2O$; combination with sulfuric acid produced the intermediate "sulfovinic acid" (the monoethyl ester of sulfuric acid), which then decomposed to form the anhydrous "ethyl oxide" (ether, $C_4H_{10}O$). Both of these schemes assumed that etherification proceeds by simple dehydration of alcohol; the point of difference was the presumed constitutions of the two substances, Dumas claiming that they were compounds of etherin, Liebig that they were compounds of the ethyl radical. Although this "chemical theory" of etherification was consistent with empirical formulas and with the recognized dehydrating ability of strong acids, it did not take into account vapor density considerations—ether was presumed to have a lower molecular weight than alcohol, despite the fact that its vapor density is nearly twice as great.

For Berzelius, by contrast, ether was the protoxide of ethyl (Ae^2O, $Ae = C^2H^5$), and alcohol the oxide of a different radical, $C^2H^6 \cdot O$. Etherification was thought to be effected by the "catalytic force" of the sulfuric acid. This "contact theory," unlike the chemical theory, denied the existence of an intermediate. Of the two theories, Liebig's was more widely held at mid-century.

Williamson began his researches, he wrote, with the intention of substituting ethyl into alcohol, to produce a higher homologous alcohol (i.e., butyl alcohol).[3] The application of a gentle heat to a mixture of "ethylate of potash" (potassium ethoxide, prepared by dissolving potassium metal in absolute alcohol) and ethyl iodide effected the desired substitution, but, he related, he was astonished to discover that the new substance was not an alcohol at all—it was nothing other than common ether. If the "higher formula" for alcohol—Dumas' and Liebig's—were correct, Williamson argued, the stoichiometry would lead one to expect the formation of a substance containing two oxygen atoms. But one of the few points of agreement between the two theories was that ether, the product obtained in Williamson's new reaction, contains but a single oxygen atom.[4]

> The alternative was evident; for having obtained aether by substituting C^2H^5 for H in alcohol, the relative composition of the two bodies is represented *by expressing that fact in our formula*. Thus alcohol is
>
> $$\begin{matrix} C^2H^5 \\ H \end{matrix} O,$$
>
> and the potassium compound is
>
> $$\begin{matrix} C^2H^5 \\ K \end{matrix} O;$$
>
> and by acting upon this by iodide of aethyl, we have
>
> $$\begin{matrix} C^2H^5 \\ K \end{matrix} O + C^2H^5I = IK + \begin{matrix} C^2H^5 \\ C^2H^5 \end{matrix} O.$$

However, Williamson noted, Liebig's formula could be salvaged by writing the reaction in the following manner:

$$\begin{matrix} C^4H^{10}O \\ K^2O \end{matrix} + C^4H^{10}I^2 = 2\,IK + 2(C^4H^{10}O),$$

whereby one molecule of ether—being a component of the potassium compound—passes through unaltered and a second is produced by the reaction between potassium oxide and ethyl iodide. But Williamson ingeniously blocked this avenue of retreat by synthesizing a mixed ether from potassium ethoxide and *methyl* iodide. Clearly, Liebig's equation would predict the formation of a mixture of two different "symmetrical" ethers, C_2H_6O and $C_4H_{10}O$, whereas in fact a single compound of the formula C_3H_8O, which Williamson formulated as the "asymmetrical" methyl ethyl ether,

$$\left.\begin{array}{l}C^2H^5\\CH^3\end{array}\right\}O,$$

was formed.[5]

Williamson also proposed a mechanism for the acid-catalyzed etherification of alcohol. He suggested that the sulfuric acid serves as an agent for transferring an ethyl group from one alcohol molecule to another, producing water and ether in the process. Thus, although siding with Berzelius on the issue of the relative molecular magnitudes of alcohol and ether, he also agreed with Liebig in ascribing a chemical role to sulfuric acid and the necessary presence of the intermediate sulfovinic acid. He wrote: "Innovations in science frequently gain ground *only* by displacing the conceptions which preceded them, and which served more or less directly as their foundation; but, if the view here presented be considered a step in our understanding of the subject, the author begs leave to disclaim for it the title of innovation; for the conclusion here deduced consists in showing the compatibility of views which have hitherto been considered contrary."[6]

On 16 June 1851, Williamson read a second paper "On Etherification" to the London Chemical Society, in which he provided experimental details absent from the earlier, largely theoretical paper.[7] An experiment offering further evidence for his mechanism of acid-catalyzed etherification was described: sulfamylic acid (amyl hydrogen sulfate, produced by the action of sulfuric acid on fusel oil) was converted into sulfovinic acid by distilling from alcohol. Clearly, Williamson noted, exchange of amyl for ethyl was occurring; moreover, a small quantity of the seven-carbon ethyl amyl ether was isolated from the reaction mixture.[8]

Williamson now expanded his views on molecular constitution to include that of acetic acid. A comparison of the empirical formulas of acetic acid and alcohol shows that the acid has two more equivalents of oxygen and two fewer equivalents of hydrogen; indeed, methods were known to effect this substitution directly, by oxidizing alcohol. But two questions required resolution: Does the acid molecule contain two or four carbon atoms? and Which hydrogen atoms of alcohol are removed and replaced by oxygen? Williamson attacked the first question in a similar manner to his investigation of the constitutions of alcohol and ether. It was known that potassium acetate decomposes thermally to form acetone and potassium carbonate. Now, if potassium acetate contains four carbon atoms, the reaction would consist of the asymmetric cleavage of a single molecule; but if it only contains two carbons, then the reaction must proceed through "double decomposition" between two identical molecules, one donating methyl in exchange for KO. Williamson heated the salts of two *different* acids — sodium acetate and potassium valerate — obtaining an asymmetric ketone as the principal product, which he formulated as

$$\begin{matrix} CH^3 \\ C^4H^9 \end{matrix} CO.$$

The subsequent argument for the smaller (atomic weight) formula for acetic acid followed the same lines as that for the smaller alcohol formula, after his earlier synthesis of the asymmetric ethers.[9]

But which hydrogen atoms of alcohol

$$\begin{pmatrix} C^2H^5 \\ H \end{pmatrix} O$$

are replaced by oxygen in conversion to acetic acid (now securely formulated $C^2H^4O^2$)? Williamson reasoned that the "basic" hydrogen atom could not be one of those removed, since both compounds share acidic behavior (for instance, in their reaction with potassium). Consequently, the substitution must take place within the ethyl radical itself.

> Viewing, therefore, alcohol as water in which half the hydrogen is replaced by ethyle,
>
> $$\begin{matrix} C^2H^5 \\ H \end{matrix} O,$$

we shall consider acetic acid as containing one equivalent [i.e., atom] of oxygen in the place of two atoms of hydrogen of that radical, or

$$\left.\begin{array}{c}C^2H^3O\\H\end{array}\right\}O.\ldots$$

[A]cetic acid differs from alcohol by containing, instead of ethyle, this other radical, differing from it by having oxygen in lieu of an equivalent [i.e., molecule] of hydrogen, and which may be called *oxygen-ethyle*, or *othyle*.

Williamson correctly formulated aldehyde as the "hydruret" of othyl and acetone as its methyl derivative. He noted in conclusion: "The method here employed of stating the rational constitution of bodies by comparison with water, seems to me to be susceptible of great extension."[10]

On 2 July 1851, at the annual meeting of the British Association in Ipswich, Williamson read his third and final paper of this series, "On the Constitution of Salts."[11] Williamson viewed the vast majority of chemical reactions as examples of double decompositions, i.e., substitutions, analogous to the formation of salts by means of double decompositions between acids and bases; hence nearly every compound was in a sense a salt. Although the definitions of "acid", "base," and "salt" would thereby have to be greatly extended, Williamson defended the use of the analogy for the light it would shed on both reaction mechanisms and chemical constitutions.

> A chemical [double] decomposition should therefore be represented by the juxtaposition of the formulae of the reacting substances, and by effecting in these formulae the change which takes place in the mixture. The adoption of such a method will of course necessitate the adoption of types, from which, by the replacement of certain elements or molecules, we can deduce the constitution of more and more complex groups. I believe that throughout inorganic chemistry, and for the best-known organic compounds, one single type will be found sufficient; it is that of water, represented as containing 2 atoms of hydrogen to 1 of oxygen, thus

$$\left.\begin{array}{c}H\\H\end{array}\right\}O.$$

And thus the official emergence of Williamson's water type. The heart of this theory was that all organic compounds containing oxygen should properly be related to two-volume atomic weight formulas, and that they are formed by double decomposition rather than by addition; "they are not compounds of water, but products of substitution in water."[12] In this paper Williamson also introduced the concept of *multiple* types to account for polybasic acids. For instance, the "sulphurous acid radical" SO^2 could replace a hydrogen atom in each of two water molecules —

$$\begin{matrix} HO \\ HO \end{matrix} SO^2$$

—to form sulfuric acid. The "bibasic" radical SO^2 "holds together" the two hydroxyl radicals, something that would be impossible for a "monobasic" radical such as ethyl.[13] (In this connection Williamson was careful to draw a sharp distinction between the conventional use of formulas as mere taxonomic aids and his own conception of formulas "as an actual image of what we rationally suppose to be the arrangement of constituent atoms in a compound."[14]) Summarizing, the following atoms and radicals were considered by Williamson to be monobasic: hydrogen, the halogens, the alkali metals, the "alcohol" (alkyl) radicals, the "othyl" (acetyl) radical, NO_2, ClO, ClO_2, and ClO_3; bibasic moieties included oxygen, CO, SO_2, and SO_4.

In this series of papers, Williamson was investigating two distinct but equally fundamental issues: relative molecular and relative atomic magnitudes. The first problem concerned the relative constitutions of alcohol and ether, as well as the very large number of derivatives of these substances. Williamson was able to settle this issue convincingly after a generation-long controversy. His elegant argument regarding the products of his new reactions—homogeneous asymmetric ethers and ketones rather than mixtures of symmetric compounds—led seemingly inexorably to the conclusion that the received formulas of alcohol and acetic acid ought to be halved. As Williamson put it, "In this experiment the two theories cross one another, and must lead to different results."[15]

Williamson also sought to apply his results to the establishment of a set of relative atomic weights that corresponded with the ideas of Gerhardt and Laurent. Despite the circumstances that he had been a student of Comte and that his chemical apprenticeship was under the "equivalentists" Gmelin and Liebig—moreover just at the time when the international shift from Berzelian atomic weights to conventional equivalents was beginning—he nonetheless became an ardent proponent of the novel theories of the French reformers from his Paris period (1846-49) onward. Williamson demonstrated that his new reactions and proposed mechanisms could be represented more simply and economically using half as many carbon and oxygen atoms (i.e., doubling their respective "equivalent" weights). He argued that, by analogies between organic and inorganic compounds the water molecule should actually be represented as H_2O_2 using conventional equivalents, or H_2O using his own atomic weights. This style of argumentation in chemistry, wherein analogies and elegance of formulation are regarded as scientifically persuasive, was derived from the work of Laurent and Gerhardt, as were the particular atomic weight values used by Williamson.

Let us pause for a moment to get our bearings. In this series of articles, Williamson was attempting to settle two troublesome issues. He succeeded in convincingly and conclusively establishing his position regarding molecular magnitudes by what I shall call his "asymmetric synthesis" argument.[16] Whether or not Williamson was justified in regarding this as a genuine *experimentum crucis*, the argument was convincing enough to his contemporaries so that the point was thereafter essentially settled.

The second issue, atomic weights, was skillfully though—perhaps necessarily—inconclusively argued by Williamson. He used two approaches. In the style of Laurent and Gerhardt, he justified the use of atomic as opposed to "equivalent" weights on the basis of analogy, simplicity, and consistency. More novelly and significantly, he sought the cause of cohesion of the ethyl radicals within the ether molecule—or the hydrogen atoms within water or sulfuric acid—in the integrity and "bibasic" character of the oxygen atom or of certain radicals such as sulfuryl or sulfate. Since this is, in modern vocabulary, nothing less than the assertion of the divalence of the oxygen atom, I shall refer to this as the "valence" argument. Williamson was well aware that his evidence touching this second issue was less than conclusive.[17]

GERHARDT'S GENERALIZATION OF TYPES

Williamson's water type was not the first of what are usually known as the "newer" types (to distinguish them from Dumas' type theory of 1839-40). In 1840 Liebig asserted that ammonia could be considered "as the type of all organic bases" and predicted the possibility of an ethyl-substituted ammonia. The prediction was fulfilled by his former student Adolphe Wurtz, who synthesized ethylamine in 1849. Another of Liebig's former students, A. W. Hofmann, developed the experimental and theoretical implications by producing a variety of secondary and tertiary amines and explaining them as the results of substitution of hydrocarbon radicals for the hydrogen in ammonia.[18]

The research of Liebig, Wurtz, and especially Hofmann thus established ammonia as the prototype for series of organic as well as inorganic compounds. This was the leading idea of the newer type theory. Nevertheless, it is to Williamson, not Hofmann, we must look for those concepts that were developed into the mature theory of types. Although the function of the nitrogen atom as linking three replaceable groups into a single molecule may have been implied by Hofmann, this idea was not stated. Hofmann was careful to limit his conclusions to a small number — a dozen or so — of the volatile organic bases, most of which had only recently been discovered. Other organic bases — the alkaloids, for example — could well have some different constitution, Hofmann cautioned.

By contrast, Williamson's water type was applicable to a huge number of substances, including the vast majority of organic compounds; any molecule containing oxygen could be subsumed under the water type. Moreover, in the third paper of the series Williamson developed his pregnant conception of the distinction between "monobasic" and "bibasic" moieties as the basis for an explanation of chemical constitution and chemical properties. This held the seed of the theories of valence and structure.

But Williamson did little to integrate these ideas into the vast body of chemical data of his day. The master synthesizer of facts and concepts was Charles Gerhardt. Williamson's first paper on etherification was first printed in the August 1850 issue of the *Philosophical Magazine*; Gerhardt translated the article for the September issue of his *Comptes rendus des travaux de chimie* and for G. A. Quesneville's *Revue scientifique* and added a commentary. Gerhardt fairly crowed at this experimental vindication of his ideas. "There can be, I think, no more palpable and conclusive proof of the principles which we have maintained [since 1842] . . . Now, perhaps, my adversaries will be a bit more humane."[19]

In the following spring Gerhardt published, in collaboration with his friend Gustave Chancel, a sketch of what he was to develop into his theory of types. The two Frenchmen proposed using "synoptic" formulas as expressions of "contracted chemical equations" and of classificatory relationships. "The determination of the constitution of a body, in our opinion, does not mean indicating the molecular arrangement of the elements which compose it, but rather identifying the series to which it belongs and the row [rang] which it occupies."[20] They wrote the recently discovered series of alkyl amines as:

$$\text{Az} \begin{cases} H \cdot CH^2 \\ H \\ H \end{cases} \qquad \text{Az} \begin{cases} H \cdot C^2H^4 \\ H \\ H \end{cases} \qquad \text{Az} \begin{cases} H \cdot CH^2 \\ H \cdot C^2H^4 \\ H \end{cases}$$

methylamine   ethylamine   methylethylamine

the "homologues of water" were written:

$$O \begin{cases} H \\ H \end{cases} \qquad O \begin{cases} H \cdot CH^2 \\ H \end{cases} \qquad O \begin{cases} H \cdot C^2H^4 \\ H \end{cases} \qquad O \begin{cases} H \cdot C^2H^4 \\ H \cdot C^2H^4 \end{cases}$$

water   methyl alcohol   ethyl alcohol   common ether

and the "homologues of hydrogen" as:

$$\begin{cases} H \\ H \end{cases} \qquad \begin{cases} H \cdot CH^2 \\ H \end{cases} \qquad \begin{cases} H \cdot CH^2 \\ H \cdot CH^2 \end{cases} \qquad \begin{cases} H \cdot CH^2 \\ H \cdot C^2H^4 \end{cases}$$

hydrogen   marsh gas   dimethyl   ethyl methyl.

This paper is a fine synthesis of Laurent's views on formula notation, Hofmann's views on the constitution of amines, Williamson's views on the constitution of alcohols and ethers, and Gerhardt's own concept of homology. In addition, the authors added the hydrogen group to the two already proposed. (Gerhardt did not use the term "type" in its new sense until 1852, by which time Liebig [1840], Hofmann [1849] and Williamson [1851] had popularized the new denotation.) In the last five years of his life, Gerhardt went on to develop a highly influential and complete type theory, a theory that became associated more and more exclusively with Gerhardt's name.[21]

It is evident that these formulas, in contrast to Williamson's and true to Gerhardt's and Chancel's disclaimer, were not intended to express the linking function of the nitrogen atom in amines or of the oxygen atom in alcohols and ethers. The new hydrogen type breaks the symmetry of the preceding two types by not having an atom outside the bracket — so there could be no thought of a single atom, analogous to oxygen in Williamson's type, holding together those atoms or groups on the other side of the bracket. This point is seen particularly clearly in Gerhardt's and L. Chiozza's conception of the constitution of sulfuric acid. Like Williamson, they regarded the molecule as derived schematically from a double water type, representing the SO^2 group as dibasic. But they conceived the two hydrogen atoms replaced by SO^2 to be derived from the *same* water molecule, so there can be no atomic linkage of the sort that is so evident in Williamson's papers:[22]

$$\left.\begin{array}{l} H \\ H \end{array}\right\} O \qquad \left.\begin{array}{l} SO^2 \\ H \\ H \end{array}\right\} O$$

$$\left.\begin{array}{l} H \\ H \end{array}\right\} O \qquad \left.\begin{array}{l} H \\ H \end{array}\right\} O$$

Gerhardt's enemies, it is true, had suffered a setback at Williamson's hands, but they could take heart at some work recently published by H. Deville. According to Laurent and Gerhardt, the customary equivalent formula for nitric acid must be halved, it is monobasic, and no monobasic acid can form an anhydride. Deville had managed to contradict this viewpoint by synthesizing nitric anhydride, which he formulated as AzO^5 ($O = 8$).[23] But Gerhardt was not perturbed. He pointed out that the formula written in his weights would be Az^2O^5 and thus could be conceived as having been formed from *two* molecules of nitric acid:[24]

$$2AzO^3H = Az^2O^5 + H^2O.$$

This was the situation when Gerhardt made what was probably his most important experimental discovery—the synthesis of anhydrides of the monobasic organic acids. Williamson had already predicted in his third paper on etherification that such compounds could be produced. He wrote at that time: "If the two atoms of hydrogen in water were replaced by othyle, we should have anhydrous acetic acid,

$$\begin{matrix}(C^2H^3O)\\(C^2H^3O)\end{matrix}O.$$

In fact, the so-called anhydrous acids are nothing else than the *aethers* of the hydrated acids."[25] Gerhardt heated a solution of sodium benzoate in benzoyl chloride, and isolated benzoic anhydride from the reaction mixture. On 5 April 1852 he wrote to Chancel in such excitement that his handwriting is barely recognizable. "Nitric acid being

$$\begin{matrix}(NO^2)\\H\end{matrix}O,$$

anhydrous nitric acid becomes

$$\begin{matrix}(NO^2)\\(NO^2)\end{matrix}O,$$

consequently benzoic acid is

$$\begin{matrix}(C^7H^5O)\\H\end{matrix}O;$$

$$\begin{matrix}(C^7H^5O)\\(C^7H^5O)\end{matrix}O$$

must be discovered—and has been! . . . *Monobasic acids are water in which H is replaced by a complex group*: they are alcohols. . . . Ethers, alcohols, acids, potash, are nothing but water. . . . Bibasic acids are two molecules of water H^4O^2 in which H is replaced by a complex group. . . . CO replaces H^2, PO replaces H^3. This could lead far."[26]

Gerhardt's first publication of his discovery was in the form of two letters to Liebig, dated 14 and 19 April, printed back-to-back in the *Annalen*.[27] He stated that water could serve as the "type" of all organic acids, as ammonia is for the organic bases, if it is written H_2O rather than HO. The "beautiful experiments" of Williamson and of Chancel are "completely favorable" to this view of the constitution of water.[28]

On 17 May Gerhardt read his first communication on acid anhydrides before the Paris Académie des Sciences. "If it is true, as is generally assumed, that alcohol and ether represent from the molecular point of view a molecule of water in which 1 or 2 atoms of hydrogen are replaced by the carbonated hydrogen C^2H^5, *ethyl*, one is led by my experiments to apply the same theory to organic acids, and to consider, for example, hydrated acetic acid and its acid anhydride as a molecule of water in which 1 or 2 atoms of hydrogen are replaced by the group C^2H^3O,"[29] whereupon followed "type formulas" for water, alcohol, ether, acetic acid, and acetic anhydride. There is no mention of Williamson.

Gerhardt had written Williamson on 28 April, expressing his delight "to see how we are in agreement on the principal question"; he said he had "just found" Williamson's article, having temporarily misplaced it.[30] The English chemist wrote his French friend a month later, informing Gerhardt that "some of my friends are shocked that you made no mention at all of my theoretical and practical researches." These "friends" had strongly urged Williamson to send a priority claim to the president of the Académie. Williamson would be sorry to have to do that, since it could have the effect of hurting Gerhardt in the eyes of the academy; he assured Gerhardt that everything would be considered settled if he would "mention the matter in a subsequent memoir."[31]

On 14 June Gerhardt read a second memoir on anhydrides

before the Académie. Now he provided what he thought was conclusive proof of his conception of the constitution of the new compounds: he synthesized a number of asymmetric anhydrides between benzoic, cuminic, cinnamic, salicylic, and acetic acids.[32] The argument for the validity of Gerhardt's anhydride formulas is the same as that for Williamson's ether and acid formulas, following the latter's synthesis of asymmetric ethers and ketones: if the older anhydride formulas were correct, one would have expected to produce mixtures of symmetrical anhydrides:

$$C_2H_3 \cdot C_2O_2 \cdot Cl + C_{12}H_5 \cdot C_2O_3 \cdot NaO \rightarrow C_2H_3 \cdot C_2O_3 + C_{12}H_5 \cdot C_2O_3,$$

| acetyl chloride | sodium benzoate | acetic anhydride | benzoic anhydride |

rather than the homogeneous asymmetric substances actually obtained. The victory was complete; even his adversaries were impressed. He wrote to Chancel: "Regnault shook my hand, Dumas nearly hugged me, even Fremy paid me the sweetest compliments . . . What a success!" Hofmann thought the work would soon "introduce very grave modifications in our viewpoints." Liebig called the discovery "one of the most brilliant of recent times," and the (Williamsonian) theoretical argument seemed to him to be "as simple as it is elegant."[33]

In this memoir Gerhardt also tried to placate Williamson: "I would point out that I was the first to insist upon the necessity of doubling the formula of ether with respect to that of alcohol; this necessity has finally been demonstrated experimentally by the excellent researches of M. Williamson and M. Chancel. The same question subsists today for the acid anhydrides; M. Williamson has already posed it himself in this sense, in an interesting article, where, adopting my notation, he considers all salts as derivatives of the water type H^2O."[34] Other references to Williamson appeared in Gerhardt's major paper on the organic anhydrides, published in the following spring, where he stated that the simplest method of formulating these new substances "is to apply the theory of ethers to them, as has been modified in recent years from the point of view of types, since the important results which have been obtained by M.

Williamson and M. Chancel." And in connection with his type formulas for the anhydrides, he remarked, "M. Williamson has already made the same comparison, whose exactitude seems to me today to be perfectly demonstrated by my experiments."[35]

These partial admissions of the English chemist's priority seem to have satisfied Williamson, who offered a modest and factual account of Gerhardt's discovery in a Friday Evening Discourse at the Royal Institution.[36] This discovery, he explained, provided important new evidence for a "familiar theory of the constitution of salts." This new theory had been based upon recent research, "especially of MM. Laurent and Gerhardt," on atomic weights: these chemists showed that acids contained no water, and they distinguished clearly between acids of different basicities. At this point "an English chemist" proved that ether is formed by substitution by ethyl of hydrogen in alcohol. "M. Gerhardt" proved "a similar fact"; acid anhydrides are to acids as ethers are to alcohols. "These results can be most simply stated in the form adopted by M. Gerhardt the discoverer, which consists in comparing the composition of these bodies with that of water, from which they are formed by the substitution of one or both atoms of hydrogen by organic radicals." Williamson thus expressed what seems to me to be an accurate assessment of relative merits, without ever mentioning his own name.[37] Gerhardt expressed his appreciation for Williamson's kind statements in a letter (4 May 1853), declaring himself "extremely flattered."[38]

FRANKLAND, KOLBE, AND THE ISOLATION OF ORGANIC RADICALS

The fourth decade of the nineteenth century was a bad time for electrochemical dualism, especially in France, where chlorine was the reagent *par excellence*. But Berzelius became a "master wriggler" in responding to theoretical challenges. For instance, he supposed a rearrangement occurred during the chlorination of acetic acid; the acid, consisting of the hydrated oxide of the acetyl radical, was transformed into a chlorocarbon radical united to "oxalic acid";

$$C_4H_6O_3 \cdot H_2O \xrightarrow{Cl_2} C_2Cl_6 \cdot C_2O_3 \cdot H_2O.$$

Berzelius remarked, in reference to the product of this reaction: "This is a type of compound of which many examples are known, which contain simple as well as compound radicals, and of which some, although not all, exhibit the property that the oxide [the acid] can combine with and separate from bases without thereby becoming separated from the chloride."[39] That is, the "sesquichloride of carbon," C_2Cl_6, was attached to the "oxalic acid," $C_2O_3 \cdot H_2O$, in such a way that the acidic properties of the acid were not affected. Berzelius' explanation had the advantage that no fundamental changes in chemical theory would be required. Moreover, the two assumed components of chloracetic acid, perchloroethane and oxalic acid, were both known and well characterized; the only new element of the hypothesis was the assumption of a distinct type of nonelectrochemical bonding.

Berzelius used this substance, chloracetic acid, as the prototype for a new theory of the constitution of certain organic compounds, which became known as the theory of copulated or conjugated compounds. Several examples were long known where the saturation capacity of a mineral acid did not seem to be affected by combination with an organic group.[40] Berzelius himself had noted a number of such cases since 1835. Finally, in the spring of 1841 he wrote to Wöhler asking for a good German equivalent for Gerhardt's "copule" (Swedish "koppling").[41] Jules Reiset had recently discovered a platinum ammine which appeared to be fully as basic as free ammonia. Berzelius wrote:

> To the degree that such substances become more numerous it will be necessary to have a specific name for this type of compound. In order to keep within the scope of published research I will use the word *copule*, proposed by Gerhardt, and translate this by the admittedly imperfect but familiar word *copulated* [gepaart], which here is to mean "fastened together." A copulated acid is thus an acid in combination with a base, but which also does not increase or decrease the saturation capacity of the acid ... The body which is in this fashion inseparably bound to an acid or base could be called a *copula* [Paarling].[42]

Thus in Reiset's ammine and in chloracetic acid the electropositive copulae (PtO and perchloroethane respectively) were presumed to be combined nonelectrochemically to the active acid or base (ammonia and oxalic acid respectively).

However ingenious Berzelius' 1839 suggestion for the constitution of chloracetic acid was, it could only be seen as a retreat in the face of Dumas' attacks. A further retrenchment of his position became necessary three years later, when Dumas' assistant Louis Melsens converted chloracetic back into acetic acid using hydrogen gas.[43] The analogy between the two compounds was now all too clear—Berzelius could not go so far as to assert that the copulated acetic acid reformed its original hydrocarbon radical under the influence of hydrogen. But Berzelius found a way out: acetic acid was also a copulated compound![44]

Beset by new examples of substitution, Berzelius was quick to capitalize on this theory. Substitution of halogen for hydrogen could occur, he now admitted, but only within the passive non-electrochemical copula. All organic acids, and all of the new amines, as well as many neutral compounds—indeed, most organic compounds—were now presumed to be copulated.[45] Kekulé later commented,

> Everything was decomposed into copulae, everywhere new radicals were assumed; almost every new substance contained a new radical. . . . Out of joy over the discovery of copulae it was forgotten that it was the substitution theory which was supposed to be controverted. . . . In short, Berzelius now developed the substitution theory himself: chlorine was capable of replacing an equal number of hydrogen atoms, but the substitution took place only in the copula. What had been absurd as long as it had been viewed without hypotheses, became "surprisingly clear and simple" when the copula hypothesis was added.[46]

But when it came to sarcasm no one could touch Laurent:

> A word let fall from the pen of Gerhardt, was thus transformed into a luminous idea for dualism. From this time everything was copulated. Acetic, formic, butyric, margaric, &c., acids,—alkaloids, ethers, amides, anilides, all became copulated bodies. So that to make acetanilide, for example, they no longer employed acetic acid and aniline, but they re-copulated a copulated oxalic acid with a copulated ammonia. I am inventing nothing—altering nothing. Is it my fault if, while writing history, I appear to be composing a romance? What then is a copula? A copula is an imaginary body, the presence of which disguises all the chemical properties of the compounds with which it is united. Thus margaric acid contains oxalic acid united to the copula $C_{32}H_{66}$; and butyric

acid, oxalic acid united to the copula C_6H_{14}. . . . Reactions are quite incapable of unravelling the mystery, nought but the penetrating spirit of dualism will suffice. . . . The dishonesty is flagrant. What have I to do with your copulae, your radicals, and your castles of cards?[47]

In spite of Berzelius' ingenuity he was in difficult theoretical straits by about 1840, and he knew it. But it was just at this time that new research on organic radicals began to be published, research that gave new life to Berzelius' theory. In a brilliant series of papers published between 1837 and 1843, Robert Bunsen described the isolation, properties, and derivatives of a new radical: "cacodyl," $C_4H_{12}As_2$ (tetramethyl diarsine).[48] This was the first isolated organic radical since Gay-Lussac's cyanogen (1815). Berzelius reported on Bunsen's work in glowing terms, as might be expected.[49] Furthermore, cacodyl was not only a beautiful organic radical, it was also one more example of a conjugated compound, for the methyl groups did not seem to affect the activity of oxygenated cacodyl ("arsenious acid") in any way.

Further evidence in favor of Berzelius' conception of the constitution of organic acids was published in the late 1840s by the English chemist Edward Frankland (1825-99) and his friend Hermann Kolbe (1818-84). Frankland, then just twenty years old, met the older German chemist in the laboratory of Lyon Playfair in 1845, and the two quickly became close friends. In an investigation begun in London in the spring of 1847 and completed by the end of the following summer at Bunsen's laboratory in Marburg, the two chemists hydrolyzed a series of alkyl cyanides to obtain the corresponding acids.[50] This they conceived to be strong evidence for Berzelius' theory that acetic acid was methyl copulated with oxalic acid. It was known that alkaline hydrolysis of cyanogen produced oxalic acid; so if, as in the Berzelian schema, the alkyl cyanides are considered to be "alcohol" (alkyl) radicals conjugated with cyanogen, it was to be expected that alkaline hydrolysis would yield a series of compounds composed of the same radicals conjugated with oxalic acid, e.g.:[51]

methyl cyanide $C_2H_3 \cdot (C_2N)$ → $C_2H_3 \cdot (C_2O_3)$ acetic acid

In an investigation carried out simultaneously with this last, Kolbe and Frankland generated what they thought was the free methyl radical (actually ethane) from the reaction of potassium metal with ethyl cyanide.[52] The following spring Kolbe announced that electrolysis of the potassium salts of acetic and valeric acids yielded the free "methyl" and "valyl" radicals.[53]

Bunsen had prepared cacodyl from the reaction of cacodyl chloride with zinc. In the spring of 1848, Frankland tried heating ethyl iodide with potassium in an attempt to discover a method for generating the free hydrocarbon radicals. He managed to isolate "ethylic hydride" (ethane), but no ethyl. He decided that he might have better luck if he replaced potassium with a less active metal; logically enough he chose Bunsen's reagent, zinc. The experiment, performed early the following year, was successful, and he published a paper "On the Isolation of the Organic Radicals" a few months after Kolbe's paper on electrolysis appeared.[54] Frankland had not only isolated "ethyl," but he had also synthesized a new organometallic substance, zinc ethyl, which he formulated C_4H_5Zn (modern Et_2Zn). The reaction of amyl iodide with zinc produced analogous substances.[55]

In the space of three years, Frankland and Kolbe had isolated what they thought were four different organic radicals — methyl, ethyl, valyl, and amyl — by means of three different methods. The radical theory in general, and Berzelius' ideas in particular, had received what seemed to be important support; it was well and good to be able to isolate a previously unknown radical such as cacodyl, but truly momentous to isolate radicals whose existence had been assumed for fifteen years. It is tragic that Berzelius died at the very inception of this new development (in August 1848).

The interpretation of the Berzelian school did not, however, remain unchallenged. Laurent and Gerhardt pointed out that the Frankland-Kolbe formulas were at odds with the even-number rule and must be doubled; if all hydrocarbons are related to the same number of volumes of vapor, the "radicals" are actually seen to be homologues of marsh gas.[56] Hofmann and Benjamin Brodie adduced additional arguments; Frankland replied.[57]

In a long paper published in the fall of 1850, Kolbe sought to consolidate, refine, and extend his views on the constitution of organic radicals.[58] He distinguished between two species of radicals: the "ether-radicals", found in ethers, alcohols and amines, for example:

$(C_4H_5)O$	$(C_4H_5)O \cdot HO$	$(C_4H_5) \cdot NH_2$;
ether	alcohol	ethyl amine

and conjugate radicals, occurring in acids and their derivatives—

$HO.(C_4H_5)C_2O_3$.
propionic acid

This distinction formed the classificatory basis for his *Ausführliches Lehrbuch der organischen Chemie*.[59]

Kolbe admitted that Liebig's acetyl radical, equivalent formula C_4H_3, really exists in acetyl compounds; "*it must not, however, be considered as a group of four equivalents of carbon and three equivalents of hydrogen, the four carbon equivalents of which possess equal functions . . . it should rather be viewed as a compound of two equivalents of carbon, and methyl as the adjunct* [Paarling, copula]:

$$\text{Acetyl} = (C_2H_3)\frown C_2,$$

in which the C_2 presents the exclusive point of action [Angriffspunkt] for the powers of affinity of oxygen, chlorine, &c."[60] The higher homologous acids were formulated by inserting the appropriate conjugate radical between the parentheses.

Kolbe's break with Berzelius consisted in the wholehearted endorsement of the substitution phenomenon; chlorine actually replaced hydrogen in *both* types of Kolbean radicals. But Berzelius' constitutional analysis of acetic acid had received both experimental support and theoretical extension. This was a result of Kolbe's dogged pursuit of his central concern—chemical constitution—a concern that formed the basis for his later brilliant contributions.

Kolbe considered contemporary French chemistry as the opposite pole of this viewpoint. He labeled his opponents dilettantes, who wished to overthrow a comprehensive and flexible theory on the basis of a few isolated observations that could not immediately be subsumed under the present scheme — perhaps out of chauvinist jealousy that the radical theory had found its roots in German rather than French soil. He criticized the sterile nonconstitutional approach of Gerhardt and Laurent, whose "phantasy-paintings" were all too reminiscent of *Naturphilosophie*, whose "pencil-and-paper laws of nature" stood in sharp contrast to the properly scientific experimental methodology.[61] Kolbe was certain that the latter methodology, his own, would eventually prevail. "The radical theory has already outlived the theory of metalepsy and the theory of types, of which at the present time scarcely mention is made in the researches of French chemists; the nucleus theory and the ingenious invention of even numbers of atoms[62] will likewise disappear from the field; for chemistry is indeed something better than a mere arithmetic problem, into which Laurent and Gerhardt endeavor to convert it."[63]

Although Kolbe here styled the even-number law "ingenious," on the next page he confessed his inability to understand the French chemists' reasoning.[64] This was indeed unfortunate. Had Kolbe better appreciated the value of the Laurent-Gerhardt style of pencil-and-paper chemistry, had he realized that more than one methodology can lead to fruitful results in science, he might have been more willing to invest the effort required to understand the simplification and unification which the new reform offered chemistry. As it was, Kolbe retained conventional equivalents in his formulas until 1870, long after atomic weights had been accepted by the majority of organic chemists, and he always maintained a cavalier attitude toward the essential question of molecular magnitudes. In the 1850s Kolbe was among the vanguard of chemists who were seeking, often successfully, to determine molecular constitutions. As theoretical organic chemistry passed him by during the next thirty years, Kolbe, like his mentor Berzelius, became increasingly bitter and unyielding and engaged in increasingly acrimonious polemics.

CONFLICT AND CONFLUENCE

When Williamson's theory of the constitution of alcohols, ethers, and salts first appeared, Kolbe must have been more than a little upset: worse than disputing Frankland's and Kolbe's experimental and theoretical work, Williamson had completely ignored them at a time when it seemed that Berzelius' radical and copula theory had received decisive confirmation through this new research. The first salvo against Williamson came from the pen of Kolbe's student Francis Wrightson.[65] Wrightson compared the two most recent suggestions for the formula for acetic acid:

$$\left. \begin{array}{c} C^2h^3O \\ h \end{array} \right\} O \qquad HO \cdot (C^2H^3) \frown C^2, O^3$$

Williamson　　　　　　　Kolbe

and pointed out that whereas Williamson's "othyl" is a purely hypothetical body, all the components of Kolbe's formula—water, methyl, and oxalic acid—had been isolated in pure states.[66] Indeed, the existence of methyl in acetic acid had been demonstrated in three different ways, a fact that Williamson had totally ignored.

Then Wrightson added the "clincher." Using Williamson's notation, the formula for acetic acid would have to be written:

$$\left. \begin{array}{c} h \\ h \end{array} \right\} O \cdot \left(\begin{array}{c} Ch^3 \\ Ch^3 \end{array} \right) \frown C^2, O^3,$$

a molecule containing two methyl groups. If this were true, one should be able to substitute an ethyl group for one of the methyls, to produce an asymmetric acetic acid (!). Wrightson wrote: "Contrary to my expectations, the experiments I have made, variously modified and conducted with the most scrupulous care, show that the acid in question does not exist," hence Williamson's theory lost credibility. Here is an example of Williamson's "asymmetric synthesis" argument, used with a negative experimental result and quite fallacious reasoning.[67]

Williamson suggested that Wrightson was laboring under an

unfortunate misapprehension: "The result which he failed in obtaining is incompatible with the othyle theory of which he conceived it a consequence, and the result he obtained is decidedly confirmatory of the theory which he expected it to upset," and he explained why this was the case.[68] Wrightson's rebuttal to Williamson shows that he had failed to understand Williamson's points.[69]

Kolbe himself now stepped into the fray.[70] He demonstrated that all of Williamson's novel reactions could just as easily be represented in his own system:

(a) $C_4H_5O \cdot HO + 2SO_3 \cdot HO \longrightarrow C_4H_5O \cdot SO_3 \cdot HO \cdot SO_3 + 2HO$
(b) $C_4H_5O \cdot SO_3 \cdot HO \cdot SO_3 + C_4H_5O \cdot HO \longrightarrow 2C_4H_5O + 2HO \cdot SO_3$
acid-catalyzed etherification

$C_4H_5O \cdot KO + C_4H_5I \longrightarrow 2C_4H_5O + KI$
symmetrical Williamson synthesis

$C_4H_5O \cdot KO + C_2H_3I \longrightarrow C_4H_5O \cdot C_2H_3O + KI$
asymmetrical Williamson synthesis

Kolbe reiterated Wrightson's arguments, and thought, for a variety of additional reasons, that Williamson's theory was "easily refuted."[71]

Kolbe's lengthy critique was first read at a meeting of the Chemical Society on 20 February 1854. There was an element of drama as Williamson had prepared a reply to be read at the same meeting. Henry Watts wrote to H. E. Roscoe at Heidelberg: "[Kolbe's] paper is to be read at the next meeting, and Williamson will answer it, so there will be a jolly row." Watts thought some of Kolbe's arguments had merit. Williamson wrote to the same correspondent at about the same time (9 February 1854): "Of course I will answer it, though from what I have seen of its contents there seems little chance of my converting him to more rational views on the subject, as he does not enter into or understand the point of view opposed to his view."[72]

In his reply Williamson first asserted that Kolbe had conceded the validity of the argument on molecular magnitudes, which was the most essential part of the new theory.[73] He then repeated, but with greater detail and clarity, his arguments touching atomic weights. He pointed out that in Kolbe's notation ethers are represented by the stable coupling of two very similar molecules, whereas in his own system the cause of the cohesion is simply the integrity of the oxygen atom (O = 16) that unites the two ethyl radicals. He turned the tables on Wrightson's and Kolbe's argument against othyl: "The radical $(C_2H_3)\frown C_2$ only exists in the imagination of Dr. Kolbe; and chemists were never before informed that it possesses a constitution analogous to that of water, to such an extent that, if its methyl cannot be halved, the hydrogen of water must be likewise indivisible." Finally, Williamson lambasted Kolbe's complicated formula notation: "The chief peculiarity of Dr. Kolbe's method of representing chemical compounds, consists in the very frequent use he makes of undefined symbols; indeed, there is hardly any other chemist who does this in the same degree as himself." He pointed out that in writing his formula for acetic acid Kolbe used five different symbols or conventions, implying the operation of five different but unnamed sorts of chemical forces.

> It would be just as reasonable to describe an oak-tree as composed of the blocks and chips and shavings to which it may be reduced by the hatchet, as by Dr. Kolbe's formula to describe acetic acid as containing the products which may be obtained from it by destructive influences. A Kolbe botanist would say that half the chips are united with some of the blocks by the force *parenthesis*; the other half joined to this group in a different way, described by a *buckle*; shavings stuck on to these in a third manner, *comma*; and finally, a compound of shavings and blocks united together by a fourth force, *juxtaposition*, is joined on to the main body by a fifth force, *full stop*. The general use of unmeaning signs has become so habitual to Dr. Kolbe, that whenever anything has to be explained, he performs the task to his own satisfaction by inventing a sign for its unknown cause. . . . Signs and words are doubtless indispensible means for the expression of facts or thoughts; but Dr. Kolbe uses them instead of facts, and as a substitute for ideas.[74]

Kolbe had not only to worry about Williamson, but soon also had to defend his ideas against those of his good friend and Williamson's countryman, Edward Frankland. A few months after his isolation of the "ethyl" radical from the reaction of ethyl iodide with zinc, Frankland was successful in producing "methyl" gas from an analogous reaction with methyl iodide. But this time another substance was also produced, a white crystalline residue deposited in the sealed reaction tube. Frankland reported his discovery of zinc methyl and zinc ethyl to the Chemical Society in November 1849.[75] He suggested that these substances could be conceived to be constituted as substitution products of the hypothetical zinc hydride.

Intent upon following up this important new research, Frankland carried out extensive experiments that produce a variety of novel organometallic compounds.[76] He later recounted: "I had not proceeded far in the investigation of these compounds before the facts brought to light began to impress upon me the existence of a fixity in the maximum combining value or capacity of saturation in the metallic elements, which had not before been suspected. . . . It was evident that the atoms of zinc, tin, arsenic, antimony &c. had only room, so to speak, for the attachment of a fixed and definite number of the atoms of other elements, or, as I should now express it, of the bonds of other elements."[77]

For instance, it ought to have been easy to oxidize or halogenate the highly electropositive zinc methyl, but all such attempts failed. The only way further combination predicted by the copula theory could take place was if the organic copulae became disattached from the metal atom. In other words, Frankland demonstrated that the combining capacity of copulated atoms or groups of atoms was indeed reduced by the addition of organic copulae. He expressed these thoughts in his most famous paper as follows:

> When the formulae of inorganic chemical compounds are considered, even a superficial observer is struck with the general symmetry of their construction; the compounds of nitrogen, phosphorus, antimony and arsenic especially, exhibit the tendency of these elements to form compounds containing 3 or 5 equivs. of other elements, and it is in these proportions that their affinities are best satisfied: thus in the ternal group we have NO_3, NH_3, NI_3, NS_3, PO_3, PH_3, PCl_3, SbO_3, SbH_3, $SbCl_3$, AsO_3, AsH_3, $AsCl_3$, &c.; and in the five-atom group, NO_5, NH_4O, NH_4I, PO_5, PH_4I, &c. Without offering any hypothesis regarding the cause of the symmetrical grouping of atoms, it is sufficiently evident, from the examples just given,

that such a tendency or law prevails, and that, no matter what the character of the uniting atoms may be, the combining power of the attracting element, if I may be allowed the term, is always satisfied by the same number of these atoms.[78]

This is usually regarded as the first statement of the law of valence. In this quotation Frankland recognized what we would call the tri- and pentavalent character of the inorganic compounds of the group V elements. In the following pages, he expanded this view to include the organometallic compounds of arsenic, antimony, zinc, tin, and mercury, the last three exhibiting "combining powers" of one and two.

Frankland's law of maximum saturation capacity illustrated clear parallels to the type theory. He wrote: "It was probably a glimpse of the operation of this law amongst the more complex organic groups, which led Laurent and Dumas to the enunciation of the theory of types."[79] Furthermore, Frankland proposed that certain inorganic compounds of the metals with which he had been working, chiefly the oxides, could serve as the "true molecular type" of each of his new organometallic substances.[80] For the third time in three years he emphasized the analogy between his new compounds and those of the type theorists: "It is obvious that the establishment of this view of the constitution of the organo-metallic bodies will remove them from the class of organic radicals, and place them in the most intimate relation with ammonia and the bases of Wurtz, Hofmann, and Paul Thenard."[81] Frankland was explicitly offering the olive branch to his "typist" opponents: "The formation and examination of the organo-metallic bodies promise to assist in effecting a fusion of the two theories which have so long divided the opinions of chemists, and which have too hastily been considered irreconcilable."[82] Now firmly convinced of the central importance of substitution phenomena,[83] Frankland eagerly began to cultivate the type theorists' garden with the aid of his new organometallic tools.

Hermann Kolbe must have been dismayed at his friend's defection. At first "strongly opposed" to Frankland's views, he expressed a "friendly but adverse" critique.[84] But Frankland's evidence was unequivocal, and he managed to persuade Kolbe of his views through correspondence during the year 1856.[85]

Early the following year, what was intended to have been a joint manifesto appeared in Liebig's *Annalen*.[86] Kolbe now agreed with Frankland in conceding, for example, that cacodylic acid was produced not by conjugation of arsenious acid with methyl, but by substitution of methyl for oxygen of arsenic acid. Kolbe thus abandoned the central tenet of the copula theory—though he continued to use the word. Further, he derived organic acids, aldehydes, and ketones in a schematic fashion from carbonic acid, by substitution of radicals or hydrogen atoms for "atoms" (conventional equivalents) of oxygen.[87]

> The foregoing considerations have led us to believe that the single atoms of oxygen in carbonic acid may, in like manner, be susceptible of substitution by hydrogen and the alcohol radicals. The exchange of an atom of oxygen in carbonic acid C_2O_4 for one of hydrogen, methyl, ethyl, &c., would result in the production of the fatty acids . . . :
>
> | | |
> |---|---|
> | carbonic acid | $2HO,C_2O_4$ |
> | acetic acid | $HO,(C_2H_3)C_2O_3$ |
> | aldehyde | $\left. \begin{array}{c} C_2H_3 \\ H \end{array} \right\} C_2O_2$ |
> | acetone | $\left. \begin{array}{c} C_2H_3 \\ C_2H_3 \end{array} \right\} C_2O_2$ |

Implicit here is a notion of the tetravalence of the carbon "radical" C_2, later recognized by all as a single atom. Frankland gave his own version of this theory in a lecture at the Royal Institution on 28 May 1858.[88]

Kolbe further developed the theory between 1857 and 1860. In the double installment 6/7 of his *Lehrbuch*, which appeared near the end of 1857,[89] Kolbe designated the C_2 group "carbonyl," and wrote: "Carbonyl combines with hydrogen and the ether radicals to form conjugate radicals, which, like carbonyl itself, unites with oxygen and other electronegative elements in various proportions, but always such that the number of positive copula atoms and of negative atoms in these compounds adds up to two or four, so that these compounds may be considered derived from the diatomic ["zweiatomig", i.e., divalent] carbonic oxide, C_2O_2, or diatomic [i.e., dibasic] carbonic acid, C_2O_4."[90] Kolbe now wrote the rational formula for acetic acid as

HO.(C_2H_3)C_2O_2,O, for it was clear that one of the three oxygen "atoms" was more loosely bound than the other two, since it could be replaced by, for example, chlorine or sulfur (as had been discovered by Gerhardt and Kekulé respectively).

So Kolbe distinguished between two radicals present in acetic acid, the more resolved radical acetyl, (C_2H_3)C_2, and the more proximate radical "acetoxyl," (C_2H_3)C_2O_2.[91] Substitution of the oxygen atom outside of the acetoxyl radical by hydrogen yields aldehyde, substitution by methyl yields acetone:

$$\left. \begin{array}{c} C_2H_3 \\ H \end{array} \right\} C_2O_2 \qquad \left. \begin{array}{c} C_2H_3 \\ C_2H_3 \end{array} \right\} C_2O_2$$
$$\text{aldehyde} \qquad\qquad \text{acetone}$$

If further substitution of oxygen "atoms" by positive elements takes place, the C_2O_2 radical (Kolbe's "Kohlenoxyd," modern "carbonyl") is itself broken up, and alcohols are the product:

$$\text{HO.} \left\{ \begin{array}{c} C_2H_3 \\ H_2 \end{array} \right\} C_2O$$
$$\text{ethyl alcohol}$$

Kolbe could see no reason why further substitution of hydrogen atoms by methyl radicals in this molecule could not take place, which would yield two as yet unknown alcohols, especially:[92]

$$\text{HO.}(C_2H_3)_3C_2O$$
"trimethyl carbonic oxide hydrate"
(*tert*-butyl alcohol)

These ideas were spelled out in more detail in the double installment 8/9, appearing in 1859.[93] Kolbe wrote:

> Our views on the manner of composition of the elements in organic chemistry have received a significant expansion by the perception that, in addition to the customary monatomic radicals, there exist also diatomic, triatomic, and tetratomic radicals. . . . Among the carbonyl radicals, the following are monatomic: methyl, H_3C_2, ethyl,
>
> $$\left. \begin{array}{c} C_2H_3 \\ H_2 \end{array} \right\} C_2,$$

and acetoxyl, $(C_2H_3)C_2O_2$, etc.; diatomic radicals include carbonic oxide, C_2O, dihydrocarbonyl (methylene), H_2C_2, and methylhydrocarbonyl (ethylene),

$$\left.\begin{array}{c} C_2H_3 \\ H \end{array}\right\} C_2;$$

formyl, HC_2, and acetyl, $(C_2H_3)C_2$, are triatomic; and finally, carbonyl itself, C_2 is tetratomic . . .

C_2O_4 or $(C_2O_2)O_2$	C_2Cl_4	H_4C_2 [94]
carbonic acid	carbon tetrachloride	marsh gas

But why, Kolbe asked, do these radicals have these particular atomicities (valences)? It all comes down to the simple fact, he stressed, that "the carbonyl of carbonic acid is equivalent to 4 atoms of hydrogen, i.e., [it is] tetratomic."[95] Acetyl is triatomic because one-fourth of the saturation capacity of carbonyl is satisfied by methyl; acetoxyl is monatomic because it is three-fourths saturated by one methyl and two oxygen atoms, and so forth.

And Kolbe went a step further: the basicities of acids are directly related to the number of oxygen "atoms" outside the radical.[96] In the following series:

$HO.[NO_4],O$	nitric acid
$2HO.[S_2O_4],O_2$	sulfuric acid
$HO.(C_2H_3)[C_2O_2],O$	acetic acid
$2HO.[C_2O_2],O_2$	carbonic acid

the radicals between the square brackets were conceived to be stable groupings whose oxygen content had no effect on basicity. What counted were the number of oxygen atoms outside these radicals, always matched by an equal number of water molecules (actually half-molecules) on the other side. Thus, nitric acid was formulated as the hydrated oxide of the monatomic radical $[NO_4]$, acetic acid as the methyl-copulated hydrated oxide of the diatomic radical $[C_2O_2]$, and carbonic acid as the dihydrated dioxide of the same radical. The oxygen within the radical could be replaced without altering the basicity, as in the conversion of acetic acid to ethyl alcohol

$$HO.\left.\begin{array}{c} C_2H_3 \\ H_2 \end{array}\right\} C_2,O,$$

but if extra-radical oxygen atoms are replaced, the basicity is correspondingly reduced, as in the following schema:

$$2HO.[C_2O_2],O_2 \rightarrow HO.H[C_2O_2],O \rightarrow H_2[C_2O_2]$$

Kolbe discussed the theory once more in a long and important article in Liebig's *Annalen*.[97] The four types of the theory of Gerhardt and his followers attempted to unify organic and inorganic compounds, Kolbe related, but the relationships they revealed were unnatural, artificial, and superficial, leading to an "empty game with formulas" and a "dead schematism"; "Nature does not limit herself to variations on four themes."[98] Kolbe's theory, on the other hand—as the title of the article proclaimed—offered a "natural" and "scientific" basis for such a unification. We need not discuss this article in detail as it is merely a somewhat expanded treatment of the theories already expressed in his *Lehrbuch*. Kolbe placed in the forefront the idea that most organic molecules could be thought of as derived from one or more molecules of carbonic acid, C_2O_4, where various oxygen equivalents have been replaced by electropositive components, hydrogen, and organic radicals.

Kolbe's theory is a remarkable synthesis of many lines of thought. The intellectual kernel from which the ideas developed was Berzelius' theory of conjugated radicals, an 1846 paper by Liebig (which Kolbe cited),[99] and Frankland's research on organometallic compounds in the early 1850s. But by 1860 we see many curious developments: oxygenated radicals (a Berzelian arch-heresy), a distinction between mono- and polybasic acids (which Kolbe had long avoided), and, above all, substitution rather than addition reactions (inconsistent with dualism). As Berzelius had been forced to seek refuge in copulae and thus provide a further development of the hated theory of substitution, so Kolbe had been forced by the chemical facts into admitting—and even extending—certain ideas of the opposing camp. In particular, Williamson's theory of polybasic (polyatomic) radicals became central to Kolbe's scheme, as did Williamson's insight of relating the basicity of an acid to the basicity of the constituent radical. And Kolbe's oxygen theory of acidity—basicity related to the number of extra-radical oxygen atoms—was exactly that which Laurent had proposed in 1836.

1. E. von Meyer, *A History of Chemistry From the Earliest Times to the Present Day*, trans. G. M'Gowan (London, 1891), p. 282; E. Divers, *Proc. Roy. Soc.* 78A (1907): xxiv–xliv; G. C. Foster, *J. Chem. Soc.* 87 (1905): 605–18; R. Anschütz, *August Kekulé*, 2 vols. (Berlin, 1929), 1:664, and 2:943–44, 950.

2. C. A. Russell, *The History of Valency* (Leicester: Leicester University Press, 1971), pp. 49–134 passim; J. Harris and W. H. Brock, "From Giessen to Gower Street: Towards a Biography of Alexander W. Williamson," *Ann. Sci.* 31 (1974): 95–130; W. H. Brock, ed., *The Atomic Debates* (Leicester University Press, 1967); W. H. Brock, "A. W. Williamson," *Dict. Sci. Biog.* 14 (1976): 394–96; E. R. Paul, "Alexander W. Williamson on the Atomic Theory," *Ann. Sci.* 35 (1978): 17–31.

3. A Williamson, "Results of a Research on Aetherification," abstract in *Rep. Brit. Assoc. Adv. Sci.* 20 (1850): pt. 2, p. 65; "Theory of Aetherification," *Phil. Mag.* [3] 37 (1850): 350–56; *J. Chem. Soc.* 4 (1851): 106–12; rpt. in *Papers on Etherification and on the Constitution of Salts* (Edinburgh: Alembic Club Rpt. no. 16, 1902), pp. 5–17 (hereafter cited as *Papers*).

4. Williamson, *Papers*, p. 8.

5. Ibid., pp. 9–10. He also synthesized methyl and ethyl amyl ethers in a similar fashion.

6. Ibid., p. 14.

7. Williamson, *J. Chem. Soc.* 4 (1851): 229–39; *Papers*, pp. 24–39.

8. Williamson, *Papers*, p. 33.

9. Ibid., pp. 37–38.

10. Ibid., p. 39.

11. Williamson, *Rep. Brit. Assoc. Adv. Sci.* 21 (1851): pt. 2, p. 54; *Chem. Gazette* 9 (1851): 334–39; *J. Chem. Soc.* 4 (1851): 350–55; *Papers*, pp. 41–49.

12. Williamson, *Papers*, p. 44.

13. Williamson actually used the phrase "holds together" only in describing the CO radical, but there can be no doubt that he meant the same idea to apply to all of his "bibasic" moieties, including SO^2. Ibid., pp. 45–47.

14. Ibid., p. 42.

15. Ibid., p. 10.

16. Obviously, I use this term to refer to skeletal asymmetry, not stereochemical asymmetry as in modern parlance.

17. Williamson, *Papers*, pp. 8–9, 32.

18. For a discussion of these developments see A. J. Ihde, *Development of Modern Chemistry* (New York: Harper & Row, 1964), pp. 209–12; and J. R. Partington, *A History of Chemistry* (London: Macmillan & Co., 1964), 4:437–41. Partington misdates Liebig's article as 1837.

19. C. Gerhardt, "Remarques sur un travail de M. Williamson relatif aux éthers," *Compt. rend. trav. chim.* 6 (1850): 361–64.

20. Gerhardt and G. Chancel, "Sur la constitution des composés organiques," *Compt. rend. trav. chim.* 7 (1851): 65–84.

21. Gerhardt, "Recherches sur les acides organiques anhydres," *Ann. chim. phys.* [3] 37 (1853): 285–342, esp. 331–42; *Traité de chimie organique*, 4 vols. (Paris, 1853–56).

22. Gerhardt and L. Chiozza, "Addition aux recherches sur les acides anhydres," *Comptes rendus Acad. Sci.* 36 (1853): 1,050-54 (1,054).

23. H. Deville, "Note sur la production de l'acide nitrique anhydre," *Compt. rend. Acad. Sci.* 28 (1849): 257-60.

24. Gerhardt, "Sur la basicité des acides," *Compt. rend. trav. chim.* 7 (1851): 129-56.

25. Williamson, *Papers*, p. 45.

26. Letter printed in E. Grimaux and C. Gerhardt, Jr., *Charles Gerhardt: sa vie, son oeuvre, sa correspondance* (Paris, 1900), pp. 230-33, 403-4.

27. Gerhardt, "Ueber wasserfreie Säuren, namentlich wasserfreie Benzoësäure und Essigsäure," *Ann.* 82 (1852): 127-32.

28. G. Chancel, "Sur l'étherification et sur une nouvelle classe d'éthers," *Compt. rend. Acad. Sci.* 31 (1850): 521-23. Chancel published experiments and conclusions very similar to Williamson's just two months after the Englishman's first etherification paper.

29. Gerhardt, "Recherches sur les acides organiques anhydres," *Compt. rend. Acad. Sci.* 34 (1852): 755-58 (758).

30. One may be permitted to express some doubt as to the veracity of this letter. By this time Gerhardt's two communications to Liebig were in press, dated before his purported rediscovery of an offprint of Williamson's paper on his disordered desk. A copy of this letter, originally prepared by Williamson's wife after his death and left to her son, was recopied by J. Harris about fifty years ago, and is now in the possession of William Brock of the University of Leicester. (I thank Professor Brock for access to his invaluable collection of Williamson materials, hereafter cited "Harris-Brock Collection," and for his generous assistance in the summer of 1979.)

31. Grimaux, *Gerhardt*, p. 412.

32. Gerhardt, "Recherches sur les acides organiques anhydres," *Compt. rend. Acad. Sci.* 34 (1852): 902-5.

33. Grimaux, *Gerhardt*, pp. 235, 239, 241.

34. Gerhardt, "Recherches" (*Compt. rend. Acad. Sci.* 34 [1852]), p. 903.

35. Gerhardt, "Recherches" (*Ann. chim. phys.*[3] 37 [1853]), pp. 332, 340.

36. Williamson, "On Gerhardt's Discovery of Anhydrous Organic Acids," *Proc. Roy. Soc.* 1 (28 January 1853): 239-42.

37. If anything, he had slighted his own role in the matter. The following year Williamson presented a less modest statement of priority: "The leading propositions [in his 1851 paper] have been adopted by several eminent chemists in this country and in France; and M. Gerhardt speedily enriched science with a series of brilliant and striking illustrations of their truth" ("Note on the Decomposition of Sulphuric Acid by Pentachloride of Phosphorus," *Proc. Roy. Soc.*, 7 [1856, recd. 23 February 1854], 11-15).

38. Grimaux, *Gerhardt*, p. 240.

39. Berzelius, *Jahresbericht* for 1838, 19 (1840): 370-71; cf. p. 350.

40. Partington, *History*, 4:372-73. Berzelius had noted such cases in his *Lehrbuch der Chemie*, 3d ed., 10 vols. (Dresden and Leipzig, 1833-41), 6:180-81 (1837, written autumn 1835).

41. O. Wallach, ed., *Briefwechsel zwischen J. Berzelius und F. Wöhler*, 2

vols. (Leipzig, 1901), 2:243. See also Berzelius' letter to Liebig, 29 April 1841, in J. Carrière, ed., *Berzelius und Liebig, Ihre Briefe von 1831-1845* (Munich, 1898), pp. 228-29.

42. Berzelius, *Jahresbericht* for 1840, 21 (1842): 105-9.

43. L. Melsens, "Note sur l'acide chloracétique," *Comptes rendus Acad. Sci.* 14 (1842): 114-17.

44. Berzelius, *Lehrbuch*, 5th ed., 5 vols. (Dresden and Leipzig, 1843-48), 1:709.

45. Ibid., 1:459-61, 2:123-24, 4:48-52, 5:15-17. An illuminating discussion of the relations between Berzelius and Laurent is J. H. Brooke, "Chlorine Substitution and the Future of Organic Chemistry," *Stud. Hist. Phil. Sci.* 4 (1973): 47-94.

46. A. Kekulé, *Lehrbuch der organischen Chemie*, 2 vols. (Erlangen, 1861-66), 1:74-75.

47. A. Laurent, *Méthode de chimie* (Paris, 1854), pp. 249-50; *Chemical Method* (London, 1855), pp. 204-5.

48. R. Bunsen, "Untersuchungen über die Kakodylreihe," *Ann.* 37 (1841): 1-57; 42 (1842): 14-46; and 46 (1843): 1-48.

49. Berzelius, *Jahresbericht* for 1837, 18 (1839): 487-501; 20 (1841): 14-25; and 21 (1842): 495-503. He wrote to Wöhler: "This is a triumphal chariot, which will overrun and smash [Dumas'] flimsy theoretical barricades;" *Briefwechsel*, 2:220 (letter of 29 January 1841).

50. E. Frankland and H. Kolbe, "Ueber die chemische Constitution der Säuren der Reihe $(C_2H_2)_nO_4$ ...," *Ann.* 65 (1848): 288-304. Reprinted (translated and edited) in Frankland's *Experimental Researches in Pure, Applied, and Physical Chemistry* (London, 1877), pp. 29-34 and 34-46.

51. Ibid., p. 294. I have slightly modified these formulas to achieve consistency with Kolbe's later formulations.

52. Kolbe and Frankland, "On the Products of the Action of Potassium on Cyanide of Ethyl," *J. Chem. Soc.* 1 (1848): 60-74; *Experimental Researches*, pp. 49-62. The cyanide was probably contaminated with alcohol or water.

53. Kolbe, "Untersuchungen über die Elektrolyse organischer Verbindungen," *Ann.* 69 (1849): 257-94. Kolbe had actually produced the dimers of these radicals, ethane and octane. In the future, radicals in quotation marks are to be understood to be the respective dimers.

54. Frankland, *J. Chem. Soc.* 2 (1849): 263-96; *Experimental Researches*, pp. 67-96.

55. Frankland, "Researches on the Organic Radicals," *J. Chem. Soc.* 3 (1850): 30-52, 322-47; *Experimental Researches*, pp. 97-118.

56. Gerhardt, "Remarques sur un travail de M. Hofmann sur les radicaux," *Compt. rend. trav. chim.* 6 (1850): 233-36.

57. Hofmann, "Note upon the Action of Heat upon Valeric Acid ...," *J. Chem. Soc.* 3 (1850): 121-34; Brodie, "Observations on the Constitution of the Alcohol-Radicals ...," ibid., pp. 405-11; Frankland, "Researches," pp. 341-47.

58. Kolbe, "Ueber die chemische Constitution und Natur der organischen Radicale," *Ann.* 75 (1850): 211-39; 76 (1850): 1-73; "On the Chemical Consti-

tution and Nature of Organic Radicals," *J. Chem. Soc.* 3 (1850): 369-405; 4 (1851): 41-79.

59. Kolbe, *Ausführliches Lehrbuch der organischen Chemie*, 1st ed, 3 vols, (appearing as vols. 3-5 of Graham-Otto's *Ausführliches Lehrbuch der Chemie*); 1 (Brunswick, 1854—actually appeared in fascicles between 1854 and 1859), and 2 (1860—actually appeared between 1860 and 1864). Volume 3 was written by others.

60. Kolbe, "Organic Radicals," p. 372.

61. Kolbe, *Lehrbuch*, 1: 40-42 (1854); "Meine Betheiligung an der Entwickelung der theoretischen Chemie," *J. prakt. Chem.* [2] 23 (1881): 305-23, 353-79, 497-517; 24 (1881): 374-425 (353).

62. The German phrase is "paare Atomzahl," which is correct; the English translation, presumably prepared by Kolbe himself, reads "whole numbers of atoms," which is incorrect, both as a translation and as a description of the law.

63. Kolbe, "Organic Radicals," 4:76.

64. Ibid., p. 77.

65. F. Wrightson, "On the Atomic Weight and Constitution of the Alcohols," *Phil. Mag.* [4] 6 (1853): 88-99.

66. Ibid., pp. 91-92. Wrightson used small letters for the small (Gerhardtian) "equivalents."

67. Ibid., pp. 92-93. Wrightson's error was one of formula translation—instead of doubling Kolbe's carbon and oxygen equivalents to obtain Williamson's atomic weights (i.e., halving the number of carbon and oxygen atoms in Kolbe's formula), he halved Kolbe's hydrogen equivalent (doubled the number of hydrogen atoms). Williamson's reply would have been clearer to Wrightson and Kolbe had he pointed this out.

68. Williamson, "Note on the Preparation of Propionic and Caproic Acids," *Phil. Mag.* [4] 6 (1853): 204-6. Williamson later stated that he had tried in vain to talk Wrightson out of publishing the paper (*J. Chem. Soc.* 7 [1854]: 122). Williamson and Wrightson had first met in Liebig's laboratory in Giessen.

69. Wrightson, "Remarks on Professor Williamson's Othyle Theory," *Phil. Mag.* [4] 6 (1853): 418-20. T. A. Hirst's impressions of Wrightson, who is otherwise little known, are recorded in his journal, now at the Royal Institution. Wrightson was "polite but insincere, assuming but in fact commonplace . . . he has an unhappy, irritable temper . . . blustering, muddled, ill-used & snappish" (Journal, vol. 2:675, 1,077, 1,081 [1850-53]).

70. Kolbe, "Critical Observations on Williamson's Theory of Water, Ethers, and Acids," *J. Chem. Soc.* 7 (1854): 111-21. Only Eqn. (b) and Kolbe's formula for ether actually appeared in this article; the other equations and formulas are inferred.

71. Ibid., pp. 114, 118.

72. Both letters—Watts' is undated—are in the H. E. Roscoe Collection at the Chemical Society in London. I thank the curator for access.

73. Williamson, "On Dr. Kolbe's Additive Formulae," *J. Chem. Soc.* 7 (1854): 122-39.

74. Ibid., pp. 123, 132-35. When Kolbe visited London a few years later, he was graciously and warmly received by Williamson. In 1881, in the context of a violent diatribe against all of his scientific foes from Dumas to Baeyer, Kolbe mentioned his respect and affection for the Englishman, claiming that their earlier disagreement had been due to a mere misunderstanding ("Meine Betheiligung," p. 311n.). In contrast to most of his colleagues, Williamson did not seem to know how to make and keep enemies.

75. Frankland, "On a New Series of Organic Bodies Containing Metals and Phosphorus," *J. Chem. Soc.* 2 (1849): 297-99.

76. The research was vividly described by Frankland in his *Sketches from the Life of Edward Frankland* (London, 1901), pp. 120-24, 187-90.

77. Frankland, *Experimental Researches*, p. 145.

78. Frankland, "On a New Series of Organic Bodies Containing Metals," *Phil. Trans. Roy. Soc.* 142 (1852): 417-44 (440); reprinted in *Experimental Researches*, pp. 160-81. The word "equivs." in the first sentence was replaced by "atoms" in the reprinted version.

79. Ibid., p. 440.

80. Ibid., p. 441.

81. Ibid., p. 442.

82. Ibid., p. 441.

83. In 1856 Frankland declared: "No generalization has, perhaps, so extensively contributed to the progress made by organic chemistry during the last fifteen years as the doctrine of substitution" (*Phil. Trans. Roy. Soc.*, 147:59).

84. Kolbe, *Lehrbuch* (1854) 1: 23. The expressions in quotation marks are Frankland's (*Sketches*, pp. 193-96; *Experimental Researches*, pp. 147-48, 150).

85. Frankland, *Experimental Researches*, pp. 147-48.

86. Kolbe, "Ueber die rationelle Zusammensetzung der fetten und aromatischen Säuren . . . ," *Ann.* 101 (1857): 257-65; English translation in *Experimental Researches*, pp. 148-53. The article appeared in Kolbe's name alone; Frankland (ibid., p. 148) said this was due to an "inadvertence." However, the pronoun "ich" appears throughout most of the paper, and Frankland is cited by name; later references by Kolbe to this paper also use "ich" ("Meine Betheiligung," pp. 363, 367).

87. Ibid., pp. 262, 264; *Experimental Researches*, p. 152.

88. Frankland, "On the Production of Organic Bodies without the Agency of Vitality," *Proc. Roy. Soc.* 2 (1858): 538-44.

89. Kolbe, *Lehrbuch*, 1:481-672. This is Kolbe's date ("Meine Betheiligung," p. 369). Kekulé, relying on Hinrich's *Bücherverzeichnis*, later asserted that it actually appeared early the following year (Anschütz, *Kekulé*, 1:560). Kayser's *Bücher-Lexikon* confirms Kolbe's date.

90. Ibid., p. 575. The passage makes sense only if "diatomic" is given the two distinct denotations in brackets.

91. Ibid., p. 569.

92. Ibid., pp. 568-69.

93. *Lehrbuch*, 1:673-864. This is Kekulé's dating (Anschütz, 1:560); Kolbe

claimed it appeared in 1858 ("Meine Betheiligung," p. 370). This time Kayser's *Bücher-Lexikon* agrees with Hinrich's *Bücherverzeichnis*.

94. Ibid., pp. 740-41.

95. Ibid., p. 742.

96. Ibid., pp. 744-49.

97. Kolbe, "Ueber den natürlichen Zusammenhang der organischen mit den unorganischen Verbindungen . . . ," *Ann.* 113 (1860): 293-332.

98. Ibid., pp. 293-94, 332.

99. In a discussion of the chemical process of respiration, Liebig had suggested that many organic substances could be represented as substitution products of carbonic acid: "Der chemische Process der Respiration," *Ann.* 58 (1846): 335-48.

9

The Emergence of Valence and Structure

RADICALS ENGULF TYPES:
THE THEORY OF POLYATOMIC RADICALS

By 1851, we have seen Williamson had developed the water type theory and had proposed double water types to account for the constitution and properties of dibasic acids. A dibasic radical, or the dibasic oxygen atom, could function either by linking together two monobasic radicals, or by linking together two other dibasic radicals in a double type, which could in turn unite two more monobasic radicals to form a single molecule consisting of five identifiable moieties. Williamson had forged a definitive "asymmetric synthesis" argument for consistent molecular magnitudes and a persuasive "valence" argument for consistent atomic magnitudes.

The first expansion of Williamson's theory was due to William Odling (1829-1921), then demonstrator in chemistry at Guy's Hospital, London, in an important paper "On the Constitution of Acids and Salts."[1] Although he claimed to offer only an "elaboration" of the views of Williamson and Gerhardt, the paper is significant in providing a generalization of Williamson's somewhat casual notion of "basicity" (valence) of atoms and radicals. Odling suggested that the "replaceable, or representative, or substitution value" of an atom could conveniently be represented by an appropriate number of dashes to the right of the symbol for the element: hence H′, K′, Sn′, Sn″, O″, N‴.[2] Complex bases as well as acids could be represented by substitution in one or more molecules of water; for instance, the formulas for the phosphoric acids and the phosphates were related to the tribasic PO‴ radical, orthophosphoric acid itself being:[3]

$$\left.\begin{array}{c} PO''' \\ 3H' \end{array}\right\} 3O''$$

Such a formula can be interpreted as representing a molecule composed of linked atoms, as in Williamson's theory. But Odling, like Gerhardt, was not always consistent in this respect. For instance, he suggested two equivalent formulas for "hyposulfuric" (dithionic) acid:[4]

$$\left.\begin{array}{c} SO_2'SO_2' \\ 2H' \end{array}\right\} 2O'' \quad \text{or} \quad \left.\begin{array}{c} S_2O_4'' \\ 2H' \end{array}\right\} 2O''.$$

The second of these can be interpreted in the "linking" sense, but not the first. Odling's criterion in establishing one of these formulas was apparently not the constructability of the molecule following the valence assignments, but merely an equalization of the sums of dashes on either side of the bracket.

Nonetheless, Odling accepted Williamson's "valence" argument and applied it to novel cases. He pointed out that chlorine could be thought to replace the oxygen in water, to produce hydrochloric acid:

$$\left.\begin{array}{c} H' \\ H' \end{array}\right\} O'' \qquad \left.\begin{array}{c} H' \\ H' \end{array}\right\} \begin{array}{c} Cl' \\ Cl' \end{array}$$

However, the latter symbol "is obviously capable of division into two equivalents of H′Cl′, so that in reality 2 atoms of hydrochloric acid are required to represent 1 atom of water, the reason being that in the latter case the separable equivalents of hydrogen are held together by the indivisible oxygen."[5] Williamson highly approved of this article. He wrote to Roscoe: "It contains some excellent formulae & some new reactions confirming them. . . . [Several] of our younger chemists are taking to them very kindly, but I very much doubt the probability of the older chemists ever adopting a change so great."[6]

Shortly after Odling's paper appeared, Williamson published an interesting article on the chlorination of sulfuric acid.[7] His intent was to contest Gerhardt's and Chiozza's recently expressed view of the constitution of sulfuric acid (see chapter 8). This view, he said, was nothing less than a return to the old Berzelian idea that ready-formed water exists as such in acids, a notion that Gerhardt himself had frequently and repeatedly contested. In support of his own viewpoint, Williamson succeeded in chlorinating the acid in two separate stages, producing a substance now known as chlorosulfonic acid,[8]

HO
SO_2 .
Cl

He wrote:

> The existence and formation of this body . . . furnishes the most direct evidence of the truth of the notion, that the bibasic character of sulfuric acid is owing to the fact of one atom of its radical SO_2 replacing, or (to use the customary expression) being equivalent to two atoms of hydrogen. Had this radical been divisible like an equivalent quantity of a monobasic acid, we would have obtained a *mixture*, not a *compound* of the chloride with the hydrate, — or, at least, the products of decomposition of that mixture.

As was the case with the 1851 "valence" argument, Williamson's reasoning is persuasive but not conclusive. The new acid could be represented in Kolbe's system just as easily ($SO_2Cl \cdot SO_3HO$), but the reason for the stable cohesion of the chloride and hydrate is in this case unclear.

Just at this time chemical circles in London were brightened by the appearance of a promising young German chemist, F. August Kekulé (1829–1896). Kekulé had an assistantship with John Stenhouse at St. Bartholomew's Hospital, but was most decisively influenced by his considerable intercourse with Williamson at University College. Kekulé later recalled: "the good fortune was alotted me, in lively and friendly conversation with Williamson, to become thoroughly familiar with the patterns of thought of this philosophical intellect. . . . Chemistry then stood at a turning point; the theory of polybasic radicals was emerging. The ingenious Odling also consorted with Williamson. Williamson insisted on clear formulas, without Kolbe's commas and buckles or Gerhardt's brackets. That was an excellent schooling for me, generating an independent spirit."[9] In a Latin curriculum vitae for the University of Heidelberg (1856), Kekulé wrote: "I must not fail to make mention of Williamson, that wisest of men and most learned of philosophers, who was not my teacher but my friend, and to whom I owe so much."[10] Two months after his arrival in London, Kekulé wrote the following lines to Planta: "I'll tell you about Williamson some other time, for once you start on him you're not soon finished, he has too many sides to be able to be characterized in a few words."[11]

Kekulé could not have chosen a better time to arrive in London. Within two months after his arrival, Odling and Williamson each published the important papers just discussed, both of which had a demonstrable influence on Kekulé's first important publication, on thiacetic acid.[12] Kekulé reasoned as follows. Since the hydrogen sulfide type (H_2S) is analogous to the water type (H_2O), and since neutral mercaptans (RSH) and sulfides (RSR), known at that time, bear analogies to alcohols (ROH) and ethers (ROR), then it is logical to suppose the possibility of a more extended parallelism between the two series, viz., to predict the possible existence of sulfur analogues of acids, acid anhydrides, and esters. To replace oxygen by sulfur in these three series Kekulé synthesized a new reagent, phosphorus pentasulfide (P_2S_5).

When Kekulé told Williamson of his idea—it must have been shortly after his arrival in London—the English chemist was enthusiastic and urged Kekulé to proceed. But Kekulé had no laboratory or time of his own and delayed the experiments. Finally, Williamson told his friend that he would do it himself if Kekulé stalled any longer. Thus it happened that Stenhouse's laboratory was blessed with the aroma of various organosulfur compounds in the early morning hours before Kekulé's official duties began. Stenhouse, who could not help noticing the remarkable odors, was displeased, but let the work continue.[13] Kekulé wrote to Planta about his work with Stenhouse: "Boredom, that's all I can say about it, and moreover the man—

otherwise very noble — is so indecent as to make a face when one wants to do something for oneself." Kekulé went on to report that fortunately Stenhouse was at the moment indisposed, so that he (Kekulé) could do "ein paar Versuche für mich."[14]

The experiments succeeded beautifully — Kekulé obtained sulfurated acids, anhydrides, and esters. The analogy to chlorination was virtually complete: "however, with this difference, that by using the chlorine compounds the product is resolved into *two* groups of atoms, while by using the sulfur compounds there is obtained only *one* group; a peculiarity, which, according to the bibasic nature of sulfur, must have been expected. Thus the new (Gerhardtian) formula notation is truly a better expression of the facts than the notation which has hitherto been customary."[15] If, using the old equivalents, alcohol and mercaptan are expressed as $C_4H_5O \cdot HO$ and $C_4H_5S \cdot HS$, then it remains mysterious why the chlorine analogue does not cohere in a similar fashion, to form a single indivisible molecule $C_4H_5Cl \cdot HCl$. The fact that it does not is simply explained, Kekulé noted, by the difference in "basicity" between hydrogen and chlorine on the one hand, and oxygen and sulfur on the other. Kekulé concluded: "It is not merely a difference in notation, but rather [is] an actual fact, that 1 atom of water contains 2 atoms of hydrogen and only 1 atom of oxygen; and that the quantity of chlorine equivalent to one indivisible atom of oxygen is divisible by 2, while sulfur, like oxygen itself, is *dibasic*, so that 1 atom is equivalent to 2 atoms of chlorine."[16]

This was the third use of Williamson's "valence" argument in as many months by three chemists working in London in early 1854: Williamson himself, Odling, and Kekulé. All three put the case in stronger terms than were justified. It was not demonstrated to be an "actual fact" that O = 16 is "dibasic," but merely more consistent and revelatory of analogies that were obscured or unexplained in the rival system of O = 8. The latter system retained the advantages — rather dubious ones to modern eyes — of successfully avoiding a confrontation with the new and puzzling idea of valence and of retaining something of the Berzelian dualism in organic chemistry.

Williamson's influence is also discernable in the work of another chemist who was intimately involved with the emergence of valence theory, Adolphe Wurtz. Building on the slightly earlier work of M. Berthelot, Wurtz investigated the "diatomic" and "triatomic" alcohols, namely derivatives of glycol and glycerine, and expressed the results in terms of the Williamson-Gerhardt type theory.[17] He wrote: "One can consider glycerine as a species of tribasic alcohol, that is, an alcohol containing 3 equivalents of hydrogen capable of being replaced by 3 groups . . . thus forming a bond [lien] between 3 molecules of conjugated water."[18]

Wurtz was also able to settle the controversy between Kolbe and Edward Frankland on the one side, and Laurent, Gerhardt, and Williamson on the other, over the monomeric or dimeric character of the hydrocarbon "radicals" isolated between 1848 and 1850 by Kolbe and Frankland. If the latter chemists' radicals, R, are actually dimers, R–R, then it should be possible, Wurtz reasoned, to produce "asymmetric" radicals, R–R'. Wurtz was able to accomplish this by fusing the iodides of two different hydrocarbon radicals with metallic sodium; for example:

$$C^4H^5I + C^8H^9I + 2Na \rightarrow C^4H^5 - C^8H^9 + 2NaI.$$

Wurtz formed five asymmetric and three symmetric hydrocarbons in this fashion (the "Wurtz reaction"). A simple application of the now familiar Williamsonian asymmetric synthesis argument — Wurtz, unlike Gerhardt, generously acknowledged his debt to Williamson — provided what Wurtz termed "conclusive proof" of the dimeric formulas for the Kolbe-Frankland hydrocarbon "radicals."[19]

Incidentally, the fact that Wurtz still retained Kolbe's "equivalents" here demonstrates that the two issues argued by Williamson were in fact independent of one another. Wurtz converted to the new atomic weights only in 1859. In his first article using these weights, he praised Williamson's classic researches that, he recognized, had formed the basis of the theory of polyatomic radicals, drawing particular attention to Williamson's idea of the linking function of radicals such as SO_2, CO, and glyceryl.[20] To Williamson personally he wrote: "I must tell you that your article on etherification has created a sensation" in

Paris.[21] His history of chemical theory contains a perceptive discussion of Williamson's influence on the development of organic chemistry theory.[22]

While Wurtz was applying Williamsonian arguments to new series of compounds, Gerhardt was completing the fourth and last volume of his monumental *Traité de chimie organique*, containing a final summary of his evolving conceptions of types, radicals, and formulas. Many substances did not fit naturally into Gerhardt's scheme of four simple types (water, ammonia, hydrogen, and hydrogen chloride), and for these he took recourse in an old idea of his: the theory of conjugated or copulated radicals. This theory had two independent and coexistent lifetimes: both versions were proposed in 1839, gained a measure of popularity, underwent various modifications, and were finally put to rest in 1857.[23] It is an odd coincidence that the demise of each version provided the occasion, in two different contexts, for the establishment of two theories of the tetravalence of the carbon atom, the fundamental axiom of the structure theory.

Gerhardt had originally suggested that conjugation was a special type of addition reaction between an acid and an organic substance, such that the union did not alter the acid's saturation capacity (in contrast to salt formation).[24] Later he used the term to apply to any organic ester of an inorganic acid; he recognized that a water molecule always split off in such reactions, reducing the basicity of the acid of one unit.[25] Gerhardt's "basicity law" stated that the basicity of the product of a conjugation reaction is equal to one plus the sum of the basicities of the reactants, less the number of reactant molecules:

$$B = b_1 + b_2 \ldots - (n - 1),$$

where the b's are basicities and n is the number of reactant molecules. The expression in parentheses represents the number of water molecules split off, i.e., the number of conjugations.

For example, in the reaction of benzene and nitric acid, we have:

$$0 + 1 - (2 - 1) = 0,$$

and nitrobenzene is indeed a neutral substance. For the formation of the monobasic sulfovinic acid, we write:

$$0 + 2 - (2 - 1) = 1$$

diethyl sulfate being:

$$(2 \times 0) + 2 - (3 - 1) = 0.$$

Monobasic picric acid is formed from carbolic acid (phenol) plus three molecules of nitric acid, hence:

$$1 + (3 \times 1) - (4 - 1) = 1.$$

This was the version Laurent adopted in his *Méthode de chimie*.[26] He cautioned, as had Gerhardt, that the basicity law must be used with circumspection, since there were instances of subtances that were on the borderline of being acidic or neutral. For example, carbolic acid must be considered monobasic when it is nitrated, but neutral when it is sulfonated:

$$0 + 2 - (2 - 1) = 1,$$

since both picric acid and phenolsulfonic acid are monobasic.

Gerhardt once more altered his definitions for his *Traité*: a "conjugate radical" was now regarded as any combination of two or more "constituent radicals" (which could also include single atoms). Radicals could be conjugated by addition—for example, organometallic substances ($M + nR$) or acyl radicals ($CO + R$); or they could be conjugated by substitution—for example, amines ($NH_3 + nR - nH$) or oxygenated radicals ($R + O - H_2$). The fact that "acetyl" (the modern denotation of the word originated with Gerhardt) could be conceived to be a member of both species of conjugate radicals ($CH_3 + CO$ or $C_2H_5 + O - H_2$) shows that Gerhardt's definitions were merely classificatory criteria—ambiguous ones, at that—and not reaction mechanisms.

Gerhardt's "hydrogen-equivalent rule" was: "Each hydrogen equivalent added to a radical diminishes the total hydrogen

The Emergence of Valence and Structure   259

equivalent of the radical by as much; and conversely, each hydrogen equivalent abstracted from a radical augments the equivalent of the remaining radical by as much."[28] Applied to additive conjugate radicals, the rule stated that the equivalent of the resulting radical is equal to the difference of the equivalents of the constituent radicals; for instance:

			equiv. in hydr.
	arseniosum radical	As	$= H_3$
	2 × methyl radical	$(CH_3)_2$	$= H_2$
	cacodyl radical	$As(CH_3)_2$	$= H$;
or:			
	carbonyl radical	CO	$= H_2$
	methyl radical	CH_3	$= H$
	acetyl radical	$CO(CH_3)$	$= H$

Applied to substitutive conjugate radicals, the rule predicted the equivalent value of the resultant radical to be equal to the *sum* of the equivalents of the constituent radicals, less the number of hydrogen atoms replaced by substitution. For example:

			equiv. in hydr.
	ammonium radical	NH_4	$= H$
	4 × ethyl radical	$(C_2H_5)_4$	$= H_4$
			$= H_5$
	less hydrogen replaced		$= H_4$
	tetraethyl ammonium radical	$N(C_2H_5)_4$	$= H$;
or:			
	benzoyl radical	C_7H_5O	$= H$
	sulfuryl radical	SO_2	$= H_2$
			$= H_3$
	less hydrogen replaced		$= H$
	sulfobenzoyl radical	$C_7H_4(SO_2)O$	$= H_2$

In a footnote, Gerhardt added that sulfobenzoic acid is "bibasic":

$$\left.\begin{array}{r}C_7H_4(SO_2)O \\ H_2\end{array}\right\} O_2.$$

The Gerhardt-Laurent system of chemistry was not popular in any European country during the 1850s, but in Germany in particular it was sedulously avoided, at time even reviled. One of the first signs of a change in attitude was the publication of H. Limpricht's *Grundriss der organischen Chemie*, which enthusiastically expounded Gerhardt's type theory. Limpricht adopted the atomic weights H = 1, C = 12, N = 14, and Cl = 35.5; however, he retained the conventional equivalents O = 8 and S = 16, because of the purported existence of half-molecules of water of hydration in certain salts. He took this step in spite of his own opinion that it was "extremely probable" that the true atomic weights of these two elements were twice these values.[29] Hence Limpricht, like Kolbe, Berthelot, and many others, assumed that the water molecule was "HO" (= ½H$_2$O), although his water *type* had to be written

$$\left.\begin{array}{r}H \\ H\end{array}\right\} O_2.$$

To compound the confusion, Limpricht always used the customary German conventional equivalents in his papers in Liebig's *Annalen*.

In the spring of 1857, Limpricht and his student von Uslar published a paper on sulfobenzoic acid,[30] a substance discovered by Mitscherlich in 1834. They regarded this compound as a conjugated acid wherein the divalent sulfuryl radical S$_2$O$_4$ substituted for one hydrogen atom of the monobasic benzoic acid

$$\left.\begin{array}{r}C_{14}H_5O_2 \\ H\end{array}\right\} O_2,$$

to yield the dibasic sulfobenzoic acid

$$\left.\begin{array}{c}C_{14}H_4(S_2O_4)O_2\\H_2\end{array}\right\}O_4.$$

Two months later O. Mendius, also a student of Limpricht, published similar views.[31] Mendius regarded the formation of sulfobenzoic acid to be truly one of addition of sulfuric anhydride to the benzoyl radical:

$$\left.\begin{array}{c}\mathsf{C}_7H_5O_2\\H\end{array}\right\}O_2 + S_2O_4\Big\}O_2 = \left.\begin{array}{c}\mathsf{C}_7H_4(S_2O_4)O_2\\H_2\end{array}\right\}O_4,$$

even though the "intrinsic" process is one of substitution.[32] He added parenthetically that chemists ought not renounce the study of the "rational, i.e., true constitution of chemical compounds. . . . Gerhardt, by ascribing only [a positivistic] significance to chemical formulas and especially to his types, has robbed them of — or perhaps failed to recognize — a large part of their value."[33] These publications by Limpricht and his student provided Kekulé the necessary inducement to publish his views on the structure theory.

KEKULÉ ON THE NATURE OF CARBON

August Kekulé (1829-1896) was raised in Darmstadt, the home town of Justus Liebig, and became the latter's student at Giessen in 1849.[34] He spent successive *Wanderjahre* in Paris (1851-52), Chur, Switzerland (1852-53), and London (1853-55). He later summarized his chemical education in the following terms: "Originally a pupil of Liebig, I had become a student of Dumas, Gerhardt and Williamson; I no longer belonged to any school. . . . From my extensive travels I got to know many different viewpoints, and learned to separate the good from the bad; I had become an eclectic."[35] This self-assessment contains a certain measure of truth, but it masks the fact that Kekulé had joined the French-English reform movement initiated by Gerhardt and Laurent and elaborated by such workers as Odling, Wurtz, and especially Williamson. Kekulé was from the early 1850s — and he always remained — inimical to the conservative school of Frankland and Kolbe.[36]

Early in 1856 Kekulé became Privatdozent at Heidelberg and began his brilliant teaching and research career. The first experimental work to emerge from Kekulé's laboratory was a series of two papers on the constitution of mercury fulminate and the hypothetical fulminic acid. He wrote:

> The following compounds can be considered as belonging to the same type [C = 6, O = 8]:
> C_2 H   H   H   H   marsh gas
> ....
> C_2 (NO_4) Hg Hg (C_2N)   mercury fulminate
> C_2 (NO_4) H   H   (C_2N)   hypothetical fulminic acid
> ... By including these bodies in the same type, I use this word not in the sense of Gerhardt's unitary theory, but rather in the sense in which it was first used by Dumas . . . All these bodies can be included in one series, one mechanical type.[37]

Thus, like Williamson and Frankland, Kekulé rejected Gerhardt's purely relativistic, "synoptic" conception of types, and turned to Dumas' "mechanical" idea. However, his (double) carbon atom united four monovalent *groups* of atoms, whereas Dumas' criterion was the conservation of the number of *equivalents* in all compounds of the series. This, of course, constituted the essence of the newer type theory since its earliest development by Williamson, Wurtz, Hofmann, and Gerhardt.[38]

Kekulé has frequently been chastized by historians for having abandoned atomic weights for the old equivalents in these two papers, the only time he ever did so. He later explained his reasons: as the new notation was virtually unknown in Germany then, and since the paper did not have much explicitly theoretical content, he made use of the commonly accepted notation in order to make himself more generally understandable.[39] A student's lecture notebook of 1857-58 shows that Kekulé followed the same practice in his lectures, i.e., he used conventional equivalents when discussion experimental facts, and atomic weights when discussing theory.[40] This was reasonable, considering the circumstances at that time.[41]

As Kekulé's structure theory probably originated in his London period, it is reasonable to suppose that here Kekulé was writing "C_2," while intending to express "C," the modern carbon atom. Seen in this light, these papers contain a clear, though implicit, statement of the tetravalence of the carbon atom, as well as a less than clear suggestion of carbon-carbon linkages.[42] In any case, we see here the first unambiguous development, if not initial suggestion, of the marsh gas type. It is ironic that all thirteen of Kekulé's formulas are properly derivable from methane *except* mercury fulminate and fulminic acid themselves, which were later found to contain neither a nitro nor a cyano group.

Between 1851 and 1857 most of the elements of the theories of valence and structure were assembled by a number of chemists: Williamson, Frankland, Odling, Kekulé, Wurtz, Gerhardt, and Kolbe, among others. What was needed at this point was a systematic development of these ideas, a clear and methodical elaboration of a unified conception of chemical constitution, with a program for its elucidation. Kekulé answered this need in two theoretical papers that appeared in late 1857 and early 1858.

Twice in later years Kekulé stated that his thoughts on structure theory dated from his stay in London. Kekulé told a now famous anecdote how the theory occurred to him one summer evening in a reverie on the upper deck of a London omnibus. This was probably in 1855.[43] In Heidelberg, by the spring or summer of 1856, he had written out a manuscript version, but only published it over a year later, in response to Limpricht's and his students' articles on sulfobenzoic acid and the former's *Grundriss der organischen Chemie*.[44] Kekulé noted that Limpricht's textbook was the first in Germany to advocate and expound Gerhardt's type theory, but Limpricht had subverted the true spirit of the theory by taking O = 8; Limpricht's water *molecule* was HO, whereas his water *type* was H_2O_2, or

$$\left.\begin{array}{c}H\\H\end{array}\right\}O_2.$$

Kekulé wrote: "For me the water type only has meaning if Limpricht's two oxygen atoms are an indivisible unit, i.e., a single atom. I do not understand how the similarities of certain organic compounds with water can lead to the view that they belong to the type H_2O_2, if water itself is considered to be HO; in a word, I do not understand the water type, if water does not belong to its own type."[45]

Here Kekulé revealed his Williamsonian prejudice. We have seen that Odling and Gerhardt, like Limpricht, had no scruples against proposing type formulas that could not be viewed as atomic linkages, whereas for Williamson and Kekulé the type theory made no sense outside of a mechanical-realist context.[46] We have noted that some of the formulas in Odling's 1854 paper cannot be viewed in a linking sense. In Kekulé's offprint copy of this paper there appear marginal penciled annotations expressing surprise every time such a structurally impossible formula is given.[47]

The stated purpose of Kekulé's paper was to show that the concept of conjugated acids was a needless additional hypothesis to account for the constitution of certain substances that could more simply be represented as "mixed" types. Limpricht's and his students' formulation of sulfobenzoic acid was portrayed as an example of this purposeless complication of compound radicals. Instead of picturing the sulfonation of benzoic acid as substitution to produce a conjugate radical, Kekulé suggested that it consisted of addition of $SO_2\}\ominus$ plus rearrangement to form a double water type molecule:[48]

$$\left.\begin{array}{c} \mathrm{C_7 \acute{H}_5 \ominus} \\ \mathrm{H} \end{array}\right\}\ominus \quad \rightarrow \quad \left.\begin{array}{c} \mathrm{H} \\ \mathrm{\acute{S}\ominus_2} \\ \mathrm{C_7 \ddot{H}_4 \ominus} \\ \mathrm{H} \end{array}\right\}\begin{array}{c}\ominus\\ \\ \ominus\end{array}$$

In this paper Kekulé declared his intention of communicating "some fragments of a method of considering chemical compounds which I have used for a long time, and which, I believe, gives a clearer conception of the relations of chemical compounds than does the hitherto customary method."[49] He emphasized, first, that his presentation was here limited only to certain selected aspects of this new theory, and second, that it was by no means entirely original to himself.

The Emergence of Valence and Structure 265

It is nothing more than a further development of the leading ideas that Williamson has from time to time communicated, and which could be called the "theory of polyatomic radicals"; ideas which Odling first further developed in his paper on the constitution of acids and salts; ideas which, since Gerhardt adopted them to some extent in the fourth volume of his *Traité* (but without conceiving them, as is easily seen, strictly in Williamson's sense) have frequently been repeated in German journals as well.[50]

It is by now clear what Kekulé meant in saying that his and Williamson's ideas were only partially shared by Gerhardt. Later in this paper, he criticized Gerhardt's nonlinking formula for glycerine

$$\left.\begin{array}{c} C_3H_5O \\ H_3 \end{array}\right\} O_2$$

as "inadmissible according to the theory of polyatomic radicals."[51]

The heart of the paper reads: "The molecules of chemical compounds consist of juxtaposed atoms. The number of atoms or radicals united with one atom (of an element, or, if one does not intend in the case of compound bodies to apply the method all the way back to the elements themselves, of a radical), is dependent on the basicity or magnitude of affinity [*Verwandtschaftsgrösse*] of the components."[52] Kekulé distinguished three "main groups" of atoms—monatomic, diatomic and triatomic—corresponding to the modern designations of mono-, di-, and trivalent atoms. In a footnote he added that carbon forms a fourth group, in that it is tetratomic. He continued: "A *unification of several molecules* of the types can only take place when a cause for the cohesion arises through the entry of a *polyatomic radical* in the place of 2 or 3 atoms of H. For example, water itself can be considered as 2 molecules of hydrogen in which 2 atoms of hydrogen are replaced by 1 atom of oxygen, etc. In addition to the schematic clarity of such a method of representation, experimental arguments are known for individual cases, particularly for the dibasic nature of Θ and S."[53] Accordingly, a monatomic radical can never serve as such a "cause for cohesion"; a diatomic radical can unite two, a triatomic radical three type molecules into a single unit; and if a polyatomic radical enters into the molecule more than once, large and complex linkages may be formed.

The last section of the paper dealt with the concept of radicals, which Kekulé defined simple as "the portion of a molecule that remains unattacked in any particular decomposition." This leads to the view that different radicals may be assumed to be present in the same molecule, depending on the reaction under consideration. For instance, sulfuric acid may be thought to contain a dibasic SO_4 radical when salt formation is considered, but when chlorination occurs the largest possible radical would be a dibasic SO_2 radical.

Kekulé's major 1858 paper is the one most frequently cited by historians as signalling the birth of the structure theory. Kekulé had already made great strides in the earlier paper in that he had generalized, systematized, and made more explicit the cautious pronouncements of his predecessors. He now continued this process of clarification and would carry it even further in the pages of his textbook (from 1859). But it should be apparent by now that each of these steps was one of degree and not of kind. Kekulé himself clearly understood this point. He wrote:

> I must repeatedly emphasize that in no way do I consider a large portion of these views as original to me, but rather am of the opinion that in addition to the earlier named chemists (Williamson, Odling, Gerhardt), from whom more detailed observations on these topics are at hand, still others share, at least, the fundamental ideas of these views; above all Wurtz, who, never feeling it necessary to develop his ideas more fully, nevertheless permitted others of us to read them between the lines of each of his classic researches, through which the development of these views first became possible.[54]

The major novelties of the paper were twofold: the extension of the principle of "atomicity" all the way down to the individual atoms; and a detailed discussion of the nature of carbon, including its ability to form bonds to itself. Regarding the first point, Kekulé wrote:

> I consider it necessary, and, in the present state of chemical knowledge, in many cases possible, to go back to the elements themselves which compose compounds, in order to account for the properties of chemical substances. I no longer consider the principal task of

The Emergence of Valence and Structure 267

the times to be the determination of atomic groupings which due to certain properties can be considered as radicals, in this way assigning compounds to a few types that thus have scarcely any more significance than those of pattern-formulas. I believe on the contrary that we must extend our considerations to the constitutions of the radicals themselves, that we must determine the relations of the radicals among each other, and that we must derive the nature of the radicals as well as that of their compounds from the nature of the elements. My earlier considerations on the nature of the elements and the basicity of the atoms serve as the point of departure for these views.[55]

For instance, the resolution of sulfuric acid to the constituent (dibasic) sulfate radical, then further to its constituent (dibasic) sulfuryl radical, can be continued all the way back to the (dibasic) atoms themselves:

$$\overset{\prime\prime}{\underset{\sim}{S}}O_4 \quad \overset{\prime\prime}{\underset{\sim}{S}}O_2 \quad \underset{\sim}{S}'' \begin{cases} \Theta'' \\ \Theta'' \end{cases} \quad \begin{matrix} Cl \\ \underset{\sim}{S} \\ Cl \end{matrix} \begin{cases} \Theta'' \\ \Theta'' \end{cases}$$

so that the last formula represents the completely resolved sulfuryl chloride molecule.[56]

Kekulé applied these concepts to organic molecules. First he discussed those compounds containing a single atom of carbon: "If one considers the simplest compounds of carbon . . . it is striking that the quantity of carbon which chemists recognize as the least possible, as the *atom*, always unites four atoms of a monatomic or two atoms of a diatomic element; that the sum of the chemical units of the elements combined with an atom of carbon is generally equal to four. This leads to the view that carbon is *tetratomic* (or tetrabasic)."[57] For compounds containing *two* carbon atoms: "The simplest and therefore most probable case of such a juxtaposition of two carbon atoms is that in which one affinity unit of the first atom is bound with one unit of the second atom."[58] So of the eight original affinity units (valences) only six remain, and C_2 is therefore a hexatomic (hexavalent) "radical." Then:[59] "If more than two carbon atoms combine in this manner, the basicity of the carbon group will increase by two units for each additional atom. Hence, the number of hydrogen atoms, for example (i.e., the number of chemical units), bound to n atoms of carbon juxtaposed in this fashion will be expressed by:

$$n(4 - 2) + 2 = 2n + 2."$$

Two more bonding patterns in organic molecules are possible. Hydrogen atoms may be only *indirectly* bonded to the carbon group; that is, they may be held only through the intermediacy of a polyatomic atom (such as oxygen or nitrogen) which is itself united to carbon with at least one affinity unit (e.g., in alcohols, acids, and primary amines). Or more than one carbon group may be indirectly held together through the intermediacy of such a polyatomic atom (e.g., in ethers, sulfides, and tertiary amines). In either case, those atoms that are indirectly or incompletely bonded to carbon, that is, those atoms in which not all of their units (valences) are bonded to one and the same carbon group, Kekulé called "typical" atoms (atoms of the type). "Typical" atoms thus included the replaceable hydrogen atoms of alcohols, acids, mercaptans, and primary and secondary amines, oxygen atoms of alcohols, ethers, acids, and esters (excluding the carbonyl oxygens), and nitrogen atoms of amines. Kekulé later defined a "radical" as a group consisting of the basic hydrocarbon skeleton plus all of the atoms directly and completely bonded to it, i.e., excluding any "typical" atoms.[60]

This definition of organic radicals formed the basis of the system of classification used in Kekulé's *Lehrbuch*.[61] Moreover, the definition was implicitly utilized in papers as early as February 1858.[62] In this paper Kekulé provided a sketch of his taxonomic system and commented that he had been using the system "for a long time" in his lectures. One may presume that the classificatory system and the definition of radicals upon which it is based is as old as the first draft of the structure theory, that is, about the summer of 1856.

This definition of radicals may be seen to be an extension of Liebig's idea of distinguishing atoms that are "inside or outside of the radical," the distinction having been provided by the type notation. Again we see a conglomeration of radicals and types into a system more comprehensive than either was able to provide alone. This definition is somewhat at odds with Kekulé's earlier definition, derived from Gerhardt, and also carried into the *Lehrbuch*, which asserted that a radical is merely the portion of a molecule that remains unattacked in any particular reaction. But the former definition is only slightly less general than the latter, for atoms "of the radical" are only occasionally attacked; that is, most chemical reactions involve only the atoms "of the type."

Kekulé reiterated his view that radicals and types are merely relative and conventional, as well as mutually complementary concepts. Different reagents can act upon one and the same molecule in different ways, revealing different aspects of its constitution. It is accordingly clear, he stated, that a substance may be thought of as belonging to more than one type, or as containing different constituent radicals (which is saying the same thing). A single reaction reveals a single such type-radical relationship and can be summarized in a rational formula, i.e., a formula exhibiting atomic groupings as revealed by chemical reactions. "Rational formulas are reaction formulas, and can be nothing else in the present state of science. . . . Therefore, every formula which expresses certain metamorphoses of a compound is *rational* . . . but the *most rational* of the various possible formulas is that which simultaneously expresses the largest number of metamorphoses."[63]

Kekulé established a pattern here that derives ultimately from Gerhardt, to which he would often return, and which many other chemists emulated: *viz.*, the distinction between the apparent atomic arrangements deduced from chemical properties ("chemical constitution" or later, "chemical structure"), and the true, actual spatial arrangement of the atoms within a molecule. The first was well within the pale of positive science, the second was not, or at least not by means of the study of chemical reactions alone; the two may in fact be identical, but we have no way of verifying this. Kekulé did not deny in principle the determinability of actual arrangements, stressing that this question was *independent* of his treatment of rational (resolved) formulas as relative, conventional, and based solely on chemical reactions.

Kekulé always sought to strike a balance between the sterility of Gerhardt's positivistic approach and the presumptuous attitude of Kolbe, who was certain that he could determine "absolute" constitutions. He wrote to his friend Lothar Meyer: "We and science quietly wend our way between the mischief of those who play with constitutional formulas, and the indolence of those who deny formulas, toward the star of a fundamental synthesis which bekons from afar."[64] For Kekulé, it was only important that a rational formula should express the chemist's ideas in a clear fashion, and that it be empirically derived.

In the autumn of 1858 Kekulé was called to be extraordinary professor of chemistry at the University of Ghent. Within two months of his arrival he had set up laboratory facilities for his students and for himself, had begun to attract large numbers of students, had initiated a research program, and, above all, had somehow found the time to complete writing the first fascicle of his soon famous textbook.[65] This portion of the work contained a "General Section" and a "Theoretical Section." The first part consisted of general introductory material together with an interesting and perceptive—if one-sided—summary of the history of organic chemical theory. The second section is basically a transcription, often verbatim but considerably expanded, of Kekulé's two theoretical papers of 1857-58.

On 1 July 1859 Kekulé wrote to Wurtz:

> Enclosed is a copy of the first fascicle of my textbook, which has just appeared. Admittedly you will find nothing very new in it, but hopefully you will recognize many unique characteristics. . . . I hope you will be convinced that in spite of the great similarities, my views go further than those of Williamson and Odling, in particular that they go back to the elements themselves, deriving the nature of both compounds and radicals from the nature of the atoms of the elements. The section concerning the process of chemical metamorphoses will hopefully show you that my viewpoint on this subject is somewhat different from that of Gerhardt. . . . If you should happen to have a free moment, I would be particularly grateful if you would tell me what you think of these theoretical considerations of mine. I think you will not consider it mere customary flattery when I tell you that I lay more value on your judgment than on that of most modern chemists.[66]

Wurtz replied:

I thank you for sending me your excellent book. I can't say I've read it all (I certainly intend to), but I have looked through it attentively and am convinced that it faithfully summarizes the current state of the science, and that it does you much credit. As you have said, I willingly recognize that you have gone further than your predecessors in seeking to consider the fundamental properties and basicity of the elements in the conceptions regarding the constitution of bodies and chemical formulas. I consider your idea of the polyatomic nature of carbon as both sound and fertile, susceptible of important development. I must also thank you for the role which you ascribed to me in your chapter on polyatomic radicals. I would not wish to claim any more than that which you have attributed to me.[67]

Wurtz's actual reaction was recorded by Beilstein: "Wurtz cut the pages open, sniffed around in it a bit, then turned to me and said, 'I think it's an excellent work!' "[68] Williamson was greatly pleased, calling the book "one of the most original and masterly productions which chemistry can boast of."[69] George Carey Foster "expected something very good, but I like it even better than I anticipated."[70] Both Odling and Brodie thought the work "admirable."[71] But not all chemists were equally impressed: Kolbe later admitted that he barely glanced through it, since, he said, he was sure he could learn nothing from it.[72]

In general the work enjoyed tremendous success. It proved to be an effective means of publicizing and propagating the views of the modern French-English school and the new developments that Kekulé himself had fathered. The general, historical, and theoretical sections, as well as the detailed "Specific Section," which comprised the majority of the work and included extensive literature references, made the work indispensible as a textbook, a practical handbook, and a guide to the literature. Friedrich Beilstein adopted Kekulé's system of classification for his monumental *Handbuch der organischen Chemie*.[73] Japp later wrote: "The effect produced by the book was enormous. The facts of organic chemistry appeared to group themselves spontaneously under the new system. Whatever might be its ultimate fate, here was a method of exposition immeasurably superior to any that had preceded it; and as a result every textbook of organic chemistry that has since appeared has shown more or less distinctly the influence of this remarkable work."[74]

It was in the pages of his textbook that Kekulé unveiled to the chemical public his unique graphic formulas, which acquired the epithet "sausages." Although he used them sparingly, there is reason to believe that he had high regard for their representational and even heuristic value.[75] This is seen, for instance, in his discussion of lactic ($CH_3CH(OH)CO_2H$) and glycolic ($HOCH_2CO_2H$) acids. He wrote:

> There are several cases where different hydrogen atoms should be equivalent according to the type theory, and are not. For instance, both glycolic acid and lactic acid behave like monobasic acids, although they contain *two* typical hydrogen atoms. The two typical hydrogen atoms of glycolic acid are non-equivalent, although they both belong to the type. One behaves just like the typical hydrogen of alcohol, the other just like the typical hydrogen of acetic acid. The different behavior of these two hydrogen atoms is clearly caused by the different positions which they occupy with respect to the other atoms, particularly with respect to oxygen. One hydrogen atom lies in the neighborhood of *two* oxygen atoms, like that of acetic acid; the other lies in the neighborhood of *one* oxygen atom, like that of alcohol.

A footnote to this paragraph reads: "If one represents the constitution of glycolic acid by the graphic mode of representation which I have repeatedly utilized, the unsymmetrical constitution of glycolic acid and the different positions of the hydrogen atoms emerge particularly clearly."[76]

Kekulé's conception of the distinction between atoms of the radical and atoms of the type, as exemplified in his discussion of the constitutions of glycolic and lactic acids, illustrates his understanding of the ability of his theory to elucidate the arrangements of the atoms in a molecule. Such illustrations are difficult to reconcile with other passages where he denied, in effect, that he was talking about constitutions at all. Some historians have taken Kekulé at his word and suggest that it was Kekulé's colleagues and students, such as Butlerov, Erlenmeyer, Couper, Crum Brown, and Ladenburg, who were the first to actually understand and develop the potential of the theory outlined by Kekulé.[77]

It was assuredly the fashion in mid-nineteenth-century physical science to keep one's feet as firmly as possible embedded in experiment and empiricism; those who strayed too far from the ideal were frequently ostracized or simply ignored. Kekulé exhibited the symptoms of a highly imaginative theoretician who feared to disclose his ideas until the circumstances were just right, even then hedging his bets by disclaiming belief in them as realistic representations. To some extent these disclaimers were sincere—for example, the lack of certainty in the identity of "chemical" and "actual" constitutions—but Kekulé often got carried away. Adolph Baeyer, who was Kekulé's student and associate from 1856 to 1860 (he followed Kekulé from Heidelberg to Ghent) said that in his textbook Kekulé was frequently untrue to his own views as expressed in lectures and conversations.[78] I believe it was this sense of caution to which Baeyer was referring. The graphic formulas are a good example: Kekulé had been using them at least since 1857 in his lectures at Heidelberg and even used models of them at Ghent; and yet virtually his only publication of them was in the pages of his own textbook, quite sparsely at that, confined to footnotes, and accompanied by positivistic disclaimers. The publication of the benzene theory followed the same cautious story as that of the structure theory.[79] Kekulé may have learned his lesson in circumspection from the examples of Laurent and Gerhardt, who, it can be argued, suffered from their premature publication and overly ardent defense of their theories.

PRIORITY CONTROVERSIES[80]

The rise of structure theory is as convoluted a subject as the discovery of energy conservation; in both cases a number of scientists simultaneously contributed bits and pieces of the picture that eventually emerged, and in both cases it was not apparent that many of the contributors were even talking about the same set of concepts until after the dust had settled.

It was well known that A. S. Couper published a paper containing thoughts astonishingly similar to Kekulé's "theory of the atomicity of the elements" just one month after the latter's landmark 1858 paper appeared.[81] But Kekulé's narrow victory and the fact that he was much better known than his Scottish rival ensured that he would receive the major share of credit for the new theory in the eyes of their contemporaries. Another chemist whose name is now associated with structural ideas,

Aleksandr Butlerov, enjoyed essentially no recognition as creator of the theory during his lifetime, nor did he even claim the honor; that claim was made only after his death by Russian historians, who have generated an extremely complex and misleading historiographic tradition.[82]

More delicate questions concern the relationship of Kekulé's work to his acknowledged predecessors, Williamson, Odling, Gerhardt, and Wurtz, and to his rivals Frankland and Kolbe. When Kekulé's 1858 paper appeared, Wurtz wrote a review of it for inclusion in his newly founded *Répertoire de chimie pure*.[83] The review was largely favorable; Wurtz agreed that one can and should go "back to the elements," using the concept of polyatomic atoms. A footnote reads: "This idea is not new to science. M. Williamson, M. Odling (1854) and I myself (1855) have already clearly expressed this view some years ago. . . . Furthermore, M. Kekulé himself recognizes in his memoir that my research served as the point of departure for the views which he developed."[84] In a private reply Kekulé pointed out that already in 1854 he had provided an experimental argument for the dibasic nature of sulfur, more than a year before Wurtz's theoretical paper of 1855.

> So if the view of the basicity of the atoms was expressed by someone before me, this was certainly not in the paper which you cited. Your elegant investigations of the glycols are even more recent [from 1856]; so they could have served just as little as the *point de départ* for my views, and I have indeed merely recognized that the *further development* of my views first became possible through your discoveries. As regards the views of Williamson and Odling, in particular the articles which you cited, I willingly recognize, and so indicated in my paper, that the ingenious considerations of those chemists formed the point of departure for my views. But whether the idea of the basicity of the atoms *as we now conceive of it* is "clearly" expressed in those papers, this is less certain. In fact it doesn't seem so to me . . .[85]

Two months later Wurtz published the following note: "In giving recently some historical indications on the theory of polyatomic elements and radicals, I forgot to mention, with the names of MM. Williamson and Odling, that of M. Kekulé. I must remark here that in his paper on thiacetic acid this chemist insisted on the bibasic nature of sulfur."[86]

Odling discussed the question of atomic and molecular magnitudes in two papers published in 1858. In the first of these he clearly distinguished between the concepts of "combining proportion" and "atomic weight."[87] The former was defined as the "least quantity of an element which can unite with, or replace, one part of hydrogen." Ammonia contains 83% nitrogen and 17% hydrogen, so the "combining proportion" is clearly $N = 4.7$ if $H = 1$. Why, then, Odling asked, does every modern chemist accept the value of the nitrogen equivalent to be 14? The answer is that the formula for ammonia is known to be not NH but NH_3. Odling cited a number of chemical arguments that had compelled chemists to assume the latter formula; for example, the quantity of ammonia which takes part in chemical reactions is invariably found to contain a weight of nitrogen divisible by 14 and a weight of hydrogen divisible by 3, never less than these amounts. Similar arguments lead to modern atomic weight values for oxygen and sulfur.[88]

These views were repeated and emphasized in a paper that Odling published a couple of months later.[89] Odling also added some historical comments. He paraphrased what he had intended to say in an 1855 lecture on radicals to the Royal Institution: each hydrocarbon contains, in a sense, as many radicals as hydrogen atoms, and the basicity of the radical is inversely proportional to the hydrogen content. He remarked: "Kekulé has recently published a paper in which these views are brought forward more definitely, and with an amount of illustration heretofore not possible."[90]

Thus Wurtz and Odling quietly and tentatively urged their own priority for certain of the ideas that Kekulé had synthesized into his structure theory. All three chemists, however, looked first and foremost to Williamson as the one who had done the most to point the way.[91] Williamson had been the first important adherent of the atomic weight reform of Laurent and Gerhardt. His work was instrumental in persuading other chemists (including, but not limited to, Odling, Kekulé, and Wurtz) of the value of that reform. He was the first to broach a limited concept of valence, though others (Frankland, Odling, Kekulé) were more active in the elaboration of the idea. He clarified the troublesome questions concerning relative molecular magnitudes, questions that had stymied the consistent development of organic chemistry. His essential innovation lay in the ingenious use of chemical experiment to decide critical issues. He knew well the art of posing unambiguous questions to nature, of closing the door to possible alternatives, of tightening all of the interpretational screws necessary for a convincing and consistent theory. Williamson never asserted a claim to any part of the structure theory; he probably felt that mention of his name by colleagues and successors was sufficient recognition.

Kekulé was an eager spectator to the 1854 debate between Williamson and Kolbe, and there can be little doubt as to where his sympathies lay. During the next thirty years, until Kolbe's death, the battle lines between the two German chemists were always clearly drawn although they restricted themselves by and large to occasional potshots. Kolbe regarded the entire direction that modern chemistry had taken to be fundamentally erroneous, even worse, pernicious. He always held strong opinions and was never afraid to express them; his increasingly vitriolic tirades could always find publication in his own *Journal für praktische Chemie*. Kolbe's conduct has been ascribed to his reaction to the Franco-Prussian war (which made him a violent nationalist and "*Franzosen-fresser*") and to his wife's death.[92]

But his students were devoted to him. Henry Armstrong gave the following sketch of Kekulé, Frankland, and Kolbe, all of whom he knew personally:

> Kekulé was a born aristocrat in manner. An intellectual of a high order, many-sided in his interests, he was too critical and cynical to be a leader of men in the way that Hofmann was, though even superior to him as an orator; he attracted through his clearcut talent, his gift of precise speech and his great command of knowledge. He had an astounding memory. Frankland was a man of eminently simple, retiring nature, with a strong practical outlook, a demon worker but in no way eloquent. Kolbe was equally simple, never a man of the world, a good lecturer and a far better writer but not an orator; the best chemist of all. Hofmann and Kekulé were cosmopolitans; Frankland was ever the plain, high-souled Lancashireman; Kolbe—just the dear old German, academic pedagogue of the highest class: there is no other way of describing him.
> . . . When I left him early in 1870, he was already peculiar: he

afterwards, in his last years, so fixed his mind upon certain grievances as to be little short of monomaniac.[93]

The gathering storm finally broke in 1881 when Kolbe published a long and violent diatribe.[94] Kolbe's general position was something like the following. The whole of modern chemistry, especially the structure theory, is a highly injurious product of the unscientific sterile French chemistry of the 1830s and 1840s, particularly that of Dumas, Laurent, and Gerhardt; Kekulé and his students had substituted flights of fantasy and the painting of pretty pictures for the exact scientific principles that Berzelius first introduced into organic chemistry, and whose further development in this field was due to men such as Liebig, Wöhler, Bunsen, and Kolbe himself. Kolbe felt that his seminal work had been systematically ignored by the new school — which is not far from the truth. He asserted that the valence theory was due to Frankland alone and that its application to carbon should be credited to Frankland and himself jointly. Kolbe did not hesitate to accuse Kekulé of both the largest and the smallest transgressions: he repeatedly proclaimed that Kekulé's conduct could only be viewed as intentional usurpation of the theories of others, then added insult to injury by filling page after page with stylistic criticisms.[95]

Certainly Kolbe was often justified in his critique. Kekulé frequently used the same word with different meanings (we have discussed at length Kekulé's two different definitions for the word "radical," for instance — and of course Kolbe found both of them incomprehensible) or different words to express the same concept. Many passages in his textbook and papers were vaguely and inconsistently worded. Kolbe wrote:

> I have been charged with rejecting and despising Kekulé's theoretical views and hypotheses *on principle*, thereby placing myself in opposition — which I would certainly rather not do — to the majority of chemists, who have come to view Kekulé's utterances as oracular. They are indeed quite similar to the sayings of the Delphic Oracle: their sense is obscure and usually so incomprehensible, that everyone can see something different in them, or even nothing at all.[96]

This quotation reveals an important point: there is no evidence that Kolbe ever *tried* to master the sense of Kekulé's theories. There was perhaps no chemist more poorly equipped by nature or inclination than Kolbe to be willing to invest the effort that was often admittedly required to perceive Kekulé's views, and many of his criticisms rest on pure misunderstandings. Kekulé was occasionally even intentionally obscure. In response to criticisms of his *Lehrbuch* by his friend Erlenmeyer, Kekulé explained that such ambiguity was on purpose and was designed to reserve priority for ideas that he intended to develop in more detail in the future.[97]

Kekulé replied to Kolbe's attack, but it was not published during Kekulé's lifetime. He wrote the rejoinder as the introduction to an experimental paper and sent the manuscript to his friend Jacob Volhard, then editor of Liebig's *Annalen*. Volhard urged Kekulé not to publish it. "Don't you see that you will only thereby legitimize his attack, which can only be properly condemned by maintaining silence. You may be sure that no one will welcome your defense more than Kolbe himself, for it would finally break this terrifying silence, and give him the opportunity for a salty reply. And, as before, you will be sure to draw the short end of the stick, for you are no match for Kolbe when it comes to coarseness and ruthlessness."[98] Kekulé took his friend's advice.

Kolbe had based his claim for the tetravalence of carbon on his and Frankland's 1857 paper, wherein organic compounds were conceived to be derivatives of carbonic acid and on certain passages of his textbook printed between 1857 and 1859. Kekulé gave examples of some of the carbonic acid formulas, and commented: "Probably no one has really understood these formulas. It is not at all clear why the attached water equivalents (i.e., half water molecules) vanish as the oxygen atoms (i.e., half oxygen atoms) are replaced. The whole derivation is only possible if, proceeding from the hypothetical hydrated carbonic acid, hydroxyls or water residues are replaced by hydrogen or radicals. In any case the procedure has nothing at all to do with the tetravalence of carbon."[99] Kolbe had spoken of his recognition of "the tetravalence of the carbon atom (or of the double carbon atom, as it was then viewed)." Kekulé pointed out that Kolbe never spoke of a double atom at that time, "but always of the *carbonyl radical* consisting of two atoms." Even if one should grant that Kolbe was the first to express the idea that the two-carbon "carbonyl radical" has a "saturation capacity" of four, "what does that have to do with the *tetravalence* of the carbon *atom*? . . . Even if he had mentioned a double atom at that time, this still would not have been a recognition of the tetravalence of carbon," since a tetravalent two-carbon group is by no means equivalent, indeed is contradictory, to the concept of a tetravalent single atom.[100]

Kolbe had attempted to forestall this line of argumentation by pointing out that Kekulé had used the "double carbon atom" $\text{\c{C}} = C_2$ until 1867, shortly before Kolbe also accepted $C = 12$. Kekulé replied that the chemists at the Karlsruhe Congress had agreed among themselves to utilize barred atoms, not in Berzelius' sense, but merely to designate the newer atomic weights. This was a subtle dig at Kolbe, for the latter had boycotted the 1860 meeting. Kekulé also explained his reasons for using $C_2 = 12$ in his two fulminate papers of 1857-58.

Finally, Kekulé showed that the passages on which Kolbe based his claim were in every case later than the corresponding publication dates of Kekulé's papers. Even the Kolbe-Frankland carbonic acid paper was published a month after Kekulé's 1857 theoretical paper, which contained the footnote explicitly asserting that carbon was "tetratomic or tetrabasic."[101]

Frankland's claims were somewhat different, and Kekulé also had to deal with them.[102] In the 1860s Frankland's services had been ignored even more unconscionably than Kolbe's; he was not even mentioned in the historical section of Kekulé's *Lehrbuch*, or in Wurtz's *Histoire*, for example.[103] Frankland tried to set the record straight in his 1877 collected papers. His 1852 paper on organometallic substances contained a paragraph that "constitutes the basis of what has been called the doctrine of atomicity or equivalence of elements; and it was, so far as I am aware, the first announcement of that doctrine." Furthermore, "it was an obvious and easy step to apply it to the compounds of carbon."[104] In support of this statement, Frankland cited his own papers from 1855 to 1858, as well as his joint work with Kolbe published in 1857. But his assertions were neither as abrasive nor as absolute as were Kolbe's. He remarked: "In thus claiming the discovery of the law of atomicity in the study of organometallic bodies, and, in conjunction with Kolbe, its application to the compounds of carbon, I do not forget how much, in its present developments, this law owes to the labours of other chemists, especially to those of Kekulé and Cannizzaro."[105]

Kekulé did not deny the significance of Frankland's 1852 paper; however, he considered its "great importance" to lie not in valence considerations, but in the fact that Frankland here first openly embraced the type theory and showed the theory of conjugate radicals to be untenable. The paragraph on "saturation capacities" is not emphasized and, "in comparison to other considerations, retreats into the background."[106] Frankland specifically eschewed "offering any hypothesis regarding the cause" of his "tendency or law," presenting it merely as an empirical observation that seemed to be general for the elements in question (N, P, As, Sb, Zn, Sn, and Hg).

Kekulé's trump card was the fact that Frankland never distinguished between atoms and equivalents; he used both terms interchangeably and accepted the lower weight for oxygen. Frankland can have no claim on the valence *theory* for he was not dealing with any kind of theory in 1852, but rather a law or empirical generalization. Thus, Kekulé asserted, the use of equivalents is itself sufficient to invalidate Frankland's claim.[107] This assertion loses its force if, as we have argued, conventional equivalents are fully as theory-laden as atomic weights.

Even when one steps back from the heat of the battle and soberly considers relative merits, an unequivocal decision as to who was the originator of valence cannot be made. Frankland was correct in saying that his law requires little modification to convert it into a modern statement of valence. Nevertheless, Kekulé was correct in asserting that the modern concept derives directly from the Williamson-Kekulé school, Frankland's work being essentially without influence, and this in fact explains the chemical world's neglect of Frankland. Valence was discovered independently and in different contexts by two theoretical schools; Frankland's statement has technical priority, but Williamson's and Kekulé's proved more fertile and influential.

As for the question of carbon tetravalence, the apportion of priority is less problematic: Kekulé's formulation has both technical priority and superiority of form. Nevertheless, the ideas of Frankland and Kolbe have much in common with Kekulé's view, and this must be seen as an independent discovery of the same phenomenon.

Seven years after writing his reply to Kolbe and Frankland, Kekulé attempted to summarize in a few sentences the history of organic chemical theory since 1840.

> Fifty years ago the river had divided into two branches; one flowed mostly on French soil through luxuriant fields of flowers, and those following it, led by Laurent and Dumas, could reap rich harvests along the entire journey almost without effort. The other followed the direction indicated by the long established and approved guidepost erected by the great Swede Berzelius. It led mostly through broken boulders, and only later returned to fruitful land. Finally, when both branches had approached quite close to each other, a thick underbrush of misunderstandings still separated them. The groups of travelers did not see each other, nor did they understand the others' language. Suddenly a loud cry of triumph rang out in the camp of the type theorists. The others had also arrived, Frankland at their head. It was now seen that both had had the same goal, even if they had traveled different paths. Experiences were exchanged; each party gained advantage from the achievements of the other, and with united forces the travelers continued their journey on the reunited river, through the most fertile fields. Only a few held themselves apart and sulked; they thought only they had followed the right path, only *they* were in navigable waters; but they still followed the current.[108]

Frankland must surely have noticed Kekulé's acknowledgment of his leadership role. In his autobiography, written in the last three years of his life, Frankland returned the compliment. He still maintained his claim on valence, but now granted that Kekulé had been the first to apply it to organic compounds; this is in fact the most commonly expressed modern view. Frankland wrote: "The application of my theory of valence to carbon compounds, however, belongs substantially to Kekulé, whose brilliant application of this theory to carbon compounds generally, and to benzol and its congeners especially, constitutes one of the most important epochs in the history of chemical science."[109] It is pleasant to be able to end this troubled section on such a note of gracious and well reasoned compromise.

1. W. Odling, "On the Constitution of Acids and Salts," *J. Chem. Soc.* 7 (1854): 1-22.

2. Ibid., pp. 2-3.

3. Ibid., p. 10.

4. Ibid., p. 9.

5. Ibid., p. 19.

6. Williamson to Roscoe, 5 December 1853, H. E. Roscoe Collection, Chemical Society, London.

7. Williamson, "Note on the Decomposition of Sulphuric Acid by Pentachloride of Phosphorus," *Proc. Roy. Soc.* 7 (1856, rec'd. 23 February 1854): 11-15.

8. Of this substance he wrote to Roscoe: "The poor thing has been badly illtreated by chemists . . . I have got the compound in *4* different ways perfectly answerable results" (Letter of 10 March 1854, H. E. Roscoe Collection, Chemical Society, London).

9. R. Anschütz, *August Kekulé*, 2 vols. (Berlin, 1929), 2:943-44, 950-51. The second volume reprints Kekulé's published papers, and I will normally cite only this collection (as Anschütz).

10. Ibid., 1:664.

11. "Von Williamson ein andermal, denn wenn man von ihm anfängt, wird man so bald nicht fertig, er hat zu viele Seiten um mit wenig Worten charakterisirt werden zu können" (Kekulé to Planta, 3 March 1854; the letter is in the August Kekulé Sammlung, Institut für Organische Chemie, Technische Hochschule, Darmstadt).

12. Kekulé, "On a New Series of Sulphuretted Acids," *Proc. Roy. Soc.* 7 (1856, rec'd. 5 April 1854): 37-40; more detailed version in *Ann. Chem.* 90 (1854): 309-16; reprinted in Anschütz, 2:54-57, 57-63.

13. Anschütz, 1:45-46.

14. "Langweilerei, das ist alles was ich darüber sagen kann und dabei ist der Mensch (sonst sehr nobel) so unanständig das Gesicht zu verziehen, wenn man etwas für sich selbst thun will" (Kekulé to Planta, 3 March 1854, August Kekulé Sammlung, Darmstadt).

15. Anschütz, 2:54, 58, 61.

16. Ibid., pp. 61-62.

17. A. Wurtz, "Théorie des combinaisons glycériques," *Ann. chim. phys.* [3] 43 (1855): 492-96; "Sur le glycol ou alcool diatomique," *Comptes rendus* 43 (1856): 199-204.

18. Ibid., pp. 492-93.

19. Wurtz, "Sur une nouvelle classe de radicaux organiques," *Ann. chim. phys.* [3] 44 (1855): 275-313 (300).

20. Wurtz, "Mémoire sur les glycols ou alcools diatomiques," *Ann. chim. phys.* [3] 55 (1859): 400-78 (468-69, 470-78).

21. Letter of 18 April 1854, Harris-Brock Collection, Leicester.

22. Wurtz, *Histoire des doctrines chimiques depuis Lavoisier jusqu'à nos jours* (Paris, 1869); H. Watts, trans., *History of Chemistry From the Age of Lavoisier to the Present Day* (London, 1869), pp. 140-57.

23. A limited version of Berzelius' theory was proposed in 1839, although he did not borrow Gerhardt's name until 1841. The year 1857 is selected as the date of Kolbe's final capitulation to Frankland's evidence against Berzelian copulae, and of Kekulé's effective critique of the Gerhardtian doctrine.

24. Gerhardt, "Sur la constitution des sels organiques à acides complexes, et leur rapports avec les sels ammoniacaux," *Ann. chim. phys.* [2] 72 (1839): 184-214.

25. Gerhardt, "Sur la loi de saturation des corps copulés," *Comptes rendus* 20 (1845): 1648-57; Laurent and Gerhardt, "Recherches sur les anilides," *Ann. chim. phys.* [3] 24 (1848): 163-207.

26. Laurent, *Méthode de chimie* (Paris, 1854), pp. 256-62; *Chemical Method* (London, 1855), pp. 212-15.

27. Gerhardt, *Traité de chimie organique*, 4 vols. (Paris, 1853-56), 4:604-10, 659-72, 827-37.

28. Ibid., pp. 601, 607-10. Gerhardt's terminology and tables are reproduced verbatim.

29. H. Limpricht, *Grundriss der organischen Chemie* (Brunswick, 1855), p. 2.

30. Limpricht and L. von Uslar, "Ueber die Sulfobenzoësäure," *Ann.* 102 (1857): 239-64.

31. O. Mendius, "Ueber gepaarte Säuren und insbesondere über Sulfosalicylsäure," *Ann.* 103 (1857): 39-80.

32. Ibid., p. 73.

33. Ibid., p. 66.

34. On Kekulé see Anschütz' excellent biography, comprising volume 1 of his *August Kekulé*, and two autobiographical and anecdotal speeches that Kekulé gave in his last years, reprinted in 2:937-47, 947-52.

35. Ibid., 2:943-44, 951.

36. For which see C. A. Russell, *History of Valency* (Leicester: University of Leicester Press, 1971), pt. 1; N. W. Fisher, "Kekulé and Organic Classification," *Ambix* 21 (1974): 29-52; and A. J. Rocke, "Origins of the Structural Theory in Organic Chemistry" (Ph.D. diss. University of Wisconsin-Madison, 1975).

37. Kekulé, "Ueber die Constitution des Knallquecksilber," *Ann.* 101 (1857): 200-13; 105 (1858): 279-86; reprinted in Anschütz, 2:64-79 (67, 77). The "marsh gas type" was used earlier, though only implicitly, by Dumas and by Odling, and simultaneously (in 1857) by Kolbe and Frankland.

38. Fisher first made this point ("Kekulé," p. 37).

39. Anschütz, 1:558n. Anschütz (1:85) suggested that Kekulé was the only German adherent of consistent atomic weights at that time. As far as I am aware this is an accurate statement.

40. Anschütz, 1:71.

41. E. Hiebert ("The Experimental Basis of Kekulé's Valence Theory," *J. Chem. Educ.* 36 [1959]: 320-27) and Russell (*History*, p. 64) also agree with Anschütz (1: 85) in this judgment.

42. Hiebert ("Experimental Basis") has offered this interpretation.

43. Kekulé spent two summers in London, and no one has hitherto attempted to date this event. Kekulé related that he was returning from a visit to his friend Hugo Müller. He did not mention his closest friend in London, R. Hoffmann (the three young chemists called themselves the "Dreibund"), and so the incident most likely occurred after the latter's return to Germany in September 1854 (Anschütz, 1:39, 50, 53; 2:941-42, 951).

44. Anschütz, 2:940-41. I have given reasons to prefer the date cited in my "Origins," pp. 270-72.

45. Kekulé, "Ueber die Constitution und die Metamorphosen der chemischen Verbindungen und über die chemische Natur des Kohlenstoffs," *Ann.* 106 (1858): 129-59; Anschütz, 2:97-119 (101).

46. Compare, for instance, Williamson's and Kekulé's 1854 papers with Odling's 1854 paper, or with Gerhardt's *Traité* 4: 629, 749.

47. Anschütz, 1: 95; I also have examined this offprint copy, in the August Kekulé Sammlung, and noted this correlation that apparently escaped Anschütz.

48. Kekulé, "Ueber die s.g. gepaarten Verbindungen und die Theorie der mehratomigen Radicale," *Ann.* 104 (1857): 129-50; Anschütz, 2:80-96 (96).

49. Ibid., p. 80.

50. Ibid., pp. 80-81.

51. Ibid., p. 84n.; the reference is to Gerhardt's *Traité*, 4: 629. Two years later Kekulé commented: "Subsequently the theory of polyatomic radicals was also adopted by Gerhardt, and used in the fourth volume of his *Traité*, admittedly not always in the full sense of Williamson's viewpoint nor always with entirely consistent execution" (*Lehrbuch der organischen Chemie*, 2 vols. [Erlangen, 1861-66] 1:94).

52. Ibid., p. 83. What we now call valence was referred to by Kekulé variously as "basicity" (from Williamson, 1851), "atomicity" (from Gerhardt and others, 1856), "affinity," "affinity units," "chemical units," and "units." Other designations included "value" ("Werthigkeit"; Erlenmeyer, 1860), "monaffin" (Wislicenus, 1863), "monad" (Odling, 1864, following Laurent), "hydrogen equivalence" (Gerhardt, 1856), "quantivalence," "monovalent" (Hofmann, 1865), and "valence" (Claus, 1866). The latter term gained currency only in the course of the 1870s. See Russell, *History*, pp. 83-89 (some of which is modified by the above data).

53. Ibid. In a footnote Kekulé cited his 1854 thiacetic acid paper as such an "experimental argument" for the dibasicity of sulfur.

54. Ibid., p. 102.

55. Ibid.

56. Ibid., p. 114.

57. Ibid., pp. 114-15.

58. Ibid., p. 115.

59. Ibid., p. 116.

60. Kekulé, "Note sur l'action du brome sur l'acide succinique . . . ," *Bull. Acad. Roy. Belg.* [2] 10 (1860): 63-72; "Note sur les acides itaconique et pyrotartrique," *Bull. Acad. Roy. Belg.* 11 (1861): 662-77; *Lehrbuch*, 2:245 (1864); Anschütz, 2:153, 204.

61. This was stated explicitly by Kekulé in 1864 (*Lehrbuch*, 2:244), but it emerges clearly from inspection of a table summarizing the classificatory scheme which appeared in the 1859 fascicle (ibid., 1: 224). For examples of the application of the definition, see discussions in ibid., 1: 131, 156–57, 164, 174.

62. Kekulé, "Bildung von Glycolsäure aus Essigsäure," *Verh. nat.-med. Vereins Heidelberg* 1 (1858): 105–7; Anschütz, 2: 138–41.

63. Anschütz, 2: 112.

64. Letter of 23 October 1860, printed in Anschütz, 1: 205.

65. Begun in 1856, the three fascicles comprising vol. 1 (1861) emerged in 1859, 1860, and 1861; the three fascicles of vol. 2 (1866) were issued in 1863, 1864, and 1866. The work was continued as *Chemie der Benzolderivate oder der aromatischen Substanzen*.

66. Letter of 1 July 1859, Anschütz, 1:157–58.

67. Letter of 21 July 1859, in Anschütz, p. 159. Kekulé (*Lehrbuch*, 1:94) said that Wurtz was the first to correctly interpret Berthelot's important work on the glycerides, and that his "ingenious conceptions and brilliant discoveries" on the di- and trialcohols have subsequently served to further develop the theory of polyatomic radicals.

68. Anschütz, 1:154.

69. From a review (*Reader*, 1 April 1865) of the first five fascicles, cited in a letter from Kekulé to Baeyer, 10 April 1865, in Anschütz, 1:368.

70. Letter to Kekulé, 7 October 1859; Anschütz, 1:167.

71. Odling to Kekulé, 22 May [1860]; Brodie to Kekulé, 4 January 1872. Both letters are preserved in the August Kekulé Sammlung.

72. Kolbe, *J. prakt. Chem.* [2] 23 (1881): 377.

73. 1st ed., 2 vols. (Hamburg, 1881–83). The scheme was finally abandoned in 1907, the year after Beilstein's death, when Paul Jacobson and Bernard Prager began preparing the fourth edition; this edition is still being published, and so far spans over a hundred volumes.

74. F. R. Japp, "Kekulé Memorial Lecture," *J. Chem. Soc.* 73 (1898): 97–138 (102).

75. Rocke, "Origins," pp. 309–21.

76. Kekulé, *Lehrbuch*, 1:174; cf. pp. 129–31 and 729–41.

77. See for example Fisher, "Kekulé," or the writings of G. V. Bykov.

78. A. Baeyer, *Gesammelte Werke* (Brunswick, 1905), 1:xv.

79. Anschütz, 1: 67; 2: 940–41.

80. See the relevant discussions in Russell, *History*, and Rocke, "Origins."

81. R. Anschütz, *Life and Chemical Work of A. S. Couper* (Edinburgh, 1909).

82. A. Rocke, "Kekulé, Butlerov, and the Historiography of the Theory of Chemical Structure," *Brit. J. Hist. Sci.* 14 (1981): 27–57.

83. Wurtz, *Rép. chim. pure* (1858), 1:20–24.

84. Ibid., p. 24n.

85. Anschütz, 1:147–48 (Kekulé to Wurtz, 15 February 1859).

86. Wurtz, *Ann. chim. phys.* [3] 55 (1859): 470n.

87. Odling, "On the Atomic Weights of Oxygen and Water," *J. Chem. Soc.* 11 (1858): 107–29.

88. Ibid., pp. 110, 127–29.

89. Odling, "Remarks on the Doctrine of Equivalents," *Phil. Mag.* [4] 16 (1858): 37–45.

90. Ibid., pp. 43–44.

91. Kekulé, in Anschütz, 1:147–48, 157–58, 554; 2:80, 102; Wurtz, *Rép. chim. pure* (1858), 1:24n.; Odling, *J. Chem. Soc.* 22 (1854): 1.

92. H. E. Armstrong, in a review of Anschütz' *Kekulé* (*Nature* 125 [1930]: 807–10 [808]).

93. Ibid., pp. 808–9.

94. Kolbe, "Meine Betheiligung an der Entwickelung der theoretischen Chemie," *J. prakt. Chem.* [2] 23 (1881): 305–23, 353–79, 497–517; 24 (1881): 374–425.

95. Kolbe was a linguistic purist, and his writing was a model of clarity and precision. Kekulé was a romantic; neologisms flowed freely from his pen, and many utterances were vague because they contained veiled hints at views that Kekulé was not ready to advocate openly.

96. Ibid., 23:516.

97. Kekulé to Erlenmeyer, 13 June 1860, August Kekulé Sammlung.

98. Anschütz found the MS in Kekulé's papers and printed it in his biography (1: 540–69). A fascimile edition of the MS was published in 1965: *Cassirte Kapitel aus der Abhandlung: Ueber die Carboxytartronsäure und die Constitution des Benzols* (Weinheim: Verlag Chemie). This edition includes the correspondence between Kekulé and Volhard, from which the above quotation is taken (letter of 25 August 1883).

99. Anschütz, 1: 556.

100. Ibid., pp. 557–58.

101. Ibid., pp. 559–60.

102. In fact, Kekulé wrote Volhard (*Cassirte Kapitel*, 2 October 1883 [n.p.]) that Frankland was the real object of the reply, since the latter's claims were considered just by many English chemists, whereas no one really paid much attention to Kolbe's rantings.

103. The neglect of Frankland is ably discussed in Russell, *History*, pp. 128–33.

104. Frankland, *Experimental Researches in Pure, Applied, and Physical Chemistry* (London, 1877), pp. 145–46.

105. Ibid., p. 154.

106. Anschütz, 1: 561–64.

107. Ibid., pp. 565–68.

108. Anschütz, 2: 939–40.

109. Frankland, *Sketches from the Life of Edward Frankland* (London, 1901), p. 191.

10

The Karlsruhe Nexus

The late eighteenth-century nonreductionist materialism that was related to imponderable fluid theories in physics as well as the Lavoisierian concept of chemical elements yielded at the beginning of the nineteenth century to a trend toward reductionistic mechanism. This trend was particularly apparent at the Royal Institution of London, where its founder Count Rumford and his youthful employee Humphry Davy unsuccessfully attempted to displace the caloric theory with a dynamical theory of heat, and Thomas Young argued for a wave theory of light against the Newtonian particle view. This incipient program was partially countered and partially assimilated by the romantic-idealist movement in science with its concern for seeking the active forces rather than the inert pieces of matter that make up the world.[1]

Concurrently, physics and chemistry began drifting apart, a trend that was accelerated with the growing specialization of science during the course of the nineteenth century. The physicalist style of such chemists as Berthollet, Gay-Lussac, and Avogadro began to be viewed as passé; the number of workers who felt at home in both physics and chemistry, such as Wollaston, Ampére, and Faraday, declined; and chemistry textbooks that included a healthy dose of physics, like Berzelius' *Lärbok*, seemed increasingly old-fashioned. The rejection of Berzelian atomic weights in favor of conventional equivalents in the 1840s was a symptom of this divorce: Berzelius had derived his weights with some regard to physical measurements—vapor densities, specific heats, and crystal forms — whereas Wollaston and Gmelin had intentionally used solely chemical data.[2]

About mid-century another sea change seems to have come over European scientists, namely a drift back towards realism and materialist-mechanist physical explanations. By the 1850s chemists were beginning to have more confidence in the process of transdiction and to experience a growing interest in the invisibly small world of atoms and molecules. Chemical atomists were approaching explanations in terms of physical atoms.[3] At the same time, physics also experienced a reawakening of interest in kinetic-molecular interpretations of the microcosm.

KINETIC THEORIES IN PHYSICS

Explanations of thermal and gaseous phenomena in terms of an imponderable caloric fluid survived in some circles past mid-century, though it is now recognized that most leading physicists had rejected the leading propositions of the theory by about 1840.[4] Alternative kinetic explanations were proposed independently by John Herapath from 1820 and J. J. Waterston from 1843. Both showed that the gas laws would result if one supposed gases to be composed of elastic molecules in motion, exerting forces upon collision with each other and with the sides of the container by simple transfers of momentum. Both submitted their work to the Royal Society, were turned down, and ultimately published shorter versions in more obscure organs such as Herapath's own *Railway Magazine* and the British Association *Reports*. Neither was influential.[5]

Brush has cogently argued that one reason for the lack of acceptance of these early kineticists was the prevailing romantic idealism with its overarching concern with forces rather than with matter and mechanisms. But, as Kuhn has urged, that very prejudice seems to have predisposed a number of workers to an emerging generalized concept of energy; in the 1840s such scientists as J. R. Mayer, J. P. Joule, and H. Helmholtz proposed the idea of energy conservation. In 1850 Rudolph Clausius formalized the principle as the first law of thermodynamics and showed by means of a second law that it was fully reconcilable with Sadi Carnot's earlier work on the rules governing interconversions between heat and work. The interchangeability of heat and work made it natural to assume that heat was nothing but molecular motion of some sort, so it was no coincidence that several prominent early thermodynamicists were also kineticists. Indeed, it was Clausius himself who, in another landmark paper (1857), initiated sustained and successful investigations into the kinetic theory of gases and of heat.[6]

This paper was a remarkable performance in many respects. In addition to showing the connection between kinetic assumptions and the gas laws more completely, generally, and correctly than his predecessors, he also applied the theory in a variety of other ways. For instance, he was able to derive the ratio of specific heats of an ideal gas at constant pressure and at constant volume, a kinetic explanation for changes of state, and an estimate of molecular velocities.

The Dutch meteorologist Buys-Ballot immediately objected that if molecules really moved as fast as Clausius calculated — hundreds of meters per second — then gaseous diffusion should take place nearly instantaneously, a phenomenon that was not observed. Clausius developed an effective rebuttal in an important paper published in 1858. He argued that if the mean free path of the gas molecules were sufficiently short, slow diffusion would result despite the high mean velocities. His argument was not fully convincing, though, since it was necessary to estimate an unknown, molecular sizes, in order to calculate mean free paths.[7]

A few years later Joseph Loschmidt provided the first convincing evidence of those sizes from experimental data. An ingenious argument using kinetic assumptions together with measurements of gas viscosity (to give mean free paths) and the ratio of condensed to vapor phases for the same substance led him to an approximate value for molecular diameters, a figure that agrees with modern measurements to within about half an order of magnitude and that supported Clausius against Buys-Ballot.[8]

Clausius' 1858 article was also influential in bringing the kinetic theory to the attention of James Clerk Maxwell, who read an English translation in the February 1859 number of the *Philosophical Magazine*. Maxwell was doubtful but sufficiently intrigued to reach a number of novel results "as an exercise in mechanics," communicating them privately to his friend G. G. Stokes on 30 May and publicly to the British Association annual meeting in Aberdeen on 21 September. A major two-part paper appeared in the *Philosophical Magazine* for January and July 1860.[9] Maxwell was the first kineticist to employ a statistical distribution of molecular velocities rather than a single mean speed, and he deduced several predictions of gaseous transport processes from the theory.

One of the most interesting of these predictions was that the viscosity of a gas ought to be independent of its density, a result that seemed to contradict both common sense and some earlier rather equivocal experiments on the question. But new experiments carried out in the 1860s by Maxwell and others confirmed the prediction;[10] the unexpected character of the phenomenon gave significant support to the kinetic assumptions used to deduce it. This is one of the most beautiful and dramatically successful examples of transdiction in the history of nineteenth-century science.

During the third quarter of the century, the kinetic theory made rapid strides toward general acceptance. Such major physicists as Boltzmann, Kelvin, van der Waals, Graham, and Tyndall contributed to formulating and propagating the new theory, and Clausius and Maxwell continued to improve their own versions, learning from each other's critiques. A number of estimates of the sizes of molecules inferred from experimental data emerged in the late 1860s and early 1870s, some of them independent of kinetic assumptions and all of them reasonably consistent among themselves. The new ability to specify molecular weights in grams and volumes in milliliters lent concreteness and credibility to the theory, even though the figures remained highly approximate.[11]

The kinetic theory of gases made conceptual contact with chemistry in the support it gave to Avogadro's hypotheses. Waterston noted as early as 1843 that identifying temperature with molecular kinetic energies and assuming equipartition of energy among all molecules in a mixture (a restricted form of the theorem of the equipartition of energy) led to the result that equal volumes of gases under similar conditions contain equal numbers of molecules. He added that the only way to reconcile this result with known chemical facts is to assume submolecularity in those physical molecules.[12]

There is no reason to think that these passages in an anonymous and obscure book were read by any prominent chemist. However, in a memoir given at the 1851 British Association meeting in Ipswich, Waterston reiterated that kinetic assumptions lead to the EVEN and the submolecularity hypotheses. There is good circumstantial evidence that reading the abstract of this paper led A. K. Krönig to his 1856 reproduction of the derivation of the gas laws from simplified kinetic assumptions.[13] Krönig noted the implication of EVEN but drew no particular attention to it. Clausius, on the other hand, laid stress on this as a highly satisfying result of the theory and proposed the submolecularity hypothesis as well, in apparent ignorance of earlier work on the idea. He was the first to observe that experimental measurements of the ratio of specific heats at constant temperature and at constant pressure differed from that predicted from the theory; the discrepancy could be explained by assuming internal energy of vibration, e.g. of a diatomic molecule. Later he properly noted that Dumas, Laurent, and Gerhardt had suggested the submolecularity hypothesis years earlier purely from chemical reasoning.[14] Maxwell emphasized even more strongly than Clausius the relation of kinetic theory to the EVEN hypothesis, drawing attention to it in his letter to Stokes, in the abstract to his British Association paper, and in both his 1860 and his 1867 memoirs.

Although evidence that these physical investigations were immediately and widely influential for chemists is not strong, it must be stressed that in the decade before the Karlsruhe Congress, five derivations of the EVEN hypothesis from the kinetic theory by four authors appeared in three of the leading science journals of the day. Counting translations and reprintings, these papers appeared no less than fourteen times, all but one between 1856 and 1860.[15] Any chemist who kept up with the literature must surely have encountered at least one of these derivations.

THE KARLSRUHE CONGRESS

The idea of an international convention to discuss significant contemporary issues in chemistry originated with Kekulé. He first mentioned the idea to an enthusiastic Wurtz during a trip to Paris in the spring of 1859. Kekulé had the highest regard for Wurtz: "Excellent chap! two people could not possibly agree more on a science in a general way than we do . . . I am convinced Wurtz is the most important personality [for the project], and if he takes a leading role the matter is already half won."[16] Serious planning for the congress began when Kekulé discussed his ideas with Karl Weltzien on a visit in the following autumn to Karlsruhe, a lovely Badische city on the Rhine in the northern fringes of the Black Forest. During the winter of 1859–60, the two friends wrote several leading European chemists and in March met in Paris with Wurtz, Adolf Baeyer, and Henry Roscoe. Conjointly, they drew up a preliminary circular to garner support for the project.

Kekulé was assigned to correspond with English chemists. He regarded the Oxford chemist B. C. Brodie as "definitely one of the most philosophical minds in chemistry,"[17] and he must have been disappointed by Brodie's lukewarm response.

> If you consider that the objects which you have in view are at all likely to be promoted by the addition of my name to the names of the distinguished Chemists who have expressed their willingness to sign the document, which I have received from you, I should certainly not withhold it. . . . It appears to me however that we should be on our guard against even the apparent attempt to force the opinion of others or to attempt a premature settlement of questions, which are by no means ripe for it, and on which opinion is so unformed.[18]

G. C. Foster was an enthusiastic collaborator, and worried about Brodie's obvious indifference: "I could not get much out of Brodie. He seems a somewhat undecided kind of being who is not much to trust for getting anything going. The last time I saw him he strongly advocated putting off the meeting till next year and holding this year a preparatory meeting only."[19]

Williamson was cordial and cooperative, but also was concerned to ensure objectivity and avoid compulsory decrees:

I shall have much pleasure in coming [to] the meeting, and will endeavor as far as I can to promote its object. . . . I think it may probably be desirable to advance during the earlier part of the proceedings the form of *merely* stating facts favorable to or unfavorable to each particular theory, without balancing them against the opposite facts & drawing a conclusion. *That* would but be done when the facts on both sides have been brought forward & carefully arranged & weighed. It will be essential to have the best exponent we can get of the older theories.[20]

Unfortunately, it proved impossible to attract many advocates of the older chemistry to Karlsruhe. Hermann Kolbe replied to Weltzien that since the chemistry of the day was dominated by the despised Gerhardt-Wurtz-Williamson-Kekulé school, any resolution reached by the congress would have to be replaced when the fraudulence of the new chemistry was finally recognized, a revolution he expected to see "in aller kürzester Zeit." It is uncertain whether he attended.[21] Marcellin Berthelot likewise dissociated himself from the project, expressing similar sentiments.[22]

A final circular printed in three languages was issued early in the summer, and the Congress took place on 3–5 September 1860. About 140 chemists from a dozen countries attended, of whom we know the names of 126.[23] The course of events at the Congress has been well treated elsewhere,[24] and here I wish to focus particularly on the relationship between the two chief protagonists, Kekulé and Stanislao Cannizzaro.

From an early date, Kekulé developed detailed views on the program to be presented. One goal was simply standardization: agreement on uniform terminology, nomenclature, and symbolism concerning such concepts as atoms, molecules, equivalents, atomicity, basicity, and so on. A second much riskier but more important goal was to debate current theories and conceptions regarding molecular magnitudes.

A crucial issue that touched on both categories was deciding on the meaning of the terms *atom* and *molecule*. At this time Kekulé distinguished chemical atoms and molecules as Laurent did from 1846, which is to say about as we do today. However, he remained cautious about identifying chemical molecules with the physical molecules of the kinetic theory of gases.[25] Caution seemed warranted because dissociation in the vapor phase was just then becoming widely recognized;[26] this phenomenon suggested that some physical molecules consist of varying numbers of chemical molecules depending on conditions. The differences in size might be particularly dramatic, Kekulé thought, if it were possible to compare the three phases of condensation. The chemical molecule—the smallest unit of a substance to enter into a chemical reaction—might never exist in nature in what later chemists called a monomeric form.

But Kekulé was curiously inconsistent. Although he expressed the above viewpoint repeatedly in the first fascicle of his textbook (issued in June 1859), he also observed in the same section that empirically speaking molecular weights are nearly always directly proportional to vapor densities, hence "the chemical molecule is identical to the physical gas molecule." He also noted explicitly that the condition of equal numbers of molecules in equal volumes that could validate the correlation was supported by the kinetic gas theory of Rudolph Clausius.[27]

Cannizzaro strongly held to the view that the chemical need not be distinguished from the physical molecule; all we need to distinguish are the physical gas molecules whose relative weights are given directly by vapor densities from the chemical atoms that form those gas molecules and that enter into the composition of compounds. Kekulé carried the ambiguity of his *Lehrbuch* presentation into his statements before the Karlsruhe Congress. In both the committee meetings and the plenary sessions he urged, against Cannizzaro, that the two sorts of molecules must always be distinguished, "at least in principle." On the other hand, he admitted that "ordinarily" the gas molecule coincides with the chemical molecule.[28]

In the session of 4 September, Wurtz interrupted the discussion between Kekulé and Cannizzaro, suggesting that the two were merely quibbling over a semantic point and that the only truly important question was not in doubt, namely the chemical meaning of the terms *atoms* and *molecules*.[29] Wurtz was right. Kekulé and Cannizzaro were much closer than it seemed, either to them at the time or to later observers. In practice the two chemists agreed on the probable magnitudes of atoms and molecules for all elements with the exception of three divalent metals.

Kekulé's attitude exemplifies the hesitancy among most chemists at this time to admit physical data on an equal level with chemical data. The "equivalentist" school of Gmelin and the "unitary" school of Gerhardt both shared this tendency (though both also seized on physical data when it supported their chemical systems). Gerhardt and Laurent advocated consistent molecular magnitudes not by building from the basis of the Avogadro-Ampère gas theory, but because they thought that the use of equal volumes created a consistent *chemical* system. Williamson's experimental support for that system was likewise derived from chemical and not physical arguments. Odling, too, reiterated several times in the 1850s the chemical rationale for reformed atomic weights.[30] And Kekulé was decisively influenced by this school of thought: hence his hesitancy to admit the identity of chemical and physical atomic arrangements, of chemical and physical molecules.

Avogadro had been a provincial northern Italian chemist who worked in an obsolescent (physicalist) tradition, two reasons why his gas theory remained unaccepted for so long. Cannizzaro was another provincial northern Italian chemist also (it would seem) out of touch with the prevailing modes of thought. In any case Cannizzaro atypically paid close attention to developments in physics. His consistent and confident use of Avogadro's theory and the Petit-Dulong law come most readily to mind. Moreover, he provided in his 1858 "Sketch of a Course in Chemical Philosophy" what was perhaps the first favorable public response by *any* scientist to Clausius' pathbreaking 1857 memoir, noting that it "confirmed" the Avogadro-Ampère theory.[31] He later reiterated the importance of the kinetic theory of gases for chemistry.[32]

The last three chapters of this book have followed the movement from equivalents back to modified Berzelian atomic weights and to consistent molecular magnitudes, a movement that began with Gerhardt and Laurent and was effectively consummated by such workers as Williamson, Odling, Wurtz, and Kekulé, nearly entirely from within organic chemistry, and mostly before 1860. This depiction of events varies from the more traditional view that locates a revolution in Cannizzaro's 1858 "Sketch" and his performance at the Karlsruhe Congress. The traditional version has recently been properly discredited by two fine analyses of the Avogadro/Cannizzaro episode.[33] J. H. Brooke stresses the essential point that there was a "conceptual interdependence" between various chemical conventions and theories, which meant that "nothing less than a reformation was required" before a reformer such as Cannizzaro could even hope for success.[34] It is this gradual reformation of the entire science of chemistry that the last three chapters have sketched.

In this light Cannizzaro's role appears somewhat less heroic than was formerly thought. But there is still much in his performance that merits our attention and respect. Cannizzaro himself emphasized perhaps most strongly his contribution in publicizing and advocating the Avogadro-Ampère gas theory. As earlier sections of this book and other recent work have shown, "Avogadro's hypothesis" was by no means unknown between 1811 and 1860, as used to be thought.[35] Indeed, many chemists, both major and minor, exhibited knowledge of the theory, and several even accepted it. I suspect that there may have been an even more prominent oral tradition of discussion of Avogadrian issues. In any case, the French-English reform school of the 1840s and 1850s asserted both the EVEN and submolecularity hypotheses (though with some reservations).

Nonetheless, Cannizzaro was virtually alone in *basing* his chemical system on his conviction of the truth of the theory, a conviction that was reinforced by the recent work in kinetic theory. He was able to show that all apparent anomalies to EVEN could be eliminated by supposing gaseous dissociation of certain labile compounds and variable submolecularity in molecules of elements. Neither of these were new proposals in 1858 — Dumas, as we have seen, had long ago asserted the probability of variable submolecularity — but neither had ever been assimilated into a straightforward and comprehensive theoretical discussion.

A second Cannizzaran contribution was what he called the "law of atoms," embodying his central operational rule for atomic weight determinations: "each element has a special [least] numerical value by means of which and of integral coefficients the composition by weight of equal volumes of the different substances in which it is contained may be expressed."[36] Cannizzaro did not claim novelty for the law itself (though he may have for the formulation); indeed, Cannizzaro's mentors Gerhardt and Laurent implicitly placed essentially the same rule at the center of their considerations of atomic weights. Cannizzaro made it explicit. He regarded the existence of this law not only as justifying the atomic theory, but also as constituting yet another argument for the EVEN assumption.

The third essential contribution of Cannizzaro was his use of the laws of atomic heats and isomorphism in a systematic way.[37] Berzelius used these physical laws, but—as with the EVEN assumption—he did not accept their full generality. For Berzelius chemical analogy was more indicative of composition than completely consistent application of any physical rule. Chemical analogy was even more important for Gerhardt and his school than for Berzelius. For example, despite the evidence from atomic heats and isomorphism, it simply did not seem reasonable to place the alkalis in a different class from the alkaline earths. Whereas from 1826 Berzelius erred by classifying basic metal oxides in the ferrous oxide (RO) group, Gerhardt put them in the alkali metal (R_2O) group. Hence, Berzelius proposed atomic weights for the monovalent metals that are twice the modern values. Gerhardt's "correction" worked well for the monovalent metals, but halved Berzelius' good divalent metal atomic weights.

Even before Gerhardt's atomic weight reform, Victor Regnault noted Berzelius' incomplete use of specific heats and isomorphism, and from this physical evidence he advocated halving Berzelius' values for silver and the alkali metals.[38] By the late 1850s, there was more general concern for determining the most appropriate multiples for atomic and molecular magnitudes[39]—brought on largely by the organic-chemical followers of Gerhardt and Laurent—and it is possible that Heinrich Rose's reiteration of the same argument in 1857 was more influential than Regnault's work.[40] But Regnault was primarily a physicist and Rose an inorganic chemist who published in a physics journal, and both were merely suggesting ad hoc adjustments to the prevailing systems.

It was Cannizzaro who first developed the fully consistent use of specific heats and isomorphism, integrated into a chemical system that also made full use of gas theories. Chemical analogy was by no means missing from Cannizzaro's work, but it no longer occupied pride of place. This, indeed, was Cannizzaro's signal and most original accomplishment. It was not a discovery in the usual sense—he introduced no new law, theory or idea. But he was able to develop the work of many predecessors—especially Avogadro, Berzelius and Gerhardt—to construct the first chemical system of atomic weights and formulas that incorporated all relevant physical data in a fully consistent fashion.

Cannizzaro's speeches before the Karlsruhe Congress were considered by more than one observer to have been the most eloquent of the meeting.[41] They appear so to the modern reader as well. Kekulé's objections seemed (and seem) as supercilious quibbles against the plain common sense of the Italian. The offprint of Cannizzaro's "Sketch" distributed at the end of the meeting also made a strong impression on at least one reader—Lothar Meyer. In a friendly critique of a paper by Kekulé five years later Meyer wrote: "Perhaps my honored friend will now decide to concede what he has so often contested in our private conversations and correspondence: that the fundamental hypotheses of chemistry should be derived not just from purely chemical data in a narrow sense, but rather, as in all investigations of the most intimate nature of matter, *all* scientific aids must be applied."[42]

As great as was Cannizzaro's accomplishment, a cautionary remark must be elaborated in more detail. The contending armies that met in Karlsruhe were the atomists and the greatly underrepresented equivalentists. The war, however, was already virtually won. Gerhardt and Laurent had been advocating since 1842 essentially the Berzelian atomic weight system modified to achieve consistent molecular magnitudes.[43] Williamson joined their camp in 1849, Odling and Kekulé in 1854. Wurtz and Hofmann both advocated the conceptual reform from about

1850, though Wurtz only began using atomic weight notation in 1859 and Hofmann in 1860. By the time of the call to Karlsruhe, the equivalentists were in retreat, as is clearly revealed in the response of Kolbe and Berthelot to that call. Even those who continued to use equivalent formulas, such as Liebig and Gmelin, believed in the truth of the atomic theory. The outcome of the vote on the resolution henceforth to use only the reformed atomic weights in formulas was a foregone conclusion.[44]

CONCEPTIONS OF VALENCE AND EQUIVALENCE AFTER KARLSRUHE

Cannizzaro, Kekulé, Wurtz, and the other members of the reform movement had won the day regarding formula notation. However, the equivalentists also celebrated a minor victory. During the second plenary session the following resolution was approved by acclamation: "The concept of equivalents is empirical and independent of the concept of molecule or atom."[45] The statement is certainly true as regards chemical equivalents in the precise sense, but it is not true as applied to conventional equivalents. That the latter were nothing other than a rival set of atomic weights had been noted by Laurent,[46] but virtually no one echoed an understanding of this until the late 1860s; even then confusion over this point has continued to the present day.

The ascendancy of the theory of valence effected a subtle transformation in the concept of chemical equivalents. The discovery that atoms had definite and limited capacities for combination with other atoms seemed to indicate that the combining power was somehow localized on different parts of the atom. This in turn gave rise to speculations on subatomic structure, since it was hard to imagine how featureless homogeneous atoms could have localized powers or forces. In particular, such prominent figures as Kekulé, Wurtz, Crum Brown, and Erlenmeyer hypothesized that atoms consist of accretions of monovalent subatoms, the absolute number of these subatoms constituting the valence of the atom. According to this idea a carbon atom, for instance, was thought to consist of a complex of four carbon subatoms, each retaining one valence and weighing three atomic weight units.[47]

Such a conception provided a convenient visualization of both valence and equivalence: the valence is the number of subatoms in an atom (its "atomicity"[48]), and the equivalent is the weight of a subatom. The shorthand definition of equivalent became "atomic weight divided by valence," a formula still used by writers of elementary chemistry textbooks.

However, as I have argued in chapter 1, this definition is identical neither with the empirical and operational (i.e., proper) definition of chemical equivalents, nor with that of conventional equivalents. Like the latter, it is theory-laden, but there is a certain degree of operational access: analysis of the simple hydrides or halides of the element in question. Analysis of methane, water, ammonia, hydrogen sulfide, phosphine, arsine, phosphorus pentachloride, salt, calomel, corrosive sublimate, etc. provide what I will call "hydride equivalents" of carbon, oxygen, nitrogen, sulfur, phosphorus, arsenic, sodium, mercury, etc. It is only the hydride equivalent that is determined by the rule "atomic weight divided by valence."

As the cited examples illustrate, the hydride equivalent, hence the valence, often varies for a given element. There was much controversy over this point in the early 1860s. Active discussion was initiated by a remark by Carey Foster. Foster had studied successively with Williamson, Kekulé, and Wurtz, and became foreign correspondent for both Erlenmeyer's *Zeitschrift für Chemie* and Wurtz's *Répertoire de chimie pure*. In 1861 he reported on one of Peter Griess' researches on diazo compounds, taking issue with Griess' utilization of variable nitrogen valence in formulating the novel substances. Foster wrote: "The idea of atomicity, or capacity of combination, is a property of elementary atoms just as fixed and unalterable as their weight, and appears to us to be the one idea which, in the present state of the science, can serve as the basis for a general theory of chemical combination."[49]

A few months later Wurtz wrote a similar passage: "The idea of types is ultimately an artifice, a pure convention. In my opinion it must be subordinated to a more fundamental notion, that of the atomicity of the elements." But, citing Foster's comment, he disagreed with the doctrine of constant valence, calling it "too absolute," and asserting that atomicity "tends to a maximum" and can vary.[50] Erlenmeyer followed Wurtz in ascribing the doctrine of constant valence to Foster. In a review of Erdmann's *Ueber das Studium der Chemie*, he asserted that two assumptions in addition to those of Dalton are necessary to explain the laws of chemical combination:

> 1) that each atom contains at the same time an unchanging number of affinities (G. C. Foster);
> 2) that not only the affinities of dissimilar atoms, but also those of atoms of the same element may enter into combination with each other (Kekulé).[51]

Although Kekulé had never stated the doctrine of constant valence in the explicit manner that Foster had, it is clear that Kekulé never had any sympathy with the defenders of variable valence. Throughout the first volume of his textbook, Kekulé assiduously avoided the assumption that carbon could be divalent and that nitrogen or phosphorus could be pentavalent. This of course caused Kekulé problems in certain cases. He was frankly at a loss to explain the constitution of carbon monoxide.[52] And for ammonium chloride and similar molecules he supposed the formation of "molecular compounds" such as $NH_3 \cdot HCl$; these consisted of two saturated molecules held together by an (unspecified) attraction weaker than that of true chemical combination.[53]

Foster, for his part, never had any pretensions of having introduced anything new into science. He wrote to Kekulé: "What do you think of Erlenmeyer's critique of Erdmann's 'Theorie.' I think it is the best statement I have seen of the essential questions at issue between the old and new schools. It was very insane of him, all the same, to mention my name in it."[54]

The French chemist Alfred Naquet was a warm advocate of the new chemistry, but like Wurtz took exception to Kekulé's idea of constant valence. In an 1864 article he pointed out that sulfur, selenium, and tellurium all form tetrahalides and must therefore be considered to possess a maximum valence of four. The analogies of these elements to oxygen would argue that the latter element, too, could perhaps be tetravalent, although no compounds containing tetravalent oxygen were as yet known (with the possible exception of carbon monoxide). Naquet suggested an illuminating analogy:

> If one imagines for example certain hook-shaped appendages fixed to the atoms which can serve for hooking onto the corresponding appendages of other atoms, thus resulting in the formation of a compound, it is clear that the number of hooks attached to a single atom would represent its absolute atomicity. Now if these hooks are not able to bond equally well with the appendages of other bodies, it is conceivable that the effective or relative atomicity of a radical could sometimes remain below its absolute or true atomicity.[55]

Erlenmeyer made a similar suggestion. In accordance with his earlier expressed speculation that atoms consist of accretions of "Grenzäquivalenten" (hydride equivalents), he suggested that the varying apparent atomicity of sulfur could be due to a difference in intensity of affinity of the six sulfur equivalents:[56]

$$SO_3 = \frac{\overbrace{S\ S\ S\ S\ S\ S}}{\underbrace{O\ O\ O\ O\ O\ O}} \qquad SCl_4 = \frac{\overbrace{S\ S\ S\ S\ S\ S}}{ClClClCl} \qquad H_2S = \frac{\overbrace{S\ S\ S\ S\ S\ S}}{H\ H}$$

Presumably the affinity of the sulfur equivalents decreases left to right, and the relative affinities for sulfur of oxygen, chlorine and hydrogen are O > Cl > H.

Kekulé published a note in the *Comptes rendus* as a reply to Naquet.[57] He wrote: "Several recently published memoirs, among them a note which M. Naquet presented to the Academy in one of its last sessions, have tended, it seems to me, to throw a certain confusion into the theory of atomicity of the elements. I feel obliged to enter into the debate, especially since, if I am not mistaken, it was I who introduced into chemistry the notion of the atomicity of the elements." Kekulé's chief aim was to uphold the concept of constant atomicity. He proclaimed: "The equivalent can vary, but not the atomicity." In order to justify this statement, Kekulé took refuge in the hypothesis of molecular compounds, which we have already briefly described. The hypothesis was supported by known cases of thermal dissociation and by water of crystallization in salts. Kekulé attempted to show that the assumption of variable valence leads to absurd complications. Erlenmeyer's translation of Kekulé's paper was equipped with exclamation points, question marks, and sarcastic footnotes; it severely strained the previously close relationship of the two men.

Naquet wrote a concise and effective rebuttal.[58] He questioned the value of Kekulé's molecular compound hypothesis; the phenomenon of dissociation represented, in Naquet's view, merely a lower than normal stability limit for certain substances, above which decomposition began to take place. Kekulé's assumption was unnecessary, and so unjustified.

Wurtz then entered the debate, joining Naquet and Erlenmeyer against Kekulé. Wurtz did not deny that molecular compounds exist—one clear example is water of hydration. But he doubted that ammonium chloride, phosphorus pentachloride, and iodine trichloride had such constitutions as Kekulé proposed. Wurtz declared that the only possible explanation for the varying atomicities of organic radicals was that of varying atomicity of carbon. In the bromination of allyl alcohol, for example, the monatomic alcohol became triatomic, and Wurtz's suggestion was that a divalent carbon becomes tetravalent in order to bond two more atoms.[59]

Williamson, too, disagreed with the doctrine of constant valence. He wrote to Kekulé: "In a recent note in the comptes rendus you refer to a theory of atomicity in your book which I have not found there and of which I am anxious to know particulars. I have only used the term atomicity in a sense strictly reciprocal to equivalence, and I believe that most chemists so understand it."[60] Here Williamson revealed his commitment to valence as the reciprocal of the hydride equivalent. A. W. Hofmann also used these definitions in his influential textbook, *Introduction to Modern Chemistry*.[61] The variability of the hydride equivalent led these chemists to accept variable valence.

Olefins also posed a problem for Kekulé and other advocates of constant carbon tetravalence: ethylene, after all, contains two carbon atoms each of which is combined with three rather than four other atoms. The idea of double bonds between atoms was used by Kekulé as early as 1858; he and others developed the theory of unsaturated compounds until by the 1870s olefinic, aromatic, and carbonyl compounds were formulated in a recognizably modern fashion. Probably the principal theoretical justification for the double bond concept, at least at first, was its role in salvaging constant valence, especially carbon tetravalence.[62]

RETROSPECT

By the early years of the nineteenth century, there were sufficient grounds for assuming the existence of chemically integral smallest parts. But the evidence for the specific elemental weights proposed by Berzelius and others—rather than some multiple or submultiple thereof—were meager and unconvincing. Arguments from chemistry alone always proved circular,[63] and even such physical approaches as isomorphism and atomic heats seemed plagued by both conceptual and empirical lacunae. As Knight has so eloquently argued, the "cash value" of the atomic theory—the totality of its solved problems—was appallingly small at first.[64] The sizable number of chemists of several countries who were converted to the theory in an astonishingly brief period (e.g., the single decade 1808-18, as documented in chapters 2-5) may by contrast seem inexplicable. Laudan has plausibly argued that "the rationality of pursuit" of the atomic theory in the early nineteenth century is explained not so much by its demonstrated scientific cash value, but rather by its scientific promise, or *relative* success rate in solving problems.[65]

Laurent and Gerhardt, who advocated returning to a version of Berzelius' weights that would harmonize organic and inorganic chemistry, did not introduce any new data in support of this step. Like Berzelius, much of their case rested on the basis of consistency, simplicity, and unity; their "paper arguments" were reasonable, but not compelling to most of their contemporaries. Until the middle of the century, in short, chemists could see no convincing "concilience of inductions"—as Whewell would put it—in favor of any particular set of atomic weights. This explains both the apparent nonchalance at the bewildering multiplicity of systems and the eagerness in the 1840s to accept a single set of weights—the Wollaston-Gmelin "equivalents"—that were explicitly *conventional* in status and moreover had the proper empirical certification.

One of two important turning points was the year 1851 when Williamson presented an experimental argument for the divalence of the oxygen atom on the basis of his synthesis of the asymmetric ethers. Williamson's efforts were followed in the next three years by those of Gerhardt, Odling, and Kekulé. The experimental discovery of valence by the followers of the French reformers was one of the events that ultimately resulted in the standardization achieved after Karlsruhe. Williamson's 1851 research was the first purely chemical evidence for the divalence of oxygen, hence for the atomic weight O = 16. Equivalents had always been experimentally accessible; when atomic valences began to become so, among the French-English school in the 1850s, Berzelius' old atomic weights received the necessary evidentiary support they had lacked.

In contrast to inorganic compounds, in nearly all organic molecules valence is perfectly constant—if one accepts such auxiliary hypotheses as chain formation, multiple bonding, and ring formation. Historians frequently draw attention to the analogy between Kolbe and Frankland's carbonic acid type and the marsh gas type of the Kekulé school. But the difference is significant: Frankland's inorganic types led to the recognition of variable valence and no further, whereas the newer type theorists' organic types led directly to constant valence and to structure theory. If one ponders the reformed formulas of the homologous saturated hydrocarbon, the regularity is readily perceivable; the one additional hypothesis required is the concept of "loss" of valence due to carbon-carbon bonding. Two factors were required for the recognition of valence in organic compounds: the Laurent-Gerhardt reform of weights and formulas; and the discovery of sufficient raw data—i.e., sufficiently accurately analyzed new organic substances—to provide a basis for generalization.[66] Valence and structure followed shortly after these conditions had been fulfilled.

When the controversy between the adherents of constant and variable valence is considered, it must not be forgotten that both groups of chemists recognized that valence was in any case much less variable than equivalent weights. At most, valence could alternate between two, or, in the case of sulfur, three values. And most of the theories of variable valence assumed some sort of constancy behind the apparent inconstancy of valence (just as variable valence was itself a lesser inconstancy behind the greater inconstancy of equivalent weights);

Frankland's "latent atomicity," Erlenmeyer's subatoms possessing unequal affinities, Hofmann's "incomplete" molecules, Kekulé's "gaps," and Naquet's unused hooks were all theories of this type.

It is not coincidence that Kekulé, the most important single figure in the origin of valence and structure, clung stubbornly, not to say irrationally, to constant valence. His suggestion that Frankland's use of (conventional) equivalents and his advocacy of variable valence was in itself sufficient to disqualify the latter's claim on valence theory, is understandable, if rather inflexible. For Kekulé, variable valence was no sort of valence at all.

Kekulé's graphic formulas, much emulated in the 1860s and much maligned in the 1880s, depict the distinction between hydride equivalent and atom in a vivid and strikingly simple manner. Briefly put, it is impossible to recognize the distinction between hydride equivalent and atom without simultaneously recognizing the phenomenon known as valence, for the former distinction *defines* the latter concept. Thus it is not coincidence that it was specifically the pioneering structural chemists who picked up on and developed the suggestion of subatomic structure that was merely implicit in Kekulé's graphic formulas. Unfortunately, to my knowledge no one ever explained the distinction between hydride equivalent and conventional equivalent; such an explanation might have done much to promote understanding of the new developments among those who did not share the emerging structural paradigm. Nor did anyone ever discuss the difference between hydride equivalent — which fundamentally is no more than a new type of conventional equivalent — and true empirical chemical equivalents. In fact, there is no evidence to suggest that this latter distinction was even perceived by the structural chemists.

The second critical turning point of the 1850s was the emergence of the kinetic theory of gases and of heat. It was critical in two ways. It provided chemists with a model of successful transdiction — the physicists' success became increasingly spectacular through the 1860s — and perhaps encouraged many of them to continue down the path of mechanism and realism. Furthermore, since both EVEN and submolecularity could be derived from kinetic-molecular assumptions, it gave reformers, especially Cannizzaro, some powerful ammunition to use in favor of the Avogadro-Ampère theory. By the end of the 1860s, when the periodic law was proposed independently by D. Mendeleev and L. Meyer, the reformed atomic weights had ceased to be controversial, at least among atomists. Mendeleev and Meyer simply assumed their validity, and the acceptance by the chemical community reflected nearly universal assent.

1. S. G. Brush, *The Temperature of History: Phases of Science and Culture in the Nineteenth Century* (New York: Franklin, 1977). See also L. P. Williams, *Michael Faraday* (London: Chapman & Hall, 1965), pp. 66-80.

2. For the rejection of Avogadro as a manifestation of chemists' avoidance of physics, see N. W. Fisher, "Avogadro, the Chemists, and Historians of Chemistry," *History of Science* (1982), 20:77-102, 212-31, esp. 90-92; for the changing styles of nineteenth-century science, see R. Fox, "The Rise and Fall of Laplacian Physics," *Hist. Stud. Phys. Sci.* 4 (1974): 89-136, and J. K. Bonner, "Amedeo Avogadro: A Reassessment of His Research and its Place in Early Nineteenth Century Science" (Ph.D. diss., Johns Hopkins University, 1974).

3. For details of these shifts, see Brush, *Temperature*, and Fisher, *Avogadro*, pp. 212-17.

4. R. Fox, *The Caloric Theory of Gases* (Oxford: Clarendon Press, 1971).

5. S. G. Brush, *The Kind of Motion We Call Heat* (Amsterdam: North Holland, 1976), chapters 2 and 3.

6. R. Clausius, "Ueber die Art der Bewegung, welche wir Wärme nennen," *Ann. Phys.* 100 (1857): 353-80.

7. C. H. D. Buys-Ballot, "Ueber die Art der Bewegung welche wir Wärme und Electricität nennen," *Ann. Phys.* 103 (1858): 240-59; Clausius, "Ueber die mittlere Länge der Wege . . . ," *Ann. Phys.* 105 (1858): 239-58.

8. J. Loschmidt, "Zur Grösse der Luftmolecüle," *Sitzungsber. Akad. Wiss. Wien* 52:2 (1865): 395-413.

9. Maxwell to Stokes, 30 May 1859, in *Memoir and Scientific Correspondence of the Late Sir George Gabriel Stokes*, ed. J. Larmor (Cambridge: Cambridge University Press, 1907), 2: 8-11; Maxwell, "On the Dynamical Theory of Gases," *Rep. Brit. Assoc. Adv. Sci.* 29:2 (1859): 9; "Illustrations of the Dynamical Theory of Gases," *Phil. Mag.* [4] 19 (1860): 19-32; 20 (1860): 21-37.

10. Maxwell, "On the Dynamical Theory of Gases," *Phil. Trans. Roy. Soc.* 157 (1867): 49-88.

11. Brush, *Kind of Motion*, pp. 76-78, 198-204.

12. Ibid., p. 139, citing Waterston's *Thoughts on the Mental Functions* (Edinburgh, 1943).

13. Edward Daub, "Waterston's Influence on Krönig's Kinetic Theory of Gases," *Isis* 62 (1971): 512-15. The abstract to Waterston's speech is "On a

General Theory of Gases," *Rep. Brit. Assoc. Adv. Sci.* 21 (1851): 6; Krönig's paper is "Grundzüge einer Theorie der Gase," *Ann. Phys.* 99 (1856): 315-22.

14. Clausius, "Ueber die Natur des Ozon," *Ann. Phys.* 103 (1858): 644-52 (645n).

15. The five papers are Waterston's 1851 abstract, Krönig's 1856 recapitulation, Clausius' 1857 paper, Maxwell's 1859 abstract and his major paper of 1860. Translations and reprints appeared in *Annales de chimie, Fortschritte der Physik, Zeitschrift für Mathematik und Physik, Bibliothèque universelle, Athenaeum, l'Institut,* and *Il nuovo cimento.* For specific citations, see Brush, *Kind of Motion,* pp. 732, 744, 747.

16. Kekulé to Erlenmeyer, 16 June 1859, and Kekulé to Weltzien, 14 March 1860; cited in R. Anschütz, *August Kekulé* (Berlin, 1929), 1:152, 183-84.

17. Kekulé to Weltzien, 14 March 1860, in Anschütz, 1:187.

18. Brodie to Kekulé, 27 May 1860. This letter is in the August Kekulé Sammlung, Technische Hochschule, Darmstadt.

19. Foster to Kekulé, 8 July 1860, August Kekulé Sammlung, Technische Hochschule, Darmstadt.

20. Williamson to Kekulé, 24 April 1860, August Kekulé Sammlung, Technische Hochschule, Darmstadt. As it turned out, at the last minute Williamson was prevented from going by unexpected business (Williamson to Weltzien, 29 August 1860, in A. Stock, *Der internationale Chemiker-Kongress... vor und hinter den Kulissen* [Berlin: Verlag Chemie, 1933], pp. 42-43).

21. Kolbe to Weltzien, 17 April 1860; Anschütz, 1:188. Anschütz states that Kolbe stayed away: pp. 197, 579, 673; Stock, *Chemiker-Kongress*, p. 10, and E. von Meyer, "Die Karlsruhe Chemiker-Versammlung im Jahre 1860," *J. prakt. Chem.* 191 (1911): 182-89 (186n), both say he *did* attend. Meyer may have had some inside information—he was Kolbe's son-in-law—but given Kolbe's resentful letter to Weltzien and his unawareness of the formula convention established at the Congress (Anschütz, 1:559), Anschütz's statement is also credible.

22. Related distastefully by Kekulé in a letter to Weltzien of 17 April 1860, in Anschütz, 1: 189.

23. Anschütz, 1: 673. The official report was written by Wurtz, who was Secretary of the Congress, but was only published in K. Engler, *Festgabe zum Jubiläum der vierzigjährigen Regierung seiner Königlichen Hoheit des Grossherzogs Friedrich von Baden* (Karlsruhe, 1892), pp. 346-55, and more accessibly in Stock, and in Anschütz, 1: 671-88.

24. Anschütz, 1: 183-209; C. deMilt, "Carl Weltzien and the Congress at Karlsruhe," *Chymia* 1 (1948): 153-69; C. deMilt, "The Congress at Karlsruhe," *J. Chem. Educ.* 28 (1951): 421-24; A. J. Ihde, "The Karlsruhe Congress: A Centennial Retrospect," *J. Chem. Educ.* 38 (1961): 83-86; as well as Stock and Engler.

25. Kekulé, *Lehrbuch der organischen Chemie* (Erlangen, 1861-66), 1:95-98, 106, 235; Anschütz, 1: 205-7, 677-78.

26. H. Deville and L. Troost, "Mémoire sur les densités de vapeur à des températures très-élevées," *Comptes rendus* 49 (1859): 239-42; H. Kopp, "Zur Erklärung ungewöhnlicher Condensationen von Dämpfen," *Ann.* 105 (1858): 390-94; Kekulé, "Ueber die Constitution und Metamorphosen der chemischen Verbindungen," *Ann.* 106 (1858): 129-59 (142).

27. Kekulé, *Lehrbuch*, 1:233, 236.
28. Anschütz, 1: 676-79.
29. Ibid., p. 679.
30. Fisher, "Avogadro," pp. 90-92, 214-17.
31. Cannizzaro, "Sunto di un corso di filosofia chimica," *Il nuovo cimento* 7 (1858): 321-66; *Sketch of a Course in Chemical Philosophy* (Edinburgh: Alembic Club Rpt. no. 18, 1910), p. 4. The papers of Krönig and Clausius had both appeared in vol. 6 (1857) of *Il nuovo cimento*.
32. Cannizzaro, "Notizie storiche e considerazioni sull'applicazione della teoria atomica alla chimica . . . ," *Gazzetta chimica italiana* 1 (1871): 1-33, passim; German translation in *Sammlung chemischer und chemisch-technischer Vorträge* 20 (1914): 164n.; "Considerations on Some Points of the Theoretic Teaching of Chemistry" [30 May 1872], in *Faraday Lectures: 1869-1928* (London: Chemical Society, 1928), pp. 17-43 (22-24).
33. Fisher, "Avogadro;" J. H. Brooke, "Avogadro's Hypothesis and Its Fate," *History of Science* 19 (1981): 235-73.
34. Brooke, "Avogadro's Hypothesis," p. 257.
35. Ibid.; Fisher, "Avogadro," pp. 85, 212; J. R. Partington, *A History of Chemistry* (London: Macmillan, 1964), 4:passim.
36. Cannizzaro, "Sketch," pp. 12-13.
37. Fisher, "Avogadro," p. 218, points out that in practice Cannizzaro relied much more on the Petit-Dulong law than on the Avogadro-Ampère gas theory.
38. V. Regnault, "Recherches sur la chaleur spécifique des corps simples et composés," *Ann. chim. phys.* [2] 73 (1840): 5-72 (61-64, 69); V. Regnault, "Note sur la chaleur spécifique du potassium," *Ann. chim. phys.* [3] 26 (1849): 261-67.
39. Fisher, "Avogadro," pp. 215-17.
40. H. Rose, "Ueber die Atomgewichte der einfachen Körper," *Ann. Phys.* 100 (1857): 270-91.
41. "The Congress of Chemists at Carlsruhe," *Chem. News* 2 (1860): 226-27; "Congrès des chimistes à Carlsruhe," *Moniteur scientifique* 2 (1860): 984-86; D. Mendeleev, cited by deMilt, "The Congress at Karlsruhe", p. 422.
42. L. Meyer, "Kritik einer Abhandlung von A. Kekulé . . . ," *Zeitschr. Chem.* 8 (1865): 250-54 (253-54).
43. Indeed, there was confusion and controversy at Karlsruhe over whether accepting the reformed weights meant returning to the system of Berzelius or of Gerhardt! (Anschütz, 1:680-82.)
44. The Congress resolved to use the barred letters for reformed atomic weights, a convention originally introduced unilaterally by Williamson in 1853 (Williamson to Roscoe, 5 December 1853, H. E. Roscoe Collection, Chemical Society, London).
45. See above, n. 41. Oddly—or perhaps not so oddly—Wurtz neglected to record this vote in his official report.
46. Laurent, *Chemical Method* (London, 1855), pp. 1-16. See also M. Donovan. *Treatise on Chemistry* (London, 1832), pp. 399-400. Donovan like Laurent was an atomist, but unlike Laurent believed that conventional equiv-

alents represented the true relative atomic weights for the elements. In this he was perhaps following Wollaston: see above, chapter 3.

47. A. J. Rocke, "Subatomic Speculations and the Origin of Structure Theory," *Ambix* 30 (1983):1-18.

48. Gaudin coined the word "atomicity," with its present signification, in 1833. A different meaning, approximately the same as "(poly-)basicity," was introduced in 1856 nearly simultaneously by Wurtz, Gerhardt, and Berthelot; the following year Kekulé began applying the word to *atoms*, hence using it as we now use "valence." But Kekulé's new meaning is only slightly different from the old if atoms are viewed as complexes of subatoms: "atomicity" (valence) was considered equal to the subatomicity of an atom, just as for Gaudin (and for us) atomicity is equal to the submolecularity of a molecule. In 1865 Hofmann criticized the term, by then the preferred word for "valence," as "vague and rather barbarous," and used the words "quantivalence," "univalent," "bivalent," "trivalent", etc. (*Introduction*, pp. 168-69). The following year A. Claus, apparently following Hofmann, introduced the word "Valenz": *Theoretische Betrachtungen und deren Anwendung zur Systematik der organischen Chemie* (Freiburg, 1866), p. 18. By the 1880s "atomicity" had regained its original (and modern) denotation. For more on this subject, which the above discussion supplements and modifies, see C. A. Russell, *History of Valency* (Leicester University Press, 1971), pp. 83-89.

49. Foster, *Rép. chim. pure* 3 (1861): 273n.

50. Wurtz, *Rép. chim. pure* 3 (1861):419-20. Wurtz appears to have gotten the idea of variable valence from Couper. In 1864 he acknowledged Couper's priority: *Leçons de philosophie chimique*, Paris, p. 136n.

51. Erlenmeyer, *Z. Chem.* 5 (1862): 27.

52. Kekulé, *Lehrbuch*, 1:162n.

53. Kekulé first discussed this theory in 1864 — see below — but a schematic presentation appeared in the second fascicle (1860) of his *Lehrbuch*, illustrated by graphic formulas (1: 444n.).

54. Foster to Kekulé, 12 February 1862, Anschütz, 1:218.

55. A. Naquet, "Sur l'atomicité de l'oxigène, du soufre, du sélenium et du tellure," *Comptes rendus* 58 (1864): 381-83 (383).

56. Erlenmeyer, "Ueber die Sättigungscapacität (Atomigkeit) der Elemente," *Z. Chem.* 7 (1864): 628-35 (631n). For his earlier speculation, see his "Betrachtungen über Aequivalent, Atom, Molekül und Volum," *Z. Chem.* 6 (1863): 65-75, 97-104, 609-20 (esp. 67-73).

57. Kekulé, "Sur l'atomicité des éléments," *Comptes rendus* 58 (1864): 510-14 (514). "He was, of course, mistaken," comments Partington (*History*, 4:537); "and he *knows* that he is mistaken," taunted Kolbe (*J. prakt. Chem.* [2] 23 [1881]:376).

58. Naquet, "Sur l'atomicité des éléments," *Comptes rendus* 58 (1864): 675-78.

59. Wurtz, "Sur l'atomicité des éléments," *Bull. Soc. Chim.* [2] 2 (1864): 247-53.

60. Williamson to Kekulé, 16 June 1864, August Kekulé Sammlung.

61. Hofmann, *Introduction*, pp. 166-69.

62. Russell, *History*, pp. 224-41; A. A. Baker, Jr., *Unsaturation in Organic Chemistry* (Boston: Houghton-Mifflin, 1968). See also G. B. Kauffman, "Werner, Kekulé, and the Demise of the Doctrine of Constant Valency," *J. Chem. Educ.* 49 (1972): 813-17.

63. Fisher, "Avogadro," pp. 79-81.

64. D. M. Knight, *Atoms and Elements: A Study of Theories of Matter in England in the Nineteenth Century* (London: Hutchinson, 1967), p. 32 and passim.

65. Larry Laudan, *Progress and Its Problems: Towards a Theory of Scientific Growth* (Berkeley: University of California Press, 1977), pp. 109-114.

66. Some of the simplest organic compounds were discovered surprisingly late; for example, propane, pentane, and hexane were first produced by Schorlemmer between 1862 and 1872.

11

Epilogue

In the conclusion to the second edition of *Die modernen Theorien der Chemie* (1872), Lothar Meyer commented that the use of equivalent notation seemed to have disappeared.[1] By then the minds of most working chemists were settled on a single best system of atomic weights, an amalgamation of those of Berzelius and Gerhardt, championed by Cannizzaro — essentially the system we use today. This is not to suggest, however, that everyone agreed on the philosophical basis of the atomic theory. Indeed, during the last third of the century there were recurrent episodes of sometimes heated discussion over the nature and even the existence of atoms.

THE ENGLISH DEBATES

In Great Britain the discussion centered on the introduction of a novel, purportedly nonatomistic chemical theory by Benjamin C. Brodie, Jr. As the subject has been treated in detail elsewhere,[2] my exposition will be brief. I wish to portray some of the flavor of those debates to further illuminate the ideas and opinions of some of the personalities who have figured prominently in earlier pages and to provide further illustration of certain of the themes of this book.

Brodie (1817–80) studied under Liebig before establishing a private laboratory in London; from 1855 he was professor of chemistry at Oxford. He had gotten to know Williamson in Giessen, and, like his younger colleague, became a warm advocate of the new chemistry of Laurent and Gerhardt. But about 1860 Brodie began to feel a growing skepticism toward the validity and usefulness of the atomic theory, and he devoted the remaining twenty years of his life to an attempt to create an alternative system. His "calculus of chemical operations" was published piecemeal during this period and was never completed.[3] The theory was highly mathematical, borrowing from George Boole's new algebra, and using ideas that P. W. Bridgman would later independently develop into the doctrine known as operationalism. It is likely that few contemporary chemists ever fully grasped the details of the theory.

In June 1867, not long after the first part of the new theory was published, Brodie gave a lecture to the Chemical Society, in which he attempted to summarize this portion of his system. He began by expressing his extreme dissatisfaction with the present system. "[T]he atomic doctrine has proved itself inadequate to deal with the complicated system of chemical facts, which has been brought to light by the efforts of modern chemists. I do not think that the atomic theory has succeeded in constructing an adequate, a worthy, or even a useful representation of those facts."[4] To illustrate his point, Brodie read to his audience the following advertisement for a molecular model kit, which had recently appeared in an English journal:

> Glyptic Formulae:
> Those teachers who think, with Dr. Frankland and Dr. Crum Brown, that the fundamental facts of chemical combination may be advantageously symbolized by balls and wires, and those practical students who require tangible demonstration of such facts, will learn with pleasure that a set of models for the construction of glyptic formulae may now be obtained for a comparatively small sum.[5]

At this point, the *Chemical News* reported, Brodie was forced to wait for the laughter to die down—the audience apparently was in complete sympathy. Brodie concluded that the science of chemistry had gotten "upon a wrong track," in fact, "altogether off the rules of philosophy," to have produced "such a bathos as this."[6]

Edward Frankland, who had championed Crum Brown's and Hofmann's graphic and glyptic formulas in his *Lecture Notes for Chemical Students* (1866), felt called upon, in the discussion period at the end of Brodie's lecture, to "protest at the outset, in the most emphatic manner, against the view which Sir Benjamin Brodie appears to have of such representations . . . I certainly do not imagine that any evil is likely to arise from such symbolic representations as have been hitherto used, even those of the very crudest kind which have been so strongly censured by Sir Benjamin Brodie."[7] Frankland insisted on the great value, even the necessity, of hypotheses in science; indeed, the characteristic of Brodie's "calculus" that had most favorably impressed Frankland was its extremely hypothetical nature. Just the same, Frankland stressed, one should not, nor did he, affirm one's hypotheses to be actual depictions of reality: "in repudiation of the notion that I regard such representations as these graphic or glyptic formulae, or even symbolic formulae by letters only, in the sense of representations of the constitution of those portions of matter called atoms, or as representations of the position of these atoms in the compound, perhaps I cannot do better than state, simply and at once, that I neither believe in atoms themselves, nor do I believe in the existence of centers of force."[8]

William Odling was thunderstruck at this disclaimer.

> The pleasure with which I have heard Sir Benjamin Brodie's lecture is, I am afraid, almost counterbalanced by the shock which my feelings received on hearing from Dr. Frankland that he questioned the positive existence of atoms. If Dr. Frankland's opinion on that subject was what he has stated, he has grossly deceived the chemical public. (Laughter.) The chairman [Williamson] has said quietly aside to me that, after all, Dr. Frankland never really believed in atoms, or he would not have ventured to take such liberties with them.[9] We have been led to believe that not only have we atoms, but that these atoms possess imaginary prongs, and that there is an imaginary clasping between them by means of these imaginary prongs, in a sort of hermaphroditism which it is scarcely possible to refer to. . . . In the ordinary use of our present symbolic language, there are some chemists who, differing from Dr. Frankland, bring the idea of atoms prominently forward. On the other hand, there are some who, like myself, do not believe in atoms, and who keep the idea of atoms in the background as much as possible.[10]

Coincidentally, the previous year Odling had had occasion to write referee reports for the Royal Society on papers by both Brodie and Frankland. Brodie's "calculus," Odling thought, "points out very forcibly the hypothetical nature of what have been heretofore received as demonstrated chemical truths." He recommended that Frankland's paper be published, but expressed his personal disapproval of the formulas used in it.[11] He subsequently commented in a letter to Brodie: "I suppose that like most of us here you have been amused by Frankland's marvellous picture book."[12]

Nor was Odling the only British chemist to register his disapproval of Frankland's graphic formulas. H. E. Roscoe asked Kekulé: "Have you seen Frankland's book? With pentatomic phosphorus and nitrogen? I think old Dalton will turn in his grave with delight at his string of atoms being again depicted in so prominent a manner!" A few months later G. C. Foster also wrote Kekulé: "But have you seen Frankland's book? it is a far more astonishing sight than its author."[13]

Kekulé had been following the debate in the British journals, and he penned a quick reply to Brodie's paper. "The question of whether atoms exist or not has but little significance in a chemical point of view; its discussion belongs rather to metaphysics. In chemistry we have only to decide whether the assumption of atoms is an hypothesis adapted to the explanation of chemical phenomena." Kekulé unhesitatingly rejected physical atoms, "taking the word in its literal signification of indivisible particles of matter," but forthrightly declared his belief in chemical atoms, namely "those particles of matter which undergo no further division in chemical metamorphoses." Any possible future investigation into subatomic constitution simply would not affect chemistry. As a consequence, Kekulé saw "but small value" in Brodie's attack on atomism, at least for chemistry. He showed that Brodie had proceeded from a number of gratuitous assumptions, many of which could have been quite different, and could in fact have led to far simpler consequences.[14]

Williamson's opinion of Brodie's theory was probably very similar to Kekule's; in any case, he firmly believed in the utility and validity of the atomic theory *qua* chemical theory and had little patience with those who sought to reject what he considered to be the basis of modern chemistry.[15] The highly sympathetic reception of Brodie's broadside attack against the atomic theory must have shocked Williamson even more than the actual content of Brodie's system. In an 1861 lecture to the Chemical Society—Brodie was in the chair—Williamson had innocently made the following statements: "Now, in chemistry, the physical basis which we all adopt for our own reasonings upon the more difficult transformations is that of the atomic hypothesis. . . . I have no intention of entering into the evidences of this theory, because, of course, they are quite as well known to you as to myself . . . [it is] the only physical theory at present given for the transformation of matter."[16] The 1867 lecture and discussion clearly showed Williamson that the current of antiatomistic skepticism ran deeper than he had imagined. Two years later he picked up the gauntlet by giving a speech "On the Atomic Theory" to the Chemical Society.[17] There can be little doubt that this was a direct response to Brodie, although Williamson never mentioned his colleague or the "chemical calculus" in the course of his presentation. His intention was to present a compact statement of the evidence for the existence of atoms, evidence that had been accumulating throughout the century.

Williamson began his talk by reading five passages from recent textbooks that denied belief in the real existence of atoms, portraying the theory as a mere unproved and unprovable hypothesis. He commented:

> It certainly does seem strange that men accustomed to consult nature by experiment so constantly as chemists do, should make use of a system of ideas of which such things can be said. I think I am not overstating the fact, when I say, that, on the one hand, all chemists use the atomic theory, and that, on the other hand, a considerable number of them view it with mistrust, some with positive dislike. If the theory really is as uncertain and unnecessary as they imagine it to be, let its defects be laid bare and examined. . . . But if the theory be a general expression of the best ascertained relations of matter in its chemical changes, the only general expression which those relations have as yet found, and be hypothetical only insofar as it presupposes among unknown substances relations analogous to those discovered among those which are known, then it must be classed among the best and most precious trophies which the human mind has earned, and its development must be fostered as one of the highest aims and objects of our science.[18]

Williamson explained at some length the difference between true equivalents and atomic weights and showed that any writing of chemical formulas entails an atomistic theory. His emphasis throughout the talk was more on the chemical than the physical evidence—he merely mentioned the kinetic theory, for instance. (Eight years earlier, however, he had unequivocally stated his adherence to that theory and had advised chemists to pay greater attention to the research of the physicists on atoms and molecules.[19]) He wrote:

> Of two things, one: either the existence of molecules is denied, or it is admitted. In the former case, the vast and consistent body of chemical evidence of the existence of molecules must be set aside and disproved, and the physical confirmations of their existence must also be proved to be erroneous. . . . when [the antiatomists] content themselves with saying that we are wrong, without either showing in what respect our evidence or our reasonings are at fault, and without showing any other evidence or any other reasonings which they consider preferable, it is difficult to know what else to do with them than to state our case and leave them to their reflections.[20]

Finally, Williamson added his own disclaimer: "The question whether our elementary atoms are in their nature indivisible, or whether they are built up of smaller particles, is one upon which I, as a chemist, have no hold whatsoever, and I may say that in chemistry the question is not raised by any evidence whatsoever . . . I know nothing about it." But the "vast body of evidence of the most various kinds and from the most various sources" that he had briefly summarized leads inexorably to the conclusion that matter consists of chemical atoms; and the theory describing their properties and relations is nothing less than "the very life of chemistry."[21]

As with the 1867 debate, a discussion followed Williamson's speech, but this time it took place five months later.[22] After Williamson and Brodie were allowed an initial scuffle, Frankland got up and attempted to clarify his earlier statement that had met with sarcasm on the part of Odling and Williamson. The atomic theory, he said, was inadmissable as absolute truth; it finds its value "as a kind of ladder to assist the chemist in progressing from one position to another in his science. . . .

He admired the atomic theory as much as Professor Williamson, and he thought no one could blame him for not making a sufficient use of it; but he did not wish to be considered a blind believer in the theory."[23]

Odling then provided a concise statement of the positivists' position: atoms are a mere hypothesis, an unnecessary accretion to the laws of chemical combination.[24] Like Davy, he compared the latter to Kepler's laws of planetary motion. Presumably Odling felt that the atomic theory was too materialistic, too "grob sinnlich," as Kolbe later put it. Years later, Odling averred that he had been wrong to question Williamson in this manner; he said that he had always been content to follow in Williamson's footsteps, but that unfortunately he had more than once lagged behind.[25]

The discussion continued at some length.[26] Brodie, who chaired the meeting, got the last word. The atomic theory, he said, was not only unnecessary, but pernicious. "This theory had, he thought, been often the means of deluding chemists into the belief that they understood things about which they knew nothing. He found the works of Kekulé and Naquet scribbled over with pictures of molecules and atoms, arranged in all imaginable ways, for which no adequate reason was given, and if there was no reason for this, it was a mischievous thing to do, for it led to a confusion of ideas, and to mixing up fictions with facts."[27]

Brodie had company in his opposition to atomic theory. The eccentric American mineralogist Thomas Sterry Hunt was particularly repelled by all forms of atomism and advocated a return to the Stoic interpretation of chemical reactions as interpenetration of masses rather than juxtaposition of constituent particles. Hunt's analogies were to solution or transmutation of matter; accordingly he went further even than the chemical followers of Kant and Schelling in early nineteenth-century Germany. His writings on this subject are diffuse and obscure. He could give no satisfactory account of stoichiometry, and his work seems to have been quite without influence.[28]

A more considered approach was taken by the English organic chemist Charles Wright, who was influenced by Brodie and held conventionalist views similar to Odling's. Wright outlined his ideas before the Chemical Society in February 1872 and was soon thereafter skillfully challenged by Williamson's assistant, R. W. Atkinson.[29] Williamson himself found an appropriate occasion for a reply in his Presidential Address to the British Association meeting in Bradford on 17 September 1873.[30] He denied that the antiatomists had provided any effective criticisms of the arguments he had outlined four years earlier: they had demonstrated no inconsistencies in the atomic theory nor did they have any rival continuum hypothesis worthy of being regarded as a comprehensive chemical theory. At least twice more before his retirement in 1887 Williamson returned to a public defense of the atomic theory.[31]

Reaction to the 1873 address was mixed. One observer in the *Athenaeum* wrote: "Poor Dr. Wright! We fear that he may yet be smarting under some of the President's incisive remarks."[32] On the other hand Henry Watts privately opined that Williamson:

> rather spoiled his address by trotting out his favorite hobby about the Atomic Theory. For my own part I value the Atomic Theory as highly as anyone can as a generalisation, and indeed the only generalisation we have got, of the laws of chemical combination; but I cannot go all the way with Williamson in asserting that the law of Multiples has no existence independent of it. To me indeed it appears that, even if we had never heard of atoms the Law of Multiples would still hold good as the most probable representation of the observed facts of combining proportions.[33]

Williamson, however, intended to suggest not so much that the law of multiple proportions is coextensive with the atomic theory, but rather that writing chemical formulas was simultaneously an expression of multiple proportions and of chemical atomism. It was Watts, not Williamson, who was conflating the "Law of Multiples" with the chemical atomic theory. Watts' characterization of atomism as a generalization—ipso facto, a law—rather than a theory is further evidence of conceptual confusion.

It was the atomic critic Wright who provided the most cogent analysis of these issues. In a letter to Brodie, Wright distinguished clearly between Daltonian physical atomism and Williamsonian chemical atomism. He wrote: "It seems to me that the terms of the Atomic Theory, e.g. Atom, are used in two senses by chemists; one as a material something, possessed of dimensions in space mass & time, the other as a simple number or ratio not possessed of such dimensions; . . . Dr. Williamson . . . is using the term Atom in the latter, & not in the Daltonian sense to which you restrict it, & with truth as it seems to me."[34] Atkinson had accused Wright of implicitly endorsing the atomic theory even in the act of refuting it. In a penetrating reply, Wright made the same distinction as he had to Brodie, but at much greater length. Essentially pleading guilty to a modified version of Atkinson's charge, he limited his advocacy to chemical rather than physical atomism.[35] He thus avoided Williamson's and Atkinson's semantic ambiguity, Watts' conceptual confusion, and Brodie's straw man. Wright's argument led to the conclusion that Brodie's system also contained an implicit chemical atomism, a criticism offered earlier and apparently independently by Williamson, Alexander Crum Brown, and James Clerk Maxwell.[36]

English physicists, including John Tyndall and George Stokes as well as Maxwell, were among the most ardent defenders of chemical and physical atomism in these debates. In the discussion period after Brodie's 1867 Chemical Society lecture, Maxwell applied his characteristic lightly sarcastic wit in commenting that he was shocked to find on entering the room that space was a chemical substance and that hydrogen and mercury were operations. Moreover, the kinetic gas theory (which by this time Maxwell fully accepted) supported the existence of atoms and molecules, as well as more detailed statements about them, such as EVEN.[37] Eight years later Maxwell himself delivered to the Chemical Society a lecture devoted to demonstrating the chemical applicability of kinetic-molecular physical theories.[38]

THE FRENCH DEBATES

From 1856 the principal French representative of the reform movement of Laurent, Gerhardt, and Williamson was Adolphe Wurtz. In the late 1850s, Wurtz carried out a brilliant series of investigations on polyfunctional organic acids and alcohols, and he was able to use his discoveries dramatically to support the Williamson-Gerhardt type theory. In 1859 he converted from the then-dominant equivalent notation to atomic weight notation. He began publicly and systematically to espouse chemical atomism in formal lectures given in 1863 and published in 1864 and returned to the subject in major works of 1869 and 1879.[39]

Wurtz's advocacy was effective, but he was unable to convince all prominent French chemists of the importance and validity of atomic weight notation. J. T. Pelouze and E. Fremy retained conventional equivalents in the third edition of their popular *Traité de chimie générale* (1862–1865; first edition from 1848). Victor Regnault, whose research — ironically — helped establish the reformed atomic weights, used equivalents as late as the sixth edition of the *Cours élémentaire de chimie* (1868–1869). Henri Deville and his student L. J. Troost sedulously avoided atomic weights. Despite J. B. Dumas' belief in chemical atomism and his youthful advocacy of modified Berzelian atomic weights, he used conventional equivalents consistently after 1840. Marcellin Berthelot's influential antiatomism will be discussed in some detail below.

Nevertheless, the prevailing opinion of historians that French opposition to atomism was almost unqualified from about 1840 to the turn of the century cannot be maintained. As a representative sample, let us examine the French chemical atomist community at the time of the 1860 Karlsruhe Congress. Among living scientists at this date were two elderly atomists, J. B. Biot and M. E. Chevreul. In their middle years were such chemists as F. Malaguti, M. A. Gaudin, A. E. Baudrimont and J. F. Persoz, for whom the atomic theory had been invaluable, at least heuristically. In the prime of life were such atomists as A. Cahours, G. Chancel, L. Pasteur, and A. de Chancourtois, as well as Wurtz himself. A number of French chemists then just at the beginning of their careers were also attracted to atomism, such as A. Naquet, C. Friedel, and P. Schützenberger.[40] I believe that what was true in 1860 was true thereafter, namely that French atomists always predominated numerically over antiatomists. For instance, by the turn of the century Berthelot had been joined by H. L. LeChatelier in opposition to the atomic theory, but in the atomist camp could be counted such newcomers as P. Sabatier, P. Langevin, V. Grignard, G. Urbain, J. Perrin, and the Curies.[41]

That the late nineteenth century French chemical community has *appeared* to be so uniformly opposed to atomic theory—sometimes even to contemporary observers—is, I think, due to three factors. The first is the failure to perceive the distinction between physical and chemical atomism, which makes every user of conventional equivalents appear to be rejecting all forms of atomic theory. In addition, the true antiatomists tended to be vocal; the atomists—with the exception of Wurtz—were quiet. The final factor is social and institutional and has to do with the power that Berthelot and others were able to exert over French higher education in the late nineteenth century; this aspect has been explored by Terry Shinn.[42]

The debate emerged into the public arena in a long series of papers read before the Paris Académie des Sciences in 1877-78. The particular occasion for the discussion and the complete cast of characters are less important than the fact that it gave Berthelot and Wurtz an opportunity to air their disagreements extensively. Wurtz argued, as he had been doing for two decades, that the use of conventional equivalents is neither more empirical nor more consistent than using atomic weights, indeed quite the opposite.[43] Berthelot responded: "In my opinion this question does not have the extreme importance that [Wurtz] seems to attach to it. The progress of chemical science is not subordinated to a change in notation which does not touch any fundamental issues." Nonetheless, Berthelot asked rhetorically: "Who has ever seen a gaseous molecule or an atom?" and inveighed against the atomists' "mystical conception of a simple body that combines with itself." He invoked the example of Newton, who had wisely refrained from feigning hypotheses, and he urged that the true task of science was the discovery of immutable laws rather than the concoction of probable theories.[44]

With a touch of asperity, Wurtz thanked Berthelot for the lesson in philosophy, but disclaimed any need for one. He scorned Berthelot's characterization of an element combining with itself as "mystical"; the formation of ozone is a ready experimental example. And he pointed out that Berthelot had failed to do his homework before criticizing the atomists: Berthelot, said Wurtz, was under the curious and erroneous impression that valence and equivalence express the same idea and that the atomists believe all elemental molecules consist of two atoms.[45]

About eight years later, Naquet had an interesting conversation with Berthelot (they were both then in the French Senate). Berthelot said, "I do not want chemistry to degenerate into a religion; I do not want the chemist to believe in the existence of atoms as the Christian believes in the presence of Christ in the communion wafer." Naquet replied that atoms are only a mental aid, that no one believes in their actual existence. Berthelot responded: "Wurtz has seen them."[46]

This dialogue highlights distinct aspects of the atomic debates that have figured prominently in this book: atomism as a system of conventions, as a chemical theory, and as an ontological physical theory.[47] There was no contradiction for Berthelot to believe that the choice between conventional equivalents and atomic weights — as conventions — was of no great importance, and simultaneously to assert that belief in atoms — as ontological entities — must be eliminated from the teaching and practice of chemistry.

Berthelot even declared himself a chemical atomist two years before his clash with Wurtz. He wrote: "All chemists agree on the assumption that all substances consist of very small particles indecomposable by present physical or chemical means . . . [This concept] appears to be a necessary consequence of the fundamental laws . . . of constant, multiple, and equivalent proportions."[48] What Berthelot refused to countenance was the specific atomistic theory based on the ideas of Avogadro and Ampère. Mary Jo Nye has well argued that the underlying basis for Berthelot's distaste was his eagerness to create an autonomous chemical science that was independent of physics;[49] we have seen that this desire was common in the chemical community, especially before 1860. Ironically, Berthelot was a pioneer in the emerging discipline of physical chemistry. However, he did not want chemistry to be *based* on uncertain and nonempirical physical ideas. He placed the science in the tradition of natural history rather than physics; the proper goal was classification and the discovery of laws, rather than the atomists' ontological physical transdiction.

Especially after the death of Wurtz, Berthelot was able to exert considerable influence in the French scientific community. Senator, Inspecteur Générale de l'Enseignement Supérieur, President of the Section des Sciences Physiques de l'École des Hautes Études and Vice President of the Conseil Supérieur de l'Instruction Publique, Berthelot helped retard the teaching of explicitly atomistic ideas in the public system of higher education until the end of the century.[50]

Berthelot was given aid and comfort in his campaign against atomism by the rise of an idealistic positivist movement in France and Germany toward the end of the century. Some of the scientific content of that movement will be discussed in the next section, but here I want to mention the associated work of two French scientist-philosophers who defended conventionalist positions and to some extent were seen as antiatomists. In his book, *La science et l'hypothèse*, Henri Poincaré labeled atomic theory an "indifferent hypothesis." By this he meant that accepting the hypothesis makes no difference to the results of investigation or calculation; atoms "need not therefore be rejected," though their actual existence could never be established. He did not discuss this point in detail.[51]

The other conventionalist was the physical chemist Pierre Duhem. In a long article published in 1892, Duhem examined the relation between atomic notation and atomic theory. The battle between equivalentists and atomists had been decisively won by the latter, he said, and those few who still employed equivalents did so with no valid scientific reasons. He argued that the reason for the superiority of atomic weights was their accordance with chemical analogy; the advocates of atomic weights were the *true* "equivalentists," and these weights should properly be renamed "equivalent weights." In some detail he then discussed substitution, types, valence, and structural formulas, but without ever mentioning the word "atom." In fact he concluded that chemical notation had been to its disadvantage tied to atomistic hypotheses. That tie should be broken and chemical formulas given their true function: classification and the expression of analogy.[52]

In his *Aim and Structure of Physical Theory*, Duhem amplified considerably on this point. True to his conventionalist tendencies, Duhem declared that the function of all physical theory is an economical representation and classification of experimental laws. It is often necessary to construct theories on the basis of existential hypotheses, he conceded, but those hypotheses in isolation are usually experimentally untestable. Following Poincaré, Duhem cited the example of pure inertial motion, which no one has ever, or can ever, see. A further example of his own devise was the law of multiple proportions. *Any* combination of the elements A, B, and C, even a nonstoichiometric mixture, can be represented by an "atomistic" formula $A_xB_yC_z$ to any desired degree of precision, if the integers x, y, and z are permitted high enough values. "Therefore, no chemical analysis, no matter how refined, will ever be able to show the law of multiple proportions to be wrong."[53]

But Duhem was no radical positivist. He averred that even hypotheses that are devoid of experimental meaning in isolation can and do "enter as essential foundations" into valuable theories that are indeed testable in toto. Furthermore, a good theory will evoke an "esthetic emotion" that indicates that the theory fulfills the role of a "natural" classification. What is a "natural" classification? It is one whereby the "logical order in which theory orders experimental laws is the reflection of an ontological order." Indeed, "the belief in an order transcending physics is the sole justification of physics."[54]

Duhem was a political conservative and a pious Catholic. He admitted that his antipathy to materialist-mechanist theories was partially due to his preference for an alternative that had analogies with the organismic Aristotelianism prevalent in the middle ages.[55] He was naturally inimical to Berthelot, who was a radical, an atheist, and a free-thinker.[56] Against Berthelot's chemical equivalents Duhem urged chemical atomic weights; against Berthelot's philosophical positivism Duhem urged a conventionalism that was vitiated by strong ontological hints. Duhem was an antiatomist only in a rather narrow sense.

THE GERMAN DEBATES

Poincaré and Duhem were participants in a general European resurgence of philosophical idealism and scientific positivism.

The ablest and most influential defender of the school of thought subsequently named empiriocriticism was the Austrian scientist-philosopher Ernst Mach. Mach regarded sense perceptions as basic, indeed as the only permissible elements with which to construct our ontology. Any belief in our ability to explain or elucidate causes in an ontological sense was illusory: "The law of causality is sufficiently characterized by saying that it is the presupposition of the mutual dependence of phenomena. . . . Those elements of an event which we call "cause and effect" are merely certain salient features of that phenomenon, which are important for its reproduction in our minds."[57] For Mach the proper goal of scientific laws and theories was the economical ordering of phenomena in the mind: "Denkökonomie." The chemists' atomic theory and the physicists' kinetic-molecular gas theory were at best but provisional tools for the economical construction of provisional theories, not as elements of an unchanging external reality. Mach even denied existential status to matter itself: material objects are nothing but "compendious mental symbols for groups of sensations," and any such symbol "has no existence outside of thought."[58]

One of those most strongly influenced by Mach was the Dresden physicist Georg Helm, who threw down the gauntlet to the atomists in a small book published in 1887: "Who would say to us that those atoms and their forces are truly the elements of the world? What if that were an error of the Newtonian view? *Energy* is the true element of the world, for everything that we know about the world we know by means of energy. . . . Thus arises the task of developing the energy law into a world view that includes the science of mechanics but also reaches beyond its borders."[59] Helm baptized the new world view with a word used decades before, in a different way, by W. J. M. Rankine: energetics.

The year 1887 was a sort of annus mirabilis for physical chemistry. In addition to Helm's energetics manifesto, there appeared in that year the first volume of the *Zeitschrift für physikalische Chemie*, in which were outlined two dramatic new theories: J. H. van't Hoff on dilute solutions and Svante Arrhenius on electrolytic dissociation. William Ostwald, van't Hoff's coeditor, also marked in 1887 a milestone in his life: he accepted a call from Riga to Leipzig, where he was to spend the most fruitful two decades of his career. Ostwald was a physical chemist who was strongly inclined toward the thermodynamic rather than the kinetic-molecular approach; his inaugural dissertation at Leipzig was entitled "Die Energie und ihre Umwandlungen."[60]

Influenced by Mach, Helm, and Josiah Willard Gibbs, Ostwald began to develop his ideas on energetics after 1890. His opponents were the physical atomist proponents of kinetic theories of heat and of gases. The pivotal event in this confrontation was a dramatic debate at the sixty-seventh annual meeting of the Gesellschaft deutscher Naturforscher und Ärzte in Lübeck in 1895. The chairman of the meeting, Johannes Wislicenus, was a colleague of Ostwald at Leipzig but a convinced atomist. In fact, it was none other than Wislicenus who had taken the leading role in the development of the new field of stereochemistry, following the pathbreaking 1874 papers of van't Hoff and J. A. LeBel.[61] In 1887 Wislicenus had published an elegant investigation of what he termed "geometric isomerism" in the unsaturated diacids.

At Lübeck Helm gave an address "Ueber den derzeitigen Zustand der Energetik" and Ostwald spoke on "Die Überwindung des wissenschaftlichen Materialismus." Wislicenus arranged that fellow stereochemist Viktor Meyer deliver a lecture on "Probleme der Atomistik" in order to counterbalance the energeticists. Besides Wislicenus and Meyer, Ostwald and Helm were vigorously opposed by many others at the meeting, among whom were Ludwig Boltzmann, Felix Klein, Arthur von Oettingen, Walther Nernst, and Arnold Sommerfeld. Ostwald later referred to the "closed antagonism" of his adversaries, and Helm complained of the bitter attacks in letters to his wife.[62]

In his speech Ostwald labeled mechanistic materialism as "hypothetical, even metaphysical"; indeed, "it is pure and simply an error." "I have been told that energy is after all an invention, an abstraction, whereas matter is real. I reply: *quite the opposite*! It is matter which is imaginary, an entity that we have rather imperfectly constructed to represent that which persists through the flux of appearances." Ostwald even ridiculed as "not far from pure nonsense" the assumption that elements persist materially in their compounds, since (for example) hematite has no properties in common with either metallic iron or oxygen gas.[63]

Ostwald and Helm, and energetics in general, were attacked not only orally in Lübeck, but also in a more formal way in the pages of the *Annalen der Physik und Chemie*. Boltzmann and Max Planck published searching critiques, and Ostwald and Helm responded in turn; Boltzmann then submitted a rebuttal.[64] In general, the energeticists rebuked atomistic ideas as sterile, and the atomists vehemently denied the charge. Boltzmann wrote: "Probably many a chemist who has deduced the possible number of isomeric compounds or the property of rotating the plane of polarization directly from the picture which he makes for himself of the arrangement of atoms would be of a different view regarding the purported unfruitfulness of the atomic theory."[65]

Boltzmann's comment was unusual in that most of the discussion pertained to kinetic theory, that is, to physical rather than chemical atomism. Ostwald only mentioned stoichiometry once in his talk, admitting that the atomic theory readily explains those laws.[66] Indeed, Ostwald never forsook *chemical* atomism, and always confessed as much in his numerous and popular textbooks.[67] Likewise, in a book published three years after the Lübeck meeting, Helm mentioned stoichiometry only twice, and then both times to concede the good service rendered by atomic theory for the explanation of combining proportions.[68] Finally, the renowned antiatomist polemic by the German-American J. B. Stallo contained only a short passage on chemical atomism, in the course of which the author granted the theory not only expository and representational, but even heuristic value.[69] Clearly chemical atomism was something of an embarrassment for the energeticists, for it had been able to prove its worth not only as far as the stoichiometric relations were concerned, but also, as Boltzmann, Wislicenus, and Meyer argued, in providing explanations and successful predictions for such arcane phenomena as structural and stereoisomerism. In 1904 the empirically-minded chemist Ida Freund commented that the study of isomerism "has supplied the most striking proof of the validity and utility of the introduction into the science of these hypothetical magnitudes, the atom and the molecule."[70]

Consequently, some energeticists attached considerable significance to the efforts of an otherwise obscure Bohemian chemist, František Wald, to liberate chemistry from chemical atomism.[71] In 1897 Wald triumphantly wrote Mach, his former teacher in Prague, that after four years of ceaseless desperate effort he had finally solved the stoichiometric puzzle, indeed "in an almost shamefully simple way." The key proved to be Willard Gibbs' phase rule, and Wald proceeded to publish his energetic solution in Ostwald's two journals (*Zeitschrift für physikalische Chemie* and *Annalen der Naturphilosophie*) and in monographic form.[72]

Ostwald himself adopted and developed Wald's views in his own papers and textbooks. He selected this topic when chosen Faraday Lecturer by the Chemical Society, delivered according to custom in the Royal Institution. Ostwald's message: "*It is possible to deduce from the principles of chemical dynamics all the stoichiometrical laws.* . . . Chemical dynamics has, therefore, made the atomic hypothesis unnecessary . . . I am quite aware that in making this assertion I am stepping on somewhat volcanic ground. I may be permitted to guess that among this audience there are only very few who would not at once answer, that they are quite satisfied with the atoms as they are, and that they do not in the least want to change them for any other conception." But Ostwald thought that the possible gain from his endeavor—placing chemical theory "on more secure ground than that furnished by a mere hypothesis"—more than justified the risks.[73]

Ostwald defined as "hylotropic" those homogeneous phases whose composition remains constant while undergoing a transition to another phase. For example, pure water in the process of evaporation at usual temperatures and pressures is hylotropic; most (though not all[74]) solutions are not. A chemical substance, then, is simply a special case of the more general class of homogeneous phases in that it forms hylotropic phases within certain limits of temperature and pressure. Beyond such limits most substances show nonhylotropic behavior. But there is a certain class of substances that form hylotropic phases under all conditions; these we call "elements."[75]

Having thus defined elements, compounds, and solutions with no reference to microphysics, Ostwald proceeded to deduce the stoichiometric relations in a similar fashion. To his own satisfaction, at least, he showed that the laws of constant, equivalent, and multiple proportions all followed as "a natural consequence of the mode of preparation and purification of chemical substances."[76] His viewpoint was remarkably reminiscent of that of Claude-Louis Berthollet as well as Wald. He, like his predecessors, had difficulty in convincing contemporary chemists that his assumptions did not already include the conclusions to be proved, that is, the laws of stoichiometry.[77]

In any case, as mentioned earlier, Ostwald continued throughout this entire period to utilize chemical atomism, at least in a formal sense. In response to a most courteous initial reception by the Chemical Society, Ostwald, groping for words in an unaccustomed tongue, granted that the atomic theory had "directed the discoveries of chemistry for a whole century, and if any hypothesis can do good, this one has done so."[78]

T. E. Thorpe opined that it was singularly appropriate for this distinguished German chemist to have given a lecture on "Elements and Compounds" in the very same lecture hall where Dalton had discoursed on atoms in 1809. The situation is even more singular than Thorpe realized, for Dalton's first public mention of his atomic theory outside of Manchester was in a course of lectures read in the Royal Institution in January 1804 — almost exactly a century earlier. Dalton had opened the curtain on the drama of the chemical atomic theory; Ostwald was attempting to write the end of the play. But Ostwald's opinions were to be completely transformed in the period immediately following his Faraday Lecture.

DÉNOUEMENT

Three months after the Lübeck meeting, Wilhelm Röntgen discovered X-rays. The following decade was filled with the thunder of scientific bombshells: Henri Becquerel's discovery of radioactivity; J. J. Thomson's measurement of the charge-to-mass ratio of the electron; the Curies' studies of uranium, thorium, polonium, and radium; Ernest Rutherford and Frederick Soddy's elucidation of decay series and the nature of radioactive emissions; Planck's quantum hypothesis; Einstein's special theory of relativity, his emission theory of light and his study of Brownian motion. Many of these novelties were relevant for the atomic debates and tended to weaken the energeticists' and positivists' position — subatomic particles, the Planck-Einstein light quantum, and the visible microscopic motion of colloidal particles due to random kinetic-molecular processes, all were turned to good account by the atomists. Physical atomism (though not of the Daltonian variety) was rapidly becoming as uncontroversial as chemical atomism.

Such a trend was in evidence as early as August 1898, when Boltzmann was writing the preface to the second volume of his *Vorlesungen über Gastheorie*. To the energeticists' charge that atomistics in general and kinetic gas theory in particular had been living a sterile existence, Boltzmann pointed to J. D. van der Waal's equation of state for nonideal gases, to William Ramsay's discovery of argon and neon on the basis of atomistic reasoning, and to M. Smoluchowski's verification of kineticists' prediction of otherwise anomalous aspects of heat conduction in rarified gases. But Boltzmann sounded a somber, almost despairing note as well: "In my opinion it would be a pity for science if the theory of gases were to be temporarily eclipsed by the momentary dominance of a hostile atmosphere, as for example happened to the wave theory of light due to Newton's authority. I recognize the impotence of a single individual against the trends of the times. But [I intend] to contribute what I can so that when the theory of gases is again pursued not very much will need to be rediscovered."[79]

From this and similar pronouncements, most historians have concluded that the atomic critics were expressing a majority view among scientists at the turn of the century. But Brush points out that there is no real evidence to support the widespread assumption that Boltzmann's suicide in 1906 was due to despair over the direction of the atomic debates, and Hiebert convincingly argues that the energeticists steadily lost ground from 1896.[80] Ostwald and Helm certainly felt beleaguered at Lübeck, Wald remained obscure, and Ostwald's autobiography paints a picture of a lonely crusade. Energetics was and ever remained a minority viewpoint.[81]

In 1907 Ostwald expanded his Faraday Lecture into "an introduction to all textbooks of chemistry." In the Preface to this work he remarked on the largely unfavorable reception to that lecture; the experience was "not very encouraging" and the matter perhaps even seemed "hopeless." Echoing Boltzmann's lament, he pointed out that one "must wait for the right time," and he averred that now (September 1907) "the time . . . has arrived."[82]

Nevertheless, it was just a year later that Ostwald finally converted to the atomic theory, following the studies on Brownian motion by Einstein, Smoluchowski, and Perrin (1905–10).[83] He wrote: "I am now convinced that we have recently become possessed of experimental evidence of the discrete or grained nature of matter, which the atomic hypothesis sought in vain for hundreds and thousands of years. [Recent investigations] justify the most cautious scientists in now speaking of the experimental proof of the atomic nature of matter."[84]

The elderly Ernst Mach apparently had a similar conversion experience. In a manner reminiscent of Berthelot, he had long been accustomed to respond to believers in atoms with the rhetorical question, "Habn S' eins gsehn?" But when he first saw the scintillations of alpha particles in a spinthariscope around 1903 he reportedly said, "Now I believe in the existence of the atom."[85] Although this anecdote appears to be authentic, there is no indication in Mach's later work that he ever publicly conceded an existential status to atoms.

The sharpest challenge to Mach's antiatomism was Planck's 1908 lecture on "Die Einheit des physikalischen Weltbildes," an impassioned plea for a unified realistic scientific Weltanschauung. Planck asked: "Is the physical world picture solely a more-or-less voluntary creation of our mind [as the Machists assert], or are we forced to the contrary notion, that it reflects real natural processes which are wholly independent of us?"[86] Clearly Planck inclined to the latter point of view. Along with this realist ontology Planck advocated an epistemology that did not take the empirical demarcation criterion too literally.

> Probably no physicist would contest the admissibility of the assertion that a creature endowed with both a knowledge of physics and a specific organ for ultraviolet light would recognize those rays as identical to the visible spectrum, even though no one has ever seen either an ultraviolet ray or such a creature; and no chemist would hesitate to ascribe the same chemical properties to sodium found in

the sun as to terrestrial sodium, although he can never hope to fill a test tube with salt from solar sodium.[87]

The same analysis applied to atomic theory:

> As little as we know about their detailed characteristics, atoms are no more and no less real than the heavenly bodies or the terrestrial objects that surround us; and the statement that a hydrogen atom weighs 1.6×10^{-24} grams contains no more inferior sort of knowledge than that the moon weighs 7×10^{25} grams. Of course, I can neither place a hydrogen atom on a balance, nor can I even see it; but I cannot place the moon on a balance either—and as for seeing it, it is well known that there are also invisible heavenly bodies whose weights are more or less accurately measured; indeed, the mass of Neptune was measured before any astronomer had directed his telescope to it.[88]

Planck's final tactic was to turn the tables on the energeticists' charge that the atomic theory was sterile; he ended his talk by exhorting his listeners to have "quiet trust in the power of those words which for over nineteen hundred years have taught us the ultimate infallible criterion for distinguishing the false prophets from the true: 'Ye shall know them by their fruits!' "[89]

Mach was angered, and replied:

> After [Planck] paid his respects to [me] with Christ-like generosity, he branded me at the end with the well known biblical phrase of "false prophet". Here we see physicists have formed their own religion . . . To this I simply answer: if the belief in the reality of atoms is so essential for you, I will withdraw myself from physical ways of thinking, I wish to be no true physicist, I renounce all scientific reputation, in short, I most kindly thank my community of believers. For freedom of thought is dearer to me.[90]

Three years later Mach explicitly rejected both the theory of relativity—which hitherto, to Einstein's great delight, he had favored—and the atomic theory, again referring to it as a religion.[91] Brush has colorfully depicted Mach thus as "a sinner on his deathbed, refusing to be converted by Father Planck to the faith which all his colleagues have accepted."[92]

With the benefit of hindsight, we can say that the problem with the position of Mach and his followers was that it was *too* empirical, too closely tied to direct sensory data; this is the difficulty that every phenomenalist philosophy of science eventually encounters. The dangers of excessive empiricism are certainly well understood, warnings having been uttered ever since the Presocratics. Boltzmann's advantage over Mach *as a scientist* was his liberal opinion of the importance of hypotheses, theories, and mental constructs in general. Along this line Hiebert has argued that Mach "entertained an abominably lean opinion of the role of theory construction in science. If you start with the facts of experience, as Mach did, you necessarily are led to fabricate an integrated picture of natural phenomena that is irresponsibly narrow in its potential pertinence for any aspects of nature not included in the original facts."[93]

This is certainly not to suggest, however, that Mach's phenomenalism was totally sterile. On the contrary, Mach's epistemological critique of Newtonian mechanics greatly influenced Einstein and may have helped him toward certain aspects of his relativity theory. But Einstein himself was no positivist, and it cannot be maintained that Mach's analysis *itself* contains any essential part of the theory of relativity.

Mach in his final years was a melancholy figure. By 1913 he was virtually alone in his rejection of atomism and was simply ignored by his colleagues. Chemists in particular had long since ceased to worry about what seemed to be fine philosophical distinctions, distinctions which in any case seemed to have little possible relevance to chemistry. It is doubtful whether any significant number of working chemists from the 1820s on ever rejected chemical atomism; certainly all of them, virtually without exception, *used* the theory. Most chemists throughout the century recognized that the roots of chemical atomism were solidly planted; they minded their own business, applied the theory with a sense of unquestioning confidence, and were daily rewarded for their faith.

1. L. Meyer, *Die modernen Theorien der Chemie*, 2d ed. (Breslau, 1872), p. 359.
2. W. H. Brock, ed., *The Atomic Debates: Brodie and the Rejection of the Atomic Theory* (Leicester: Leicester University Press, 1967).
3. B. Brodie, "Calculus of Chemical Operations," *Phil. Trans. Roy. Soc.* 156 (1866): 781-859; ibid., 167 (1877), 35-116; "On the Mode of Representa-

tion Afforded by the Chemical Calculus, as Contrasted with the Atomic Theory," *Chemical News* 15 (1867): 295-305; *Ideal Chemistry* (London, 1880). See D. M. Dallas, "The Chemical Calculus of Sir Benjamin Brodie," in Brock, *Atomic Debates*, pp. 31-90.

4. Brodie, "Mode of Representation", p. 296.
5. Ibid. The advertisement appeared in the *Laboratory* 1 (1867): 78.
6. Ibid.
7. Ibid, pp. 302-3.
8. Ibid., p. 302.
9. Williamson was a firm believer in atoms, but he had a strong aversion to complicated symbolic systems. We recall his masterly lampoon of Kolbe's "parenthesis, comma, buckle, juxtaposition and full stop." We should also note that Williamson studied with Auguste Comte during the time he spent in Paris. For Williamson's positivism (or, rather, lack of it), see Brock's Appendix, "Comte, Williamson and Brodie," in his *Atomic Debates*, pp. 145-52.
10. Brodie, "Mode of Representation", pp. 303-4.
11. Royal Society, London, MSS RR.6.30 and RR.6.128, 23 October 1866 and 18 January [1866].
12. Odling to Brodie, 6 November 1866, printed in Brock, *Atomic Debates*, p. 97.
13. Roscoe to Kekulé, 13 December 1866; Foster to Kekulé, 15 April 1867; both in the August Kekulé Sammlung, Darmstadt.
14. A. Kekulé, "On Some Points of Chemical Philosophy," *Laboratory* 1 (1867): 303-6; reprinted in R. Anschütz, *August Kekulé*, 2 vols. (Berlin, 1929), 2:364-70 (366-67). Kekulé's arguments are described in more detail by Dallas, "Calculus," pp. 57-58; Dallas concludes: "It is clear that Kekulé had taken time and trouble to understand the Calculus, and all his criticisms are searching."
15. See letters from Williamson to Brodie in Brock, *Atomic Debates*, pp. 95-96 and 119-20; also see Williamson's "Note on the Calculus of Chemical Operations," *Chem. News* 16 (1867): 3-4. Like Kekulé, Williamson had invested a great deal of effort to gain insight into and to reinterpret Brodie's symbolism.
16. Williamson, "On Thermo-Dynamics in Relation to Chemical Affinity," *Chem. News* 3 (1861): 234-39, 246-47 (235).
17. Williamson, "On the Atomic Theory," *J. Chem. Soc.* 22 (1869): 328-65.
18. Ibid., pp. 330-31.
19. Williamson, "On Thermo-Dynamics," pp. 234-35.
20. Williamson, "On the Atomic Theory," p. 351.
21. Ibid., p. 365.
22. *J. Chem. Soc.* 22 (1869): 433-41.
23. Ibid., pp. 434-35.
24. Ibid., p. 436.
25. *Chem. News* 78 (1898): 286.

26. For a discussion, see the essay by Brock and Knight, in Brock, *Atomic Debates*, pp. 19-26.

27. *J. Chem. Soc.* 22 (1869): 440.

28. T. S. Hunt, *A New Basis for Chemistry* (Boston, 1887). See Brock, *Atomic Debates*, pp. 24-25, and Brock, "T. S. Hunt," *Dict. Sci. Biog.* (1972), 6:564-66.

29. C. A. Wright, "On the Relations between the Atomic Hypothesis and the Condensed Symbolic Expressions of Chemical Facts," *Phil. Mag.* [4] 43 (1872): 241-64; R. W. Atkinson, "An Examination of the Recent Attack upon the Atomic Theory," *Phil. Mag.* [4] 43 (1842): 428-33. For a discussion, see Brock, *Atomic Debates*, pp. 26-29.

30. "Address of A. W. Williamson, President," *Rep. Brit. Assoc. Adv. Sci.* 43 (1873): lxx-xci, esp. lxxi-lxxvi.

31. Williamson, "On the Growth of the Atomic Theory," *Chem. News* 44 (1881): 123-29; "Dr. Williamson's Farewell Address," *Chem. News* 56 (1887): 5-9.

32. *Athenaeum*, 20 September 1873, p. 360.

33. H. Watts to H. E. Roscoe, 24 November [1873], H. E. Roscoe Collection, Chemical Society, London.

34. Wright to Brodie, 7 March 1872, in Brock, *Atomic Debates*, pp. 141-42.

35. Wright, "Reply . . . ," *Phil. Mag.* [4] 43 (1872): 503-14.

36. All cited in Brock, *Atomic Debates*, pp. 28, 49, 53, 60, 95, 129. All three critics made this point in 1867, and Williamson repeated it in his 1873 speech. Dallas, who has carried out the most thorough modern analysis of the "chemical calculus," agrees with the assertion that Brodie's system implicitly assumed a form of (chemical) atomism (Brock, *Atomic Debates*, pp. 13, 50, 53).

37. *Chem. News* 15 (1867): 303.

38. Maxwell, "On the Dynamical Evidence of the Molecular Constitution of Bodies," *Nature* 11 (1875): 357-59, 374-77.

39. A. Wurtz, *Leçons de chimie professées en 1863* (Paris, 1864), also appearing as *Leçons de philosophie chimique* (Paris, 1864); *Histoire des doctrines chimiques depuis Lavoisier jusqu'à nos jours* (Paris, 1869); *La théorie atomique* (Paris, 1879). For discussion, vide supra; also Akira Yoshida, "Charles Adolphe Wurtz et la théorie atomique," *Jap. Stud. Hist. Sci.* 16 (1977): 129-35.

40. Thus I count roughly twice as many French atomists as equivalentists in 1860. I have not included such atomists peripheral to the French scientific community as Gerhardt's Italian collaborator L. Chiozza, the Swiss chemist J. C. Marignac, the Belgian J. S. Stas, the minor organic chemist P. Havrez, or the botanist C. Delavaud, who Charles Grimaux believes to have been the first formally to teach Gerhardt's theories.

41. M. J. Nye, "Nonconformity and Creativity: A Study of Paul Sabatier, Chemical Theory, and the French Scientific Community," *Isis* 68 (1977): 375-91; M. J. Nye, "The Nineteenth-Century Atomic Debates and the Dilemma of an 'Indifferent Hypothesis'," *Stud. Hist. Phil. Sci.* 7 (1976):

245-68 (261); M. J. Nye, "Berthelot's Anti-Atomism: A 'Matter of Taste'?" *Ann. Sci.* 38 (1981): 585-90; M. J. Nye, *Molecular Reality: A Perspective on the Scientific Work of Jean Perrin* (New York: American Elsevier, 1972), pp. 60-63.

42. Terry Shinn, "The French Science Faculty System. . . ," *Hist. Stud. Phys. Sci.* 10 (1979): 271-332; Terry Shinn, *Savoir scientifique et pouvoir social: École polytechnique et les polytechniciens: 1794-1914* (Paris, 1980).

43. Wurtz, "Sur la loi des volumes de Gay-Lussac," *Comptes rendus* 84 (1877): 1,183-89 (1,186-87).

44. M. Berthelot, "Résponse. . . ," *Comptes rendus* 84:1,189-95.

45. Wurtz, "Sur la notation atomique," *Comptes rendus* 84:1,264-68, 1,349-52.

46. Carl Graebe, "Marcellin Berthelot," *Ber. Deutsch. Chem. Ges.* 41 (1908): 4,805-72 (4,855).

47. Nye, "Debate," pp. 250-51; Nye, "Taste," p. 586.

48. Berthelot, *Die chemische Synthese* (Leipzig, 1877), pp. 164-66; 1st ed. *La synthèse chimique* (Paris, 1875). In fact, *all* the equivalentists mentioned above were chemical atomists; their antiatomism applied only to physical ideas.

49. Nye, "Taste," pp. 588-90.

50. See cited works by Shinn and Nye.

51. H. Poincaré, *Science and Hypothesis* (London, 1905), pp. 152-53; 1st ed. *La science et l'hypothèse* (Paris, 1902), pp. 180-81. See Nye, "Debates," for a modern discussion.

52. P. Duhem, "Notation atomique et hypothèses atomistiques," *Rev. quest. scient.* [2] 31 (1892): 391-454 (404-5, 439, 452).

53. Duhem, *The Aim and Structure of Physical Theory* (1st ed., Paris, 1906; New York: Princeton University Press, 1954), pp. 19-24, 212-16.

54. Ibid., pp. 24-26, 335.

55. Ibid., pp. 305-11; Nye, "Debates," p. 264.

56. Duhem, "Thermochimie, à propos d'un livre récent de M. Marcelin Berthelot," *Rev. quest. scient.* [2] 42 (1897): 361-92 (390, 392); cited by Nye, in "Taste," p. 590.

57. E. Mach, *Die Geschichte und Wurzel des Satzes von der Erhaltung der Arbeit* (Prague, 1872), pp. 35-36; *The History and Root of the Principle of Conservation of Energy*, trans. P. Jourdain (Chicago, 1911), p. 61; *Popular Scientific Lectures*, trans. T. J. McCormack, 3d ed. (Chicago, 1898), p. 198.

58. *Popular Scientific Lectures*, p. 201. For Mach and the atomic theory, see E. N. Hiebert, "The Genesis of Mach's Early Views on Atomism," in R. S. Cohen and R. J. Seeger, eds., *Ernst Mach: Physicist and Philosopher* (Dordrecht: Reidel, 1970), pp. 79-106; and E. N. Hiebert, "The Energetics Controversy and the New Thermodynamics," in D. H. D. Roller, ed., *Perspectives in the History of Science and Technology* (Norman: University of Oklahoma Press, 1971), pp. 67-86.

59. G. Helm, *Die Lehre von der Energie, historisch-kritisch entwickelt, nebst Beiträgen zu einer allgemeinen Energetik* (Leipzig, 1887), pp. 56-57.

60. For Ostwald, see the biography by Hiebert and H. G. Körber, in *Dict. Sci. Biog.* (1978), 15:455-69, and sources cited therein. For Ostwald's relation to energetics see the cited works by Hiebert, and Ostwald's *Lebenslinien: Eine Selbstbiographie*, 2 vols. (Leipzig, 1927), 2:149-88.

61. Useful entrées into the enormous literature on the history of stereochemistry are provided by O. B. Ramsay, ed., *van't Hoff-LeBel Centennial* (Washington, D.C.: American Chemical Society, 1975), and by relevant articles in the *Dict. Sci. Biog.*

62. See Hiebert, "Energetics Controversy," for a detailed discussion. Ostwald's speech was reprinted in his *Abhandlungen und Vorträge allgemeinen Inhaltes (1887-1903)* (Leipzig, 1904), pp. 220-40, and illuminating correspondence between Ostwald, Boltzmann, and Helm is found in Körber, ed., *Aus dem wissenschaftlichen Briefwechsel Wilhelm Ostwalds* 1 (Berlin: Akademie-Verlag, 1961): 9-11, 21-22, 118-20; see also Ostwald's *Lebenslinien*, p. 179.

63. Ostwald, *Abhandlungen*, pp. 224, 226, 230, 234.

64. For citations and a discussion, see Hiebert, "Energetics Controversy," pp. 69-78.

65. Boltzmann, "Zur Energetik," *Ann. Phys. Chem.* 58 (1896): 595-98 (598).

66. Ostwald, *Abhandlungen*, p. 224.

67. E.g., *Lehrbuch der allgemeinen Chemie*, 1st ed. (Leipzig, 1885), 1:16; 2d ed. (Leipzig, 1891), 1:16-17; *Grundriss der allgemeinen Chemie*, 3d ed. (Leipzig, 1899), p. 12; *Grundlinien der anorganischen Chemie*, 1st ed. (Leipzig, 1900), p. 155. All these passages assert the hypothetical and provisional nature of the atomic theory, but concede that it is so useful for representing chemical composition that it will be formally retained.

68. Helm, *Die Energetik nach ihrer geschichtlichen Entwickelung* (Leipzig, 1898), pp. 138, 361.

69. J. B. Stallo, *Concepts and Theories of Modern Physics* (New York: Appleton, 1881; rpt. Cambridge, Mass.: Harvard University Press, 1960), pp. 124-28.

70. I. Freund, *The Study of Chemical Composition: An Account of its Method and Historical Development* (Cambridge: Cambridge University Press, 1904), p. 559.

71. See M. Teich, "F. Wald," in *Dict. Sci. Biog.* 14 (1976): 123-24, and J. Thiele, "Franz Walds Kritik der theoretischen Chemie," *Ann. Sci.* 30 (1973): 417-33.

72. See citations in Thiele, pp. 420-22; the monograph is *Chemie fázi* (Prague, 1918), and the letter to Mach is printed by Thiele, p. 423.

73. Ostwald, "Elements and Compounds," *J. Chem. Soc.* 85 (1904): 506-22 (508-9).

74. E.g., constant-boiling azeotropes.

75. Ibid., pp. 511-17.

76. Ibid., pp. 516-19.

77. Hiebert and Körber, *DSB*, p. 462.

78. *Chem. News* 89 (1904): 220-22 (222).

79. Boltzmann, *Vorlesungen über Gastheorie*, 2 vols. (Leipzig, 1896-1898), 2:v-vi.

80. S. Brush, *The Kind of Motion We Call Heat* (New York: North Holland, 1976), p. 247; Hiebert, "Energetics Controversy," pp. 68, 78; E. Broda *Ludwig Boltzmann: Mensch, Physiker, Philosoph* (Vienna: Deutike, 1955).

81. Ostwald, *Lebenslinien*, 2:178-88. He said that even his closest scientific friends, such as William Ramsay, refused to follow him, as they found atomic conceptions too indispensable (p. 179). Russell McCormmach has characterized energetics as having only a temporary vogue (*Night Thoughts of a Classical Physicist* [Cambridge, Mass.: Harvard University Press, 1982], p. 21).

82. Ostwald, *Prinzipien der Chemie: Eine Einleitung in alle chemischen Lehrbücher* (Leipzig, 1907). The passages cited are from H. W. Morse's translation, *The Fundamental Principles of Chemistry* (London, 1909), p. vi.

83. Nye, *Molecular Reality*, passim.

84. Ostwald, *Grundriss der allgemeinen Chemie*, 4th ed. (Leipzig, 1909); trans. by W. W. Taylor as *Outlines of General Chemistry* (London, 1912), p. vi (Preface dated November 1908).

85. The story is recounted and referenced in Brush, *Kind of Motion*, pp. 294-95; see also Hiebert, "Mach's Early Views."

86. The lecture was given at the University of Leiden, 9 December 1908, and appeared in *Phys. Zeitschr.* 10 (1909): 62-75 (73) reprinted in Planck's *Physikalische Abhandlungen und Vorträge*, 3 vols. (Brunswick: Vieweg, 1958), 3:6-29 (25).

87. Ibid., p. 25.

88. Ibid., p. 26.

89. Ibid., p. 29; the quotation is from the Sermon on the Mount, Matthew 7:16

90. Mach, "Die Leitgedanken meiner naturwissenschaftlichen Erkenntnislehre," *Phys. Zeitschr.* 11 (1910): 599-606 (603). For a discussion, see E. N. Hiebert, *The Conception of Thermodynamics in the Scientific Thought of Mach and Planck* (Freiburg: Ernst-Mach-Institut, 1968), p. 85.

91. In the preface to Mach's *Principien der physikalischen Optik* (Leipzig, 1921), preface dated July 1913; cited and discussed in G. Holton's fine article, "Mach, Einstein, and the Search for Reality," in Cohen and Seeger, *Mach*, pp. 165-99 (176).

92. Brush, *Kind of Motion*, p. 296.

93. Hiebert, "Boltzmann's Conception of Theory Construction," in J. Hintikka et al., eds., *Pisa Conference Proceedings* 2 (Dordrecht: Reidel, 1980): 175-98 (193-94). For a case study purporting to illustrate the nonheuristic and barren quality of positivist science, see Marjorie Malley, "The Discovery of Atomic Transmutation: Scientific Styles and Philosophies in France and Britain," *Isis* 70 (1979): 213-23.

Bibliography

Included are all sources cited in the notes, plus a few other uncited relevant studies. Full journal titles are given following their standardized abbreviations. Works by a single author are arranged chronologically, except for collected papers or letters which are grouped at the end. Unpublished dissertations are classified as printed secondary literature. Publishers are in general not specified for books over fifty years old. The secondary bibliography reflects the literature through the first half of 1983, but no claim of completeness is implied.

MANUSCRIPT COLLECTIONS

August Kekulé Sammlung, Institut für Organische Chemie, Technische Hochschule, Darmstadt.

L. F. Gilbert Collection, D. M. S. Watson Library, University College, London.

Harris-Brock Collection, Dr. W. H. Brock, Victorian Studies Centre, University of Leicester.

H. E. Roscoe Collection, Chemical Society, London.

Royal Society, London.

Royal Institution, London.

PRIMARY JOURNALS

Abh. Akad. Wiss. Berlin. Abhandlungen der königlich preussischen Akademie der Wissenschaften Berlin, physikalisch-mathematische Klasse

Afh. fys. kemi mineral. Afhandlingar i fysik, kemi och mineralogi, ed. J. Berzelius and W. Hisinger

Ann. Annalen der Pharmacie (1832–39); *Annalen der Chemie und Pharmacie* (1840–73); ed. J. Liebig and others

Ann. Phys. Annalen der Physik (1799–1824), ed. L. W. Gilbert; *Annalen der Physik und Chemie* (1824–76), ed. J. C. Poggendorff; (1877–99), ed. F. Wiedemann

Ann. chim. [phys.] Annales de chimie [et de physique]

Ann. Phil. Annals of Philosophy; or, Magazine of Chemistry, Mineralogy, Mechanics, Natural History, Agriculture, and the Arts, ed. T. Thomson

Archiv für die gesamte Naturlehre, ed. K. W. G. Kastner

Athenaeum; a Journal of Literature, Science, the Fine Arts, Music, and the Drama, London

Ber. Berichte der Deutschen Chemischen Gesellschaft, Berlin

Bull. Acad. Roy. Belg. Bulletin de l'Académie Royale des Sciences, des Lettres, et des Beaux-Arts de Belgique, Brussels

Bull. Soc. Chim. Bulletin de la Société Chimique de Paris

Bull. Soc. Philomatique Bulletin des sciences par la Société Philomatique de Paris

Chem. Gazette Chemical Gazette, or Journal of Practical Chemistry, ed. W. Crookes; superseded by:

Chem. News Chemical News and Journal of Industrial Science, ed. W. Crookes

Comptes rendus Comptes rendus hebdomadaires des séances de l'Académie des Sciences, Paris

Compt. rend. trav. chim. Comptes rendus mensuels des travaux de chimie, ed. A. Laurent and C. Gerhardt

Gazzetta chimica italiana, Rome

Glasgow Med. J. Glasgow Medical Journal

Jahresbericht Jahresbericht über die Fortschritte der physischen Wissenschaften (title varies), ed. J. Berzelius

J. pharm. Journal de pharmacie et de chimie, Paris

J. phys. Journal de physique, de chimie, d'histoire naturelle et des arts, Paris

J. Pharm. Journal der Pharmacie für Aerzte und Apotheker, ed. J. B. Trommsdorff

J. Chem. Phys. Journal für Chemie und Physik, ed. J. S. C. Schweigger

J. prakt. Chem. Journal für praktische Chemie

J. Nat. Phil. Chem. Arts Journal of Natural Philosophy, Chemistry and the Arts, ed. William Nicholson

J. Sci. Arts [Quarterly] Journal of Science and the Arts, London

J. Chem. Soc. Journal of the Chemical Society [of London]

Kungl. Sv. Vet. Handl. Kungliga Svenska Vetenskapsakademien Handlingar, Stockholm

Laboratory The Laboratory: A Weekly Record of Scientific Research, London

Manchester Mem. Memoirs and Proceedings of the Manchester Literary and Philosophical Society

Mém. Acad. Sci. Mémoires de l'Académie des Sciences, Paris

Mém. Inst. Mémoires de la classe des sciences mathématiques et physiques [de l'Institut de France]
Mém. Soc. Arcueil Mémoires de physique et de chimie de la Société d'Arcueil
Moniteur scientifique Le moniteur scientifique: Journal des sciences pures et appliquées, ed. G. A. Quesneville
Nature
Neues allg. J. Chem. Neues allgemeines Journal der Chemie, ed. A. F. Gehlen
Nicholson's J. [see *J. Nat. Phil. Chem. Arts*]
Il nuovo cimento
Phil. Mag. The London, Edinburgh and Glasgow Philosophical Magazine and Journal of Science
Phil. Trans. Roy. Soc. Philosophical Transactions of the Royal Society [of London]
Phys. Z. Physikalische Zeitschrift, Leipzig
Proc. Roy. Inst. Proceedings of the Royal Institution, London
Proc. Roy. Phil. Soc. Glasgow Proceedings of the Royal Philosophical Society of Glasgow
Proc. Roy. Soc. Proceedings of the Royal Society [of London]
Proc. Roy. Soc. Edin. Proceedings of the Royal Society of Edinburgh
Procès-verbaux des séances de l'Académie des Sciences, Paris
Quart. Rev. Quarterly Review, London
Reader The Reader: A Review of Literature, Science, and Art
Rép. chim. pure Répertoire de chimie pure, ed. C. A. Wurtz
Rep. Brit. Assoc. Adv. Sci. Reports of the British Association for the Advancement of Science
Rev. quest. sci. Revue des questions scientifiques
Rev. sci. Revue scientifique et industrielle, ed. G. A. Quesneville
Sitzungsber. Akad. Wiss. Wien Sitzungsberichte der kaiserlichen Akademie der Wissenschaften, Wien, Mathematisch-naturwissenschaftliche Klasse
Trans. Roy. Soc. Edin. Transactions of the Royal Society of Edinburgh
Verh. nat.-med. Ver. Heidelberg Verhandlungen des naturhistorisch-medicinischen Vereins Heidelberg
Z. ang. Chem. Zeitschrift für angewandte Chemie [from 1931 *Angewandte Chemie*]
Z. Chem. Zeitschrift für Chemie [title varies], ed. E. Erlenmeyer and others
Z. phys. Chem. Zeitschrift für physikalische Chemie

SECONDARY JOURNALS

Ambix
Am. J. Phys. American Journal of Physics
Ann. Sci. Annals of Science
Arch. int. hist. sci. Archives internationales d'histoire des sciences

Brit. J. Hist. Sci. British Journal for the History of Science
Brit. J. Phil. Sci. British Journal for the Philosophy of Science
Chymia: *Annual Studies in the History of Chemistry*
Erkenntnis
Hist. Stud. Phys. Sci. Historical Studies in the Physical Sciences
History of Science
Isis
Jap. Stud. Hist. Sci. Japanese Studies in the History of Science
J. Chem. Educ. Journal of Chemical Education
Lychnos: *Lärdomshistoriska samfundet*, Uppsala
Phil. Sci. Philosophy of Science
Physis: *Revista di storia della scienza*, Florence
Stud. Hist. Phil. Sci. Studies in the History and Philosophy of Science
Stud. Rom. Studies in Romanticism
Sudhoffs Archiv für Geschichte der Medizin und der Naturwissenschaften
Texas Quarterly, Austin
Trans. Am. Phil. Soc. Transactions of the American Philosophical Society
Vop. ist. est. tekh. Voprosy istorii estestvoznaniia i tekhniki, Akademiia Nauk SSSR, Moscow

PRIMARY BOOKS AND ARTICLES

Ampère, A. M. "Lettre . . . sur la détermination des proportions dans lesquelles les corps se combinent . . ." *Ann. chim.* 90 (1814):43–86.

———. "Note . . . sur la chaleur et sur la lumière considerées comme resultant de mouvemens vibratoires." *Ann. chim. phys.* [2] 58 (1835): 432–44.

Atkinson, R. W. "An Examination of the Recent Attack upon the Atomic Theory." Phil. Mag. [4] 43 (1872):428–33.

Avogadro, A. "Essai d'une manière de déterminer les masses relatives des molécules élémentaires des corps . . ." *J. phys.* 73 (1811): 58–76. Tr. *Foundations of the Molecular Theory* (Edinburgh: Alembic Club rpt. no. 4, 1923), pp. 28–51.

Baeyer, Adolf. *Gesammelte Werke.* Brunswick, 1905.

Baudrimont, A. E. *Introduction à l'étude de la chimie par la théorie atomique.* Paris, 1833.

———. *Traité de chimie générale et expérimentale.* 2 vols. Paris, 1844–46.

Bérard, J. E. "Essai sur l'analyse des substances animales." *Ann. chim. phys.* [2] 5 (1817): 290–98.

Berthelot, M. *La synthèse chimique.* Paris, 1875.

———. *Die chemische Synthese.* Leipzig, 1977.

———. "Réponse . . . " *Comptes rendus* 84 (1877): 1,189–95.

Berthollet, C. L. *Recherches sur les lois de l'affinité.* Paris, 1801.

———. *Ueber die Gesetze der Verwandtschaft.* Translated by E. G. Fischer. Berlin, 1802.

———. *Essai de statique chimique.* 2 vols. Paris, 1803.

Berzelius, J. J. *Lärbok i kemien.* 6 vols. Stockholm, 1808–1830. 2d ed., first 2 vols. only. Stockholm, 1817–1822.

———. "Vermischte Notizen." *Neues allg. J. Chem.* 9 (1810), 585–89.

———. "Försök rörande de bestämde proportioner, hvari den oorganiska naturens beståndsdelar finnas förenade." *Afh. fys. kem. min.* 3 (1810): 162–275.

———. "Versuch, die bestimmten und einfachen Verhältnisse aufzufinden, nach welchen die Bestandtheile der unorganischen Natur miteinander verbunden sind." *Ann. Phys.* 37 (1811): 249–334, 415–72; 38 (1811), 161–226; 40 (1812), 162–208, 235–330. Rpt. as *Ostwalds Klassiker der exakten Wissenschaften*, nr. 35. Leipzig, 1892.

———. "Schreiben des Herrn Prof. Berzelius . . . über ein zweites neues Gesetz . . . Stockholm, d. 1. Oct. 1810." *Ann. Phys.* 37 (1811): 208–20.

———. "Ueber das Verhältniss der Sauerstoffmengen etc. zu einander . . . " *J. Chem. Phys.* 1 (1811): 257–62.

———. "Ueber die bestimmten und einfachen Verhältnisse, nach welchen die Bestandtheile der unorganischen Natur verbunden sind." *J. Chem. Phys.* 2 (1811): 299–326.

———. "Lettre de M. Berzelius à M. Berthollet." *Ann. chim.* 77 (1811): 63–84.

———. "Zwei Schreiben des Hrn. Prof. Berzelius . . . " *Ann. Phys.* 42 (1812): 276–98.

———. "Versuch, die chemischen Ansichten, welche die systematische Aufstellung der chemischen Nomenclatur begründen, zu rechtfertigen." *J. Chem. Phys.* 6 (1812): 119–44, 284–322; 7 (1813): 43–78.

———. "Aus Berzelius' Tagebuch während seines Aufenthaltes in London im Sommer 1812." Translated by E. Wohler. *Z. ang. Chem.* 18 (1905): 1,946–48; 19 (1906): 187–90, 571–76.

———. "Experiments on the Nature of Azote, of Hydrogen, and of Ammonia, and of the Degrees of Oxidation of Which Azote is Susceptible." *Ann. Phil.* 2 (1813): 276–84, 357–68.

———. "Essay on the Cause of Chemical Proportions, and on Some Circumstances Relating to Them; Together with a Short and Easy Method of Expressing Them." *Ann. Phil.* 2 (1813): 443–54; 3 (1814): 51–62, 93–106, 244–57, 353–64.

———. "Neue Untersuchungen über die Natur des Stickstoffs, des Wasserstoffs, und des Ammoniaks; und einige Bemerkungen über die wahre Zusammensetzung der salpetrige Säure und der Salpetersäure . . . Erster Nachtrag . . . " *Ann. Phys.* 46 (1814): 131–75.

———. *Försök att genom användandet af den electrokemiska theorien och de kemiska proportionerna grundlägga ett rent vettenskapligt system för mineralogien.* Stockholm, 1814. Translated by J. Black as *An Attempt to Establish a Pure Scientific System of Mineralogy.* London, 1814.

———. "Experiments to Determine the Definite Proportions in which the Elements of Organic Nature are Combined." *Ann. Phil.* 4 (1814): 323-31, 401-9; 5 (1815): 93-101, 174-84, 260-75.

———. "An Address to those Chemists who wish to examine the Laws of Chemical Proportions, and the Theory of Chemistry in General." *Ann. Phil.* 5 (1815): 122-31.

———. "Sur la composition des acides phosphorique et phosphoreux, et sur leurs combinaisons avec les bases salifiables." *Ann. chim. phys.* [2] 2 (1816): 151-76, 217-40, 329-39.

———. "Untersuchungen über die Zusammensetzung der Phosphorsäure, der phosphorige Säure, und ihrer Salze: Zweiter Nachtrag . . ." *Ann. Phys.* 53 (1816): 393-446; 54 (1816): 31-55.

———. "Nouvelles recherches sur les proportions chimiques." *Ann. chim. phys.* [2] 5 (1817): 174-81.

———. "Lettre . . . sur la combinaison de l'oxigène avec le fer, le manganèse et l'étain." *Ann. chim. phys.* [2] 5 (1817): 149-60.

———. "Försök att närmare bestämma åtskilliga oorganiska kroppars sammansättning, till vinnande af en närmare utveckling af läran om de kemiska proportionerna." *Afh. fys. kem. min.* 5 (1818): 379-520.

———. "Versuche die Zusammensetzung verschiedener unorganischer Körper näher zu bestimmen . . ." *J. Chem. Phys.* 23 (1818): 98-122, 129-202, 277-99.

———. *Essai sur la théorie des proportions chimiques, et de l'influence chimique de l'electricité.* Paris, 1819.

———. "Om sammansättningen af svafvelhaltiga blåsyrade salter . . ." *Kungl. Sv. Vet. Handl.* 1820: 82-99.

———. *Lehrbuch der Chemie.* Translated by F. Wöhler. 2d ed., 4 vols. in 8, 1825-31; 3d ed., 10 vols., 1833-41; 4th ed., 10 vols., 1835-41; 5th ed., 5 vols., 1843-48 (all at Dresden and Leipzig).

———. "Ueber die Bestimmung der relativen Anzahl von einfachen Atomen in chemischen Verbindungen." *Ann. Phys.* [2] 7 (1826): 397-416; 8 (1826): 1-24, 177-90.

———. "Ueber die Zusammensetzung der Weinsäure und Traubensäure . . . nebst allgemeinen Bemerkungen über solche Körper, die gleiche Zusammensetzung, aber ungleiche Eigenschaften haben." *Ann. Phys.* [2] 19 (1830): 305-35.

———. "Proportions, determinate." D. Brewster, ed. *Edinburgh Encyclopedia* 17 (Edinburgh, 1830).

———. "Ueber die Constitution organischer Zusammensetzungen." *Ann.* 6 (1833): 173-76.

———. "Lettre . . . à M. Pelouze." *Comptes rendus* 6 (1838): 629-44.

———. "Lettre . . . à M. Pelouze." *Ann. chim. phys.* [2] 71 (1839): 137-46.

———. *Autobiographical Notes.* Translated by O. Larsell. Baltimore: Williams & Wilkins, 1934.

———. *Jac. Berzelius Bref.* Edited by H. G. Söderbaum. 6 vols., suppl. Uppsala: Almqvist & Wiksell, 1912-1961.

---. *Jahresberichte über die Fortschritte der physischen Wissenschaften* (title varies). Tübingen, 1822–48.

---. *Reiseerinnerungen aus Deutschland*. Edited by G. Klingemann. Weinheim: Verlag Chemie, 1948.

Berzelius, J. J., and Hisinger, W. "Versuch, betreffend die Wirkung der elektrischen Säule auf Salze und auf einige von ihren Basen." *Neues allg. J. Chem.* 1 (1803): 115–49.

Berzelius, J. J., and Liebig, J. *Berzelius und Liebig: Ihre Briefe von 1831–1845*. Edited by J. Carrière. Munich, 1898.

Berzelius, J. J., and Pontin, M. "Elektrisch-chemische Versuche mit der Zerlegung der Alkalien und Erden." *Ann. Phys.* 36 (1810): 247–80.

Berzelius, J. J., and Wöhler, F. *Briefwechsel zwischen J. Berzelius und F. Wöhler*. Edited by O. Wallach. 2 vols. Leipzig, 1901.

Biot, J. B., and Arago, F. "Mémoire sur les affinités des corps pour la lumière . . . " *Mem. Inst.* 7 (1806): 301–87.

Biot, J. B., and Thenard, L. J. "Sur l'analyse comparée de l'arragonite et du carbonate de chaux rhomboïdal." *Bull. Soc. Philomatique* 1 (1807): 32–35.

Bischof, C. G. *Lehrbuch der Stöchiometrie*. Erlangen, 1819.

Boltzmann, L. "Zur Energetik." *Ann. Phys.* 58 (1896): 595–98.

---. *Vorlesungen über Gastheorie*. 2 vols. Leipzig, 1896–98.

Bostock, J. "Remarks on Mr. Dalton's Hypothesis of the Manner in Which Bodies Combine with Each Other." *Nich. J.* 28 (1811): 280–92.

Brodie, B. C., Jr. "Observations on the Constitution of the Alcohol-Radicals . . . " *J. Chem. Soc.* 3 (1850): 121–34.

---. "The Calculus of Chemical Operations." *Phil. Trans. Roy. Soc.* 156 (1866): 781–859; 167 (1877): 35–116.

---. "On the Mode of Representation afforded by the Chemical Calculus, as contrasted with the Atomic Theory." *Chem. News* 15 (1867): 295–305.

---. *Ideal Chemistry*. London, 1880.

Brown, A. Crum. "On the Theory of Chemical Combination." M.D. thesis, University of Edinburgh, 1861. Printed in Edinburgh, 1879.

---. "On the Theory of Isomeric Compounds." *Trans. Roy. Soc. Edinburgh* 23 (1864): 707–19.

---. "On the Use of Graphic Representations of Chemical Formulae." *Proc. Roy. Soc. Edinburgh* 5 (1865): 429–31.

Buff, H. *Versuch eines Lehrbuchs der Stöchiometrie*. Nuremburg, 1829. 2d ed. (*Lehrbuch der Stöchiometrie*), Nuremburg, 1842.

Bunsen, R. "Untersuchungen über die Kakodylreihe." *Ann.* 37 (1841): 1–57; 42 (1842): 14–46; 46 (1843): 1–48.

Butlerov, A. M. "Einiges über die chemische Structur der Körper." *Z. Chem.* 4 (1861): 549–60.

Buys-Ballot, C. H. D. "Ueber die Art der Bewegung welche wir Wärme und Electricität nennen." *Ann. Phys.* 103 (1858): 240–59.

Cannizzaro, S. "Sunto di un corso di filosofia chimica." *Il nuovo cimento* 7 (1858): 321–66. Translated as *Sketch of a Course in Chemical Philosophy*. Edinburgh: Alembic Club Rpt. no. 18, 1961.

---. "Considerations on Some Points of the Theoretic Teaching of Chemistry." In C. S. Gibson and A. J. Greenaway, eds., *Faraday Lectures*. London: Chemical Society, 1928, pp. 17-43.

Chancel, G. "Sur l'étherification et sur une nouvelle classe d'éthers." *Comptes rendus* 31 (1850): 521-23.

Chenevix, R. "Kritische Bemerkungen, Gegenstände der Natur betreffend, geschrieben während seines Aufenthalts in Deutschland." *Ann. Phys.* 20 (1805): 417-96.

Chevreul, M. E. *Recherches chimiques sur les corps gras d'origine animale*. Paris, 1823.

Choron, F. *Théorie des atomes et des équivalents chimiques*. 2d ed. Paris, 1839.

Claus, Adolf. *Theoretische Betrachtungen und deren Anwendung zur Systematik der organischen Chemie*. Freiburg/Br., 1866.

Clausius, R. "Ueber die Art der Bewegung, welche wir Wärme nennen." *Ann. Phys.* 100 (1857): 353-80.

---. "Ueber die Natur des Ozon." *Ann. Phys.* 103 (1858): 644-52.

---. "Ueber die mittlere Länge der Wege . . . " *Ann. Phys.* 105 (1858): 239-58.

Comte, A. *Cours de philosophie positive*. 6 vols. Paris, 1830-42.

Couper, A. S. "Sur une nouvelle théorie chimique." *Comptes rendus* 46 (1858): 1,157-60.

---. "On a New Chemical Theory." *Phil. Mag.* [4] 16 (1858): 104-16.

Dalton, J. *Meteorological Observations and Essays*. London, 1793.

---. "A New Theory of the Constitution of Mixed Aeriform Fluids and particularly of the Atmosphere." *J. Nat. Phil. Chem. Arts* 5 (1801): 241-44.

---. "On the Constitution of Mixed Gases." *Manchester Mem.* 5 (1802): 535-50.

---. "Experimental Enquiry into the Proportion of the Several Gases or Elastic Fluids, Constituting the Atmosphere." *Manchester Mem.* [2] 1 (1805): 244-58.

---. "On the Absorption of Gases by Water and other Liquids." *Manchester Mem.* [2] 1 (1805): 271-87. Reprinted in *Foundations of the Atomic Theory*. Edinburgh: Alembic Club Rpt. no. 2, 1923, pp. 15-26.

---. *A New System of Chemical Philosophy*. Pt. 1, Manchester, 1808; Pt. 2, 1810; Vol. 2, 1827; 2d ed., Pt 1, 1842.

---. *Ein neues System des chemischen Theiles der Naturwissenschaften*. Translated by F. B. Wolff. Berlin, 1812-13.

---. "Observations on Dr. Bostock's Review of the Atomic Principles of Chemistry." *J. Nat. Phil. Chem. Arts* 29 (1811): 143-51.

---. "Remarks on the Essay of Dr. Berzelius on the Cause of Chemical Proportions." *Ann. Phil.* 3 (1814): 174-80.

---. "On the Constitution of the Atmosphere." *Phil. Trans. Roy. Soc.* 116 (1826): 174-87.

---. *On the Phosphates and Arseniates*. Manchester, 1840.

———. *On a New and Easy Method of Analyzing Sugar*, bound with the preceding. Manchester, ca. 1842.

Daubeny, C. *Introduction to the Atomic Theory*. Oxford, 1831.

Davy, H. "On Some Chemical Agencies of Electricity." *Phil. Trans. Roy. Soc.* 97 (1807): 1-56.

———. "On some new Phenomena of Chemical Changes Produced by Electricity, particularly the Decomposition of the Fixed Alkalies." *Phil. Trans. Roy. Soc.* 98 (1808): 1-44.

———. "An Account of some New Analytical Researches . . . "*Phil. Trans. Roy. Soc.* 99 (1809): 39-104.

———. "On some New Electrochemical Researches . . . " *Phil. Trans. Roy. Soc.* 100 (1810): 16-74.

———. "Researches on the Oxymuriatic Acid, its Nature and Combinations . . . " *Phil. Trans. Roy. Soc.* 100 (1810): 231-57.

———. "On Some of the Combinations of Oxymuriatic Gas and Oxygene . . . " *Phil. Trans. Roy. Soc.* 101 (1811): 1-35.

———. "On a Combination of Oxymuriatic Gas and Oxygene Gas." *Phil. Trans. Roy. Soc.* 101 (1811): 155-62.

———. *Elements of Chemical Philosophy*. London, 1812.

———. "On Some Combinations of Phosphorus and Sulphur . . . " *Phil. Trans. Roy. Soc.* 102 (1812): 405-15.

———. *Elements of Agricultural Chemistry*. London, 1813. 2d ed., London, 1827.

———. "Some Experiments on the Combustion of the Diamond . . . " *Phil. Trans. Roy. Soc.* 104 (1814): 557-70.

———. "Some Experiments on a solid Compound of Iodine and Oxygen, and on its chemical Agencies." *Phil. Trans. Roy. Soc.* 105 (1815): 203-13.

———. "On the Action of Acids on the Salts usually called Hyperoxymuriates, and on the Gases produced from them." *Phil. Trans. Roy. Soc.* 105 (1815): 214-19.

———. "On the Analogies between the undecompounded Substances." *J. Sci. Arts* 1 (1816): 283-88.

———. *Consolations in Travel, or The Last Days of a Philosopher*. London, 1830.

———. *Collected Works of Sir Humphry Davy*. Edited by J. Davy. 9 vols. London, 1839-40.

Deville, H. "Note sur la production de l'acide nitrique anhydre." *Comptes rendus* 28 (1849): 257-60.

———, and Troost, L. "Mémoire sur les densités de vapeur à des températures très-élevées." *Comptes rendus* 49 (1859): 239-42.

Döbereiner, J. W. "Einige stöchiometrische Untersuchungen." *J. Chem. Phys.* 14 (1815): 206-23.

———. *Grundriss der allgemeinen Chemie: Anfangsgründe der Chemie und Stöchiometrie*. Jena, 1816. 2d ed., 1819; 3d ed., 1826.

———. *Darstellung der Verhältnisszahlen der irdischen Elemente zu chemischen Verbindungen*. Jena, 1816.

———. *Neueste stöchiometrische Untersuchungen: Beiträge zur chemischen Proportionslehre.* Jena, 1816.

———. "Ueber die thierische Kohle." *J. Chem. Phys.* 16 (1816): 86–91.

———. "Ueber die Pflanzenkohle und die metallische Grundlage derselben." *J. Chem. Phys.* 16 (1816): 86–91.

———. "Das Daseyn einer Zusammensetzung aus Kohlensäure und Kohlenoxyd bewiesen." *J. Chem. Phys.* 16 (1816): 105–10.

———. "Ueber die Zusammensetzung des Zuckers und des Alkohols." *J. Chem. Phys.* 17 (1816): 188–89.

———. "Stöchiometrische Untersuchungen." *J. Chem. Phys.* 17 (1816): 241–57.

———. "Ueber die Anwendung des Kupferoxyds zur Zerlegung organischer Substanzen und über die Zusammensetzung und Sättigungs-Capacität der Weinsäure." *J. Chem. Phys.* 17 (1816): 369–75.

———. "Aus einem Schreiben des Herrn Prof. und Bergrath Döbereiner an den Prof. Gilbert." *Ann. Phys.* 57 (1817): 435–38.

———. "Chemische Bemerkungen und Versuche." *Ann. Phys.* 59 (1818): 318–27.

———. "Vermischte chemische Bemerkungen." *J. Chem. Phys.* 23 (1818): 67–97.

———. "Nachtrag zu den vermischten chemischen Bemerkungen." *J. Chem. Phys.* 23 (1818): 219–33.

———. "Versuch zu einer Gruppirung der elementaren Stoffe nach ihrer Analogie." *Ann. Phys.* 91 (1829): 301–7.

———. *Briefwechsel zwischen Goethe und Johann Wolfgang Döbereiner (1810–1830).* Edited by J. Schiff. Weimar: Böhlaus, 1914.

Donovan, M. *Treatise on Chemistry.* London, 1832.

Duhem, Pierre. "Notation atomique et hypothèses atomistiques." *Rev. quest. sci.* [2] 31 (1892): 395–454.

———. "Thermochimie, à propos d'un livre récent de M. Marcelin Berthelot." *Rev. quest. sci.* [2] 42 (1897): 361–92.

———. *The Aim and Structure of Physical Theory.* New York: Princeton University Press, 1954 (1st ed., Paris, 1906).

Dulong, P. L. [untitled secondhand extract] *Mem. Inst.* 1815 (1818): cxcviii–cc.

———. "Extrait d'un mémoire sur les combinaisons du phosphore avec l'oxigène." *Ann. chim. phys.* [2] 2 (1816): 141–50.

———. "Mémoire sur les combinaisons du phosphore avec l'oxigène." *Mém. Soc. Arcueil* 3 (1817): 405–52.

Dumas, J. B. "Sur quelques points de la théorie atomistique." *Ann. chim. phys.* [2] 33 (1826): 337–91.

———. *Traité de chimie appliquée aux arts.* 8 vols. Paris, 1828–46.

———. "Sur la densité de la vapeur du phosphore." *Ann. chim. phys.* [2] 49 (1832): 210–14.

———. "Dissertation sur la densité de la vapeur de quelques corps simples." *Ann. chim. phys.* [2] 50 (1832): 170–78.

———. "Recherches sur les combinaisons de l'hydrogène et du carbone." *Ann. chim. phys.* [2] 50 (1832): 182-97.

———. "Sur la composition de l'acide pyrocitrique." *Ann. chim. phys.* [2] 52 (1833): 295-303.

———. "Recherches de chimie organique." *Ann. chim. phys.* [2] 56 (1834): 113-50.

———. "Considérations générales sur la compositon théorique des matières organiques." *J. pharm.* 20 (1834): 261-94.

———. *Leçons de philosophie chimique.* Paris, 1837.

———. "Réplique." *Comptes rendus* 6 (1838): 646-48.

———. "Réponse . . . à la lettre de M. Berzelius." *Comptes rendus* 6 (1838): 689-702.

———. "Acide produit par l'action du chlore sur l'acide acétique." *Comptes rendus* 7 (1838): 474.

———. "Mémoire sur la constitution de quelques corps organiques et sur la théorie des substitutions." *Comptes rendus* 8 (1839): 609-22.

———. "Note sur la constitution de l'acide acétique et de l'acide chloracétique." *Comptes rendus* 9 (1839): 813-15.

———. "Premier mémoire sur les types chimiques." *Ann. chim. phys.* [2] 73 (1840): 73-100.

———. "Mémoire sur la loi des substitutions et la théorie des types." *Comptes rendus* 10 (1840): 149-78.

Dumas, J. B.,and Boullay, P. "Mémoire sur la formation de l'éther sulfurique." *Ann. chim. phys.* [2] 36 (1827): 294-310.

———. "Mémoire sur les éthers composés." *Ann. chim. phys.* [2] 37 (1828): 15-53.

Dumas, J. B., and Liebig, J. "Note sur l'état actuel de la chimie organique." *Comptes rendus* 5 (1837): 567-72.

———. "Note sur la constitution de quelques acides." *Comptes rendus* 5 (1837): 863-66.

Dumas, J. B., and Pelletier, P. J. "Recherches sur la composition élémentaire et sur quelques propriétés caracteristiques des bases salifiables organiques." *Ann. chim. phys.* [2] 24 (1823): 163-91.

Dumas, J. B., and Stas, J. S. "Second mémoire sur les types chimiques." *Ann. chim. phys.* [2] 73 (1840): 113-66.

Erlenmeyer, E. "Betrachtungen über Aequivalent, Atom, Molekül und Volum." *Z. Chem.* 6 (1863): 65-75, 97-104, 609-20.

———. "Ueber die Sättigungscapacität (Atomigkeit) der Elemente." *Z. Chem.* 7 (1864): 628-35.

Faraday, M. "On two new compounds of Chlorine and Carbon . . . " *Phil. Trans. Roy. Soc.* 111 (1821): 47-74.

Fourcroy, A. F., Guyton de Morveau, L. B., Berthollet, C. L., and Vauquelin, L. N. "Rapport sur une prétendue découverte de M. Vinterl, professeur de chimie à Pest." *Ann. chim.* 71 (1809): 225-53.

Frankland, E. "On the Isolation of the Organic Radicals." *J. Chem. Soc.* 2 (1849): 263-96.

———. "On a New Series of Organic Bodies Containing Metals and Phosphorus." *J. Chem. Soc.* 2 (1849): 197-99.

———. "Researches on the Organic Radicals," *J. Chem. Soc.* 3 (1850): 30-52, 322-47.

———. "On a New Series of Organic Bodies Containing Metals." *Phil. Trans. Roy. Soc.* 142 (1852): 417-44.

———. "On the Production of Organic Bodies without the agency of Vitality." *Proc. Roy. Inst.* 2 (1858): 538-44.

———. *Lecture Notes for Chemical Students*. London, 1866. 2d ed. 2 vols., London, 1870-72.

———. *Experimental Researches in Pure, Applied, and Physical Chemistry*. London, 1877.

———. *Sketches fron the Life of Edward Frankland*. London, 1901. Another ed., London, 1902.

Frankland, E., and Kolbe, H. "Ueber die chemische Constitution der Säuren der Reihe $(C_2H_2)_nO_4$ . . . ." *Ann.* 65 (1848): 288-304.

Fuchs, J. N. "Ueber den Gehlenit, ein neues Mineral aus Tyrol." *J. Chem. Phys.* 15 (1815): 377-86.

Gaudin, M. A. "Recherches sur la structure intime des corps inorganiques définis . . ." *Ann. chim. phys.* [2] 53 (1833): 113-22.

———. *L'architecture du monde des atomes*. Paris, 1873.

Gay-Lussac, J. L. "Mémoire sur la combinaison des substances gazeuses, les unes avec les autres." *Mém. Soc. Arcueil* 2 (1809): 207-34.

———. "Mémoire sur l'iode." *Ann. chim.* 91 (1814): 5-160.

———. "Recherches sur l'acide prussique." *Ann. chim.* 95 (1815): 126-231.

———. "Lettre . . . à M. Clément, sur l'analyse de l'alcool et de l'éther sulfurique, et sur les produits de la fermentation." *Ann. chim.* 95 (1815): 311-18.

———. "Pesanteurs spécifiques des fluides elastiques." *Ann. chim. phys.* [2] 1 (1816): 218-21.

———. "Sur l'acide des prussiates triples." *Ann. chim. phys.* [2] 22 (1823): 320-23.

———. *Cours de chimie*. 2 vols. Paris, 1828.

———. "Considérations sur les forces chimiques." *Ann. chim. phys.* [2] 70 (1839): 407-34.

Gay-Lussac, J. L., and Dulong, P. L. "Rapport . . . sur un mémoire de M. Dumas." *Ann. chim. phys.* [2] 34 (1827): 326-31.

Gay-Lussac, J. L., and Thenard, L. J. *Recherches physico-chimiques*. 2 vols. Paris, 1811.

Gehlen, A. F. "Ueber das electro-chemische System und den Grund der bestimmten Verhältnissmengen." *J. Chem. Phys.* 12 (1814): 403-11.

Gerhardt, C. "Sur la constitution des sels organiques à acides complexes, et leur rapports avec les sels ammoniacaux." *Ann. chim. phys.* [2] 72 (1839): 184-214.

———. "Recherches sur la classification chimique des substances organiques." *Comptes rendus* 15 (1842): 498-500; *Rev. sci.* 10 (1842): 145-218.

———. "Considérations sur les équivalents de quelques corps simples et composés." *Ann. chim. phys.* [3] 7 (1843): 129–43, 238–45.

———. *Précis de chimie organique.* 2 vols. Paris, 1844–46.

———. "Sur la loi de saturation des corps copulés." *Comptes rendus* 20 (1845): 1648–57.

———. *Introduction à l'étude de la chimie par le système unitaire.* Paris, 1848.

———. "Remarques sur un travail de M. Hofmann relatif aux radicaux." *Compt. rend. trav. chim.* 6 (1850): 233–36.

———. "Remarques sur un travail de M. Williamson relatif aux éthers." *Compt. rend. trav. chim.* 6 (1850): 361–64.

———. "Sur la basicité des acides." *Compt. rend. trav. chim.* 7 (1851): 129–56.

———. "Ueber wasserfreie Säuren, namentlich wasserfreie Benzoësäure und Essigsäure." *Ann.* 82 (1852): 127–32.

———. "Recherches sur les acides organiques anhydres." *Comptes rendus* 34 (1852): 755–58, 902–5.

———. "Recherches sur les acides organiques anhydres." *Ann. chim. phys.* [3] 37 (1853): 285–342.

———. *Traité de chimie organique.* 4 vols. Paris, 1853–56.

———. *Correspondance de Charles Gerhardt.* Edited by Marc Tiffeneau. 2 vols. Paris, 1918–25.

Gerhardt, C., and Chancel, G. "Sur la constitution des composés organiques." *Compt. rend. trav. chim.* 7 (1851): 65–84.

Gerhardt, C., and Chiozza, L. "Addition aux recherches sur les acides anhydres." *Comptes rendus* 36 (1853): 1050–54.

Gilbert, L. W. "Einige kritische Aufsätze über die in München wieder erneuerten Versuche mit Schwefelkies-Pendeln, Wünschelruthen, u.d.m." *Ann. Phys.* 26 (1807): 370–449.

———. *Dissertatio historico-critica de mistionum chemicarum simplicibus et perpetuis rationibus earumque legibus nuper detectis.* Inaugural dissertation, Leipzig University, 1811.

———. "Historische-kritische Untersuchung über die festen Mischungs-Verhältnisse in den chemischen Verbindungen, und über die Gesetze, welche man in ihnen in den neuesten Zeiten entdeckt hat." *Ann. Phys.* 39 (1811): 361–428.

Gmelin, L. *Handbuch der theoretischen Chemie.* 3 vols. Frankfurt, 1817–19. 3d ed., 2 vols. 1827–30; 4th ed. (*Handbuch der Chemie*), 13 vols. 1843–70.

———. "Ueber ein besonderes Cyaneisenkalium." *J. Chem. Phys.* 34 (1822): 325–46.

Graham, T. *Researches on the Arseniates, Phosphates, and Modifications of Phosphoric Acid.* Edinburgh: Alembic Club Rpt. no. 10, 1961.

Haüy, René. *Essai d'une théorie de la structure des cristaux.* Paris, 1784.

Helm, Georg. *Die Lehre von der Energie, historisch-kritisch entwickelt, nebst Beiträgen zu einer allgemeinen Energetik.* Leipzig, 1887.

——. *Die Energetik nach ihrer geschichtlichen Entwickelung.* Leipzig, 1898.

Henry, W. "Experiments on the Quantity of Gases absorbed by Water, at different Temperatures, and under different Pressures." *Phil. Trans. Roy. Soc.* 93 (1803): 29–42.

——. *Elements of Experimental Chemistry.* 11th ed. 2 vols. London, 1829.

Hildebrandt, G. F. *Anfangsgründe der allgemeinen dynamischen Naturlehre.* 2 vols. Erlangen, 1807.

Hofmann, A. W. "Note upon the Action of Heat upon Valeric Acid . . ." *J. Chem. Soc.* 3 (1850): 121–34.

——. *Introduction to Modern Chemistry, Experimental and Theoretical.* London, 1865.

Hunt, T. S. *A New Basis for Chemistry.* Boston, 1887.

Kane, Robert. *Elements of Chemistry.* London, 1841.

Kastner, K. *Materialen zur Erweiterung der Naturkunde.* Jena, 1805.

——. *Einleitung in die neuere Chemie.* Halle, 1814.

——. "Bemerkungen." *J. Chem. Phys.* 26 (1819): 253–61.

Klaproth, H., and Wolff, F. B., eds. *Chemisches Wörterbuch.* 5 vols. Berlin, 1807–10. *Supplemente zu dem chemischen Wörterbuch*, 4 vols. Berlin, 1815–19.

Kekulé, August. "On a New Series of Sulphuretted Acids." *Proc. Roy. Soc.* 7 (1854): 37–40.

——. "Notiz über eine neue Reihe schwefelhaltiger organischer Säuren." *Ann.* 90 (1854): 309–16.

——. "Ueber die Constitution des Knallquecksilbers." *Ann.* 101 (1857): 200–13; *Ann.* 105 (1858): 279–86.

——. "Ueber die s.g. gepaarten Verbindungen und die Theorie der mehratomigen Radicale." *Ann.* 104 (1857): 129–50.

——. "Ueber die Constitution und die Metamorphosen der chemischen Verbindungen und über die chemische Natur des Kohlenstoffs." *Ann.* 106 (1858): 129–59.

——. "Remarques de M. A. Kekulé à l'occasion d'une Note de M. Couper sur une nouvelle théorie chimique." *Comptes rendus* 47 (1858): 378.

——. *Lehrbuch der organischen Chemie.* 2 vols. Erlangen, 1861–66.

——. "Sur l'atomicité des éléments." *Comptes rendus* 58 (1864); 510–14.

——. "Sur la théorie atomique et la théorie de l'atomicité." *Comptes rendus* 60 (1865): 174–77.

——. "On Some Points of Chemical Philosophy." *Laboratory* 1 (1867): 303–6.

——. "Aequivalent und Aequivalenz." In H. von Fehling, ed., *Neues Handwörterbuch der Chemie.* Brunswick, 1871.

——. *August Kekulé*, vol. 2: *Abhandlungen, Berichte, Kritiken, Artikel, Reden.* Edited by R. Anschütz. Berlin, 1929.

Kolbe, H. "Untersuchungen über die Elektrolyse organischer Verbindungen." *Ann.* 69 (1849): 257–94.

———. "Ueber die chemische Constitution und Natur der organischen Radicale." *Ann.* 75 (1850): 211-39; 76 (1850): 1-73.

———. "On the Chemical Constitution and Nature of Organic Radicals." *J. Chem. Soc.* 3 (1850): 369-405; 4 (1851): 41-79.

———. "Critical Observations on Williamson's Theory of Water, Ethers, and Acids." *J. Chem. Soc.* 7 (1854): 111-21.

———. *Ausführliches Lehrbuch der organischen Chemie*. 2 vols. Brunswick, 1854-64.

———. "Ueber den natürlichen Zusammenhang der organischen mit den unorganischen Verbindungen . . ." *Ann.* 113 (1860): 292-332.

———. "Meine Betheiligung an der Entwicklung der theoretischen Chemie." *J. prakt. Chem.* [2] 23 (1881): 305-23, 353-79: 479-517; 24 (1881): 375-425.

———. *Zur Entwicklungsgeschichte der theoretischen Chemie*. Leipzig, 1881.

Kolbe, H., and Frankland, E. "On the Products of the Action of Potassium on Cyanide of Ethyl." *J. Chem. Soc.* 1 (1848): 60-74.

Kolbe, H., [and Frankland, E.] "Ueber die rationelle Zusammensetzung der fetten und aromatischen Säuren, Aldehyden, Acetone u.s.w., und ihre Beziehung zur Kohlensäure." *Ann.* 101 (1857): 257-65.

Kopp, H. "Zur Erklärung ungewöhnlicher Condensationen von Dämpfen." *Ann.* 105 (1858): 390-94.

Krönig, A. K. "Grundzüge einer Theorie der Gase." *Ann. Phys.* 99 (1856): 315-22.

Laurent, A. "Sur la nitronaphtalase, la nitronaphtalèse, et la naphtalase." *Ann. chim. phys.* [2] 59 (1835): 376-97.

———. "Théorie des combinaisons organiques." *Comptes rendus* 2 (1836): 130-32; *Ann. chim. phys.* [2] 61 (1836): 125-46.

———. "Essai sur l'action du chlore sur la liqueur des Hollandais." *Ann. chim. phys.* [2] 63 (1836): 377-89.

———. "Recherches diverses de chimie organique." *Ann. chim. phys.* [2] 66 (1837): 136-213, 314-36.

———. "Thèse de chimie et de physique." University of Paris, 1837. J. Jacques, ed., "La thèse de doctorat d'Auguste Laurent." *Bull. Soc. Chim.* [5]1 (1954): D31-39.

———. "Classification chimique." *Comptes rendus* 19 (1844): 1,089-1,100.

———. "Sur les combinaisons organiques azotées." *Comptes rendus* 20 (1845): 850-55.

———. "Recherches sur les combinaisons azotées." *Ann. chim. phys.* [3] 18 (1846): 266-98.

———. *Méthode de chimie*. Paris, 1854. *Chemical Method*, translated by W. Odling. London, 1855.

Laurent, A., and Gerhardt, C. "Recherches sur les anilides." *Ann. chim. phys.* [3] 24 (1848): 163-207.

Lavoisier, A. L. *Traité élémentaire de chimie*. Paris, 1789. *Elements of Chemistry*, translated by R. Kerr. Edinburgh, 1790.

LeRoyer, A., and Dumas, J. A. "Essai sur la volume d l'atome des corps." *J. phys.* 92 (1821): 401-11.

Liebig, J. "Neue Analyse von Wöhlers Cyansäure." *Archiv für die gesamte Naturlehre* 6 (1825): 145-53.

———. "Ueber Cyan- und Knallsäure." *J. Chem. Phys.* 48 (1826): 376-81.

———. "Ueber die Zusammensetzung der Camphersäure und des Camphers." *Ann. Phys.* [2] 20 (1830): 41-47.

———. "Ueber einen neuen Apparat zur Analyse organischer Körper . . ." *Ann. Phys.* [2] 21 (1831): 1-43.

———. "Ueber die Constitution des Aethers und seiner Verbindungen." *Ann.* 9 (1834): 1-18, 31-39.

———. "Ueber die Constitution der organischen Säuren." *Ann.* 26 (1838): 113-89.

———. "Lettre . . . á M. le Président." *Comptes rendus* 6 (1838): 823-28.

———. "Der Zustand der Chemie in Preussen." *Ann.* 34 (1840): 97-136.

———. *Familiar Letters on Chemistry.* New York, 1843. *Chemische Briefe.* 2d ed. Heidelberg, 1845.

———. "Berzelius und die Probabilitätstheorien." *Ann.* 50 (1844): 295-335.

———. "Der chemische Process der Respiration." *Ann.* 58 (1846): 335-48.

Liebig, J., and Gay-Lussac, J. L. "Analyse du fulminate d'argent." *Ann. chim. phys.* [2] 25 (1824): 285-311.

Liebig, J., and Pelouze, T. J. "Vermischte Notizen." *Ann.* 19 (1836): 241-90.

Liebig, J., Poggendorff, J. C., and Wöhler, F., eds. *Handwörterbuch der reinen und angewandten Chemie.* 9 vols. Brunswick, 1842-62.

Liebig, J., and Wöhler, F. *Aus Justus Liebig's und Friedrich Wöhler's Briefwechsel.* Edited by A. W. Hofmann. 2 vols. Brunswick, 1888.

Limpricht, H. *Grundriss der organischen Chemie.* Brunswick, 1855.

Limpricht, H., and von Uslar, L. "Ueber die Sulfobenzoësäure." *Ann.* 102 (1857): 239-64.

Loschmidt, J. "Zur Grösse der Luftmoleküle." *Sitzungsber. Akad. Wiss. Wien.* 52:2 (1865): 395-413.

Mach, Ernst. *Die Geschichte und Wurzel des Satzes von der Erhaltung der Arbeit.* Prague, 1872. *History and Root of the Principle of the Conservation of Energy.* Chicago, 1911.

———. *Popular Scientific Lectures.* Translated by T. J. McCormack. Chicago, 1898.

———. "Die Leitgedanken meiner naturwissenschaftlichen Erkenntnislehre." *Phys. Z.* 11 (1910): 599-606.

———. *Principien der physikalischen Optik.* Leipzig, 1921.

Maxwell, J. C. "On the Dynamical Theory of Gases." *Rep. Brit. Assoc. Adv. Sci.* 29:2 (1859): 9.

———. "Illustrations of the Dynamical Theory of Gases." *Phil. Mag.* [4] 19 (1860): 19-32; 20 (1860): 21-37.

———. "On the Dynamical Theory of Gases." *Phil. Trans. Roy. Soc.* 157 (1867): 49-88.

———. "On the Dynamical Evidence of the Molecular Constitution of Bodies." *Nature* 11 (1875): 357–59, 374–77.

Meinecke, J. L. G. *Die chemische Messkunst, oder Anleitung, die chemischen Verbindungen nach Maass und Gewicht auf eine einfache Weise zu bestimmen und zu berechnen.* Halle, 1815.

———. "Anleitung zur chemischen Messkunst." *J. Pharm.* 25, ii (1816): 56–214; [2] 1, i (1817): 72–130; [2] 1, ii (1817): 3–98; [2] 2, ii (1818): 3–41; [2] 3, i (1819): 21–59; [2] 3, ii (1819): 475–534.

———. "Ueber die Producte der Weingährung." *J. Chem. Phys.* 17 (1816): 177–87.

———. "Das specifische Gewicht der elastischen Flüssigkeiten nach stöchiometrischen Berechnungen." *Ann. Phys.* 54 (1816): 159–75.

———. *Erläuterungen zur chemischen Messkunst.* Halle, 1817.

———. "[Auszug aus einem Briefe] vom Herrn Prof. Meinecke in Halle." *J. Pharm.* 25, ii (1816): 241–43.

———. "Ueber die Bestandtheile und die Dichtigkeit der vorzüglichsten Gase und Dunste." *J. Pharm.* [2] 2, i (1818): 325–32.

———. "Ueber die Dichtigkeit der elastisch-flüssigen Körper im Verhältniss zu ihren stöchiometrischen Werthen." *J. Chem. Phys.* 22 (1818): 137–59.

———. "Ueber die Entwicklung der Salze aus den gediegenen Verbindungen." *J. Chem. Phys.* 25 (1819): 269–89.

———. "Beilage." *J. Chem. Phys.* 26 (1819): 296.

———. "Ueber den stöchiometrischen Werth der Körper, als ein Element ihrer chemischen Anziehung." *J. Chem. Phys.* 27 (1819): 39–47.

Meissner, P. T. *Chemische Aequivalenten- oder Atomenlehre.* 2 vols. Vienna, 1834.

Melsens, L. "Note sur l'acide chloracétique." *Comptes rendus* 14 (1842): 114–17.

Mendius, O. "Ueber gepaarte Säuren und insbesondere über Sulfosalicylsäure." *Ann.* 103 (1857): 39–80.

Meyer, Lothar. *Die modernen Theorien der Chemie.* Breslau, 1864. 2d ed., 1872.

———. "Kritik einer Abhandlung von A. Kekulé . . ." *Z. Chem.* 8 (1865): 250–54.

Mitscherlich, E. "Ueber die Krystallization der Salze, in denen das Metall der Basis mit zwei Proportionen Sauerstoff verbunden ist." *Abh. Akad. Wiss. Berlin* 1818–1819: 427–37.

———. "Sur la rélation qui existe entre la forme crystalline et les proportions chimiques." *Ann. chim. phys.* [2] 14 (1820): 172–90.

———. "Om förhållandet emellan kemiska sammansättningen och krystallformen hos arseniksyrade och phosphorsyrade salter." *Kungl. Sv. Vet. Handl.* 1822: 4–79.

———. *Ueber das Verhältniss zwischen der chemischen Zusammensetzung und der Krystallformen arseniksaurer und phosphorsaurer Salze.* Ostwalds Klassiker Nr. 94, Leipzig, 1898.

———. "Ueber allgemeine Körperanziehung, mit Hinsicht auf die Theorie der Krystallelektricität." *J. Chem. Phys.* 39 (1823): 231-50.

Stallo, J. B. *Concepts and Theories of Modern Physics.* New York: Appleton, 1882.

Thenard, L. J. "Observations sur le phosphore." *Ann. chim.* 85 (1813): 326-28.

———. *Traité de chimie élémentaire, théorique et pratique.* 4 vols. Paris, 1813-16. 2d ed., 4 vols., 1817-18.

Thomson, T. *System of Chemistry.* 4 vols. Edinburgh, 1802. 2d ed., 1804. 3d ed., 1807. *Système de chimie.* Translated by J. Riffault. Paris, 1809. *System der Chemie.* Translated by F. B. Wolff. Berlin, 1811.

———. "On Oxalic Acid." *Phil. Trans. Roy. Soc.* 98 (1808): 63-95.

———. "On the Daltonian Theory of Definite Proportions in Chemical Combinations." *Ann. Phil.* 2 (1813): 32-52, 109-15, 167-71, 293-301; 3 (1814): 134-40, 375-78; 4 (1814): 11-18, 83-89.

———. "On the Discovery of the Atomic Theory." *Ann. Phil.* 3 (1814): 329-38.

———. "A Sketch of the Latest Improvements in the Physical Sciences." *Ann. Phil.* 5 (1815): 1-53.

———. "Some Observations on the Relations between the Specific Gravity of Gaseous Bodies and the Weights of their Atoms." *Ann. Phil.* 7 (1816): 343-46.

———. *An Attempt to Establish the First Principles of Chemistry by Experiment.* 2 vols. London, 1825.

Ure, Andrew. *Dictionary of Chemistry.* London, 1821.

Vogel, F. C. "Beiträge zu der Lehre von den bestimmten chemischen Mischungs-Verhältnissen." *J. Chem. Phys.* 7 (1813): 1-42, 175-250.

Wald, F. *Chemie fázi.* Prague, 1918.

Waterston, J. J. *Thoughts on the Mental Functions.* Edinburgh, 1843.

———. "On a General Theory of Gases." *Rep. Brit. Assoc. Adv. Sci.* 21 (1851): 6.

Williamson, A. W. "Results of a Research on Aetherification." *Rep. Brit. Assoc. Adv. Sci.* 20:2 (1850): 65.

———. "Theory of Aetherification." *Phil. Mag.* [3] 37 (1850): 350-56; *J. Chem. Soc.* 4 (1851): 106-12.

———. "On Etherification." *J. Chem. Soc.* 4 (1851): 229-39.

———. "On the Constitution of Salts." *Rep. Brit. Assoc. Adv. Sci.* 21:2 (1851): 54; *J. Chem. Soc.* 4 (1851): 350-55. Reprinted (along with the 3 preceding entries) in *Papers on Etherification and on the Constitution of Salts.* Edinburgh: Alembic Club Rpt. no. 16, 1902.

———. "Suggestions for the Dynamics of Chemistry derived from the Theory of Etherification." *Proc. Roy. Inst.* 1 (1851): 90-94.

———. "Note on the Preparation of Propionic and Caproic Acids." *Phil. Mag.* [4] 6 (1853): 204-6.

———. "On Gerhardt's Discovery of Anhydrous Organic Acids." *Proc. Roy. Inst.* 1 (1853): 239-42.

———. "Note on the Decomposition of Sulphuric Acid by Pentachloride of Phosphorus." *Proc. Roy. Soc.* 7 (1856): 11-15.

———. "On Dr. Kolbe's Addition Formulae." *J. Chem. Soc.* 7 (1854): 122-39.

———. "Thermo-dynamics in Relation to Chemical Affinity." *Chem. News* 3 (1861): 234-39, 246-47.

———. *Chemistry for Students.* London, 1865.

———. "Note on the Calculus of Chemical Operations." *Chem. News* 16 (1867): 3-4.

———. "On the Atomic Theory." *J. Chem. Soc.* 22 (1869): 328-65, 433-41; *Chem. News* 20 (1869): 19.

———. "Address of A. W. Williamson, President." *Rep. Brit. Assoc. Adv. Sci.* 43 (1873): lxx-xci.

———. "On the Growth of the Atomic Theory." *Chem. News* 44 (1881): 123-29.

———. "Dr. Williamson's Farewell Address." *Chem. News* 56 (1887): 5-9.

Winterl, J. J. *Prolusiones ad chemiam saeculi decimi noni.* Budapest, 1800.

Wöhler, F. "Ueber die eigenthümliche Säure, welche entsteht, wenn Cyan (Blaustoff) von Alkalien aufgenommen wird." *Ann. Phys.* 71 (1822): 95-103.

———. "Bildung der Cyansäure auf neuem Wege, und fernere Untersuchungen über die Cyansäure und deren Salze." *Ann. Phys.* 73 (1823): 157-72.

———. "Zerlegung der Cyansäure." *Ann. Phys.* [2] 1 (1824): 117-24.

———. "Recherches analytiques sur l'acide cyanique." *Ann. chim. phys.* [2] 27 (1824): 196-99.

———. "Ueber die Zusammensetzung der Cyansäure." *Ann. Phys.* [2] 5 (1825): 385-88.

Wöhler, F., and Liebig, J. "Untersuchung über das Radikal der Benzoësäure." *Ann.* 3 (1832): 249-87.

Wollaston, W. H. "On Super-acid and Sub-acid Salts." *Phil. Trans. Roy. Soc.* 98 (1808): 96-102; *Nich. J.* 21 (1808): 164-69.

———. "On the Elementary Particles of Certain Crystals." *Phil. Trans. Roy. Soc.* 103 (1813): 51-63.

———. "A Synoptic Scale of Chemical Equivalents." *Phil. Trans. Roy. Soc.* 104 (1814): 1-22.

———. "On the Finite Extent of the Atmosphere." *Phil. Trans. Roy. Soc.* 112 (1822): 89-98.

Wright, C. A. "On the Relations between the Atomic Hypothesis and the Condensed Symbolic Expressions of Chemical Facts." *Phil. Mag.* [4] 43 (1872): 241-64.

———. "Reply . . ." *Phil. Mag.* [4] 43 (1872): 503-14.

Wrightson, F. "On the Atomic Weight and Constitution of the Alcohols." *Phil. Mag.* [4] 6 (1853): 88-89.

———. "Remarks on Professor Williamson's Othyle Theory." *Phil. Mag.* [4] 6 (1853): 418-20.

Wurtz, C. A. "Théorie des combinaisons glycériques." *Ann. chim. phys.* [3] 43 (1855): 492-96.

———. "Sur une nouvelle classe de radicaux organiques." *Ann. chim. phys.* [3] 44 (1855): 275-313.

———. "Sur le glycol ou alcool diatomique." *Comptes rendus* 43 (1856): 199-204.

———. "Mémoire sur les glycols ou alcools diatomiques." *Ann. chim. phys.* [3] 55 (1859): 400-478.

———. "Sur l'atomicité des éléments." *Bull. Soc. Chim.* [2] 2 (1864): 247-53.

———. *Leçons de philosophie chimique*. Paris, 1864. (Identical to *Leçons de chimie professees en 1863*. Paris, 1864.)

———. *Dictionnaire de chimie pure et appliquée*. 3 vols. in 5. Paris, 1869-78.

———. "Sur la loi des volumes de Gay-Lussac." *Comptes rendus* 84 (1877): 1,183-89.

———. "Sur la notation atomique." *Comptes rendus* 84 (1877): 1,264-68, 1,349-52.

Wurzer, F. "Auszug eines Briefes von Hofrath Wurzer, Prof. der Chemie zu Marburg." *Ann. Phys.* 56 (1817): 331-34.

Young, T. *Introduction to Medical Literature*. London, 1813.

———. *Miscellaneous Works of the Late Thomas Young*. Edited by G. Peacock. 2 vols. London, 1855.

SECONDARY BOOKS AND ARTICLES

Allgemeine deutsche Biographie. 55 vols. Leipzig, 1871-1910.

"The Congress of Chemists at Carlsruhe." *Chem. News* 2 (1860): 226-27.

———. "Congrès des chimistes à Carlsruhe." *Moniteur scientifique* 2 (1860): 984-86.

Anschütz, Richard. *Life and Chemical Work of A. S. Couper*. Edinburgh, 1909.

———. *August Kekulé*. Vol. 1: *Leben und Wirken*. Berlin: Verlag Chemie, 1929.

Armstrong, Henry E. "The Doctrine of Atomic Valency." *Nature* 125 (1930): 807-10.

Bachelard, Gaston, *Les intuitions atomistiques*, 2d ed. Paris: Vrin, 1975.

Baker, A. Albert, J. *Unsaturation in Organic Chemistry*. Boston: Houghton-Mifflin, 1968.

Benfey, O. Theodor. "Prout's Hypothesis." *J. Chem. Educ.* 29 (1952): 78-81.

———, ed. *Classics in the Theory of Chemical Composition*. New York: Dover, 1963.

———. *From Vital Force to Structural Formulas*. Boston: Houghton-Mifflin, 1964.

Bernatowicz, A. J. "Dalton's Rule of Simplicity," *J. Chem. Educ.* 47 (1970): 577-79.

Bonner, J. K. "Amedeo Avogadro: A Reassessment of His Research and its Place in Early Nineteenth Century Science." Ph.D. diss., Johns Hopkins University, 1974.

Brescia, F. "Equivalents—A Winner or a Dead Horse." *J. Chem. Educ.* 53 (1976): 362-65.

Brittan, Gordon G. *Kant's Theory of Science*. Princeton: Princeton University Press, 1978.

Brock, William H., ed. *The Atomic Debates: Brodie and the Rejection of the Atomic Theory*. Leicester: Leicester University Press, 1967.

_____. "Dalton versus Prout: The Problem of Prout's Hypotheses," in Cardwell, ed., *John Dalton*, pp. 240-58.

_____. "Lockyer and the Chemists: The First Dissociation Hypothesis." *Ambix* 16 (1969):81-99.

_____, and Knight, David M. "The Atomic Debates: 'Memorable and Interesting Evenings in the Life of the Chemical Society'." *Isis* 56 (1965):5-25.

Broda, Engelbert. *Ludwig Boltzmann: Mensch, Physiker, Philosoph*. Vienna: Deutike, 1955. English translation, *Ludwig Boltzmann*. Woodbridge, Conn.: Ox Bow Press, 1983.

Brooke, John H. "Wöhler's Urea, and its Vital Force?—A Verdict From the Chemists." *Ambix* 15 (1968): 84-114.

_____. "Organic Synthesis and the Unification of Chemistry—A Reappraisal." *Brit. J. Hist. Sci.* 5 (1971):363-92.

_____. "Chlorine Substitution and the Future of Organic Chemistry: Methodological Issues in the Laurent-Berzelius Correspondence (1843-1844)." *Stud. Hist. Phil. Sci.* 4 (1973):47-94.

_____. "Laurent, Gerhardt, and the Philosophy of Chemistry." *Hist. Stud. Phys. Sci.* 6 (1975):405-29.

_____. "Davy's Chemical Outlook: The Acid Test." In Forgan, ed., *Humphry Davy*.

_____. "Avogadro's Hypothesis and its Fate: A Case Study in the Failure of Case-Studies." *Hist. Sci.* 19 (1981): 235-73.

Brush, Stephen G. *The Kind of Motion We Call Heat: A History of the Kinetic Theory of Gases in the Nineteenth Century*. 2 vols. Amsterdam: North-Holland, 1976.

_____. *The Temperature of History: Phases of Science and Culture in the Nineteenth Century*. New York: Franklin, 1977.

Buchdahl, Gerd. "Sources of Scepticism in Atomic Theory." *Brit. J. Phil. Sci.* 10 (1959): 120-34.

Bykov, G. V. "The Origin of the Theory of Chemical Structure." *J. Chem. Educ.* 39 (1962): 220-24.

Campbell, Norman. *Physics, the Elements*. Cambridge: Cambridge University Press, 1920.

_____. *What is Science?* London: Methuen, 1921.

Caneva, Kenneth. "From Galvanism to Electrodynamics: The Transformation of German Physics and Its Social Context." *Hist. Stud. Phys. Sci.* 9 (1978):63-159.

Cannizzaro, Stanislao. "Notizie storiche e considerazioni sull'applicazione della teoria atomica alla chimica . . ." *Gazzetta chimica italiana* 1 (1871):1-33, 213-30, 293-314, 389-97, 567-86, 629-83. German trans. in *Sammlung chemischer und chemisch-technischer Vorträge*. Edited by W. Herz. Vol. 20. Stuttgart: Enke, 1914.

Cardwell, D. S. L., ed. *John Dalton and the Progress of Science*. Manchester: Manchester University Press, 1968.

Causey, R. L. "Avogadro's Hypothesis and the Duhemian Pitfall." *J. Chem. Educ.* 48 (1971):365-67.

Chemnitius, Fritz. *Die Chemie in Jena von Rolfinck bis Knorr*. Jena: Fromm, 1929.

Cohen, I. Bernard. *Franklin and Newton*. Philadelphia: American Philosophical Society, 1956.

———. *The Newtonian Revolution: With Illustrations of the Transformation of Scientific Ideas*. Cambridge: Cambridge University Press, 1980.

Cole, Theron M., Jr. "Early Atomic Speculations of Marc Antoine Gaudin: Avogadro's Hypothesis and the Periodic System." *Isis* 66 (1975):334-60.

———. "Dalton, Mixed Gases, and the Origin of the Chemical Atomic Theory." *Ambix* 25 (1978):117-30.

Colmant, P. "Querelle à l'Institut entre équivalentistes et atomistes." *Rev. quest. sci.* 33 (1972):493-519.

Crosland, Maurice P. "The Origins of Gay-Lussac's Law of Combining Volumes of Gases." *Ann. Sci.* 17 (1961):1-26.

———. *The Society of Arcueil: A View of French Science at the Time of Napoleon I*. Cambridge: Harvard University Press, 1967.

———. "The First Reception of Dalton's Atomic Theory in France." In Cardwell, ed., *John Dalton*.

———. "Lavoisier's Theory of Acidity." *Isis* 64 (1973):306-25.

———, ed. *The Emergence of Science in Western Europe*. New York: Science History, 1976.

———. *Gay-Lussac: Scientist and Bourgeois*. Cambridge: Cambridge University Press, 1978.

Daub, Edward. "Waterston's Influence on Krönig's Kinetic Theory of Gases." *Isis* 62 (1971):512-15.

Daumas, Maurice. *Lavoisier: Théoricien et expérimentateur*. Paris: Presses Universitaires de France, 1955.

Davy, John. *Memoirs of the Life of Sir Humphry Davy*. 2 vols. London, 1836.

Debus, Heinrich. *Ueber einige Fundamentalsätze der Chemie, insbesondere das Dalton-Avogadro'sche Gesetz*. Kassel, 1894.

———. "Die Genesis von Daltons Atomtheorie." *Z. phys. Chem.* 20 (1896):359-76; 24 (1897):325-52; 29 (1899):266-94.

deMilt, Clara. "Carl Weltzien and the Congress at Karlsruhe." *Chymia* 1 (1948):153-69.

———. "Auguste Laurent—Guide and Inspiration of Gerhardt." *J. Chem. Educ.* 28 (1951):198-204.

_____. "The Congress at Karlsruhe." *J. Chem. Educ.* 28 (1951): 421–24.

_____. "Auguste Laurent, Founder of Modern Organic Chemistry." *Chymia* 4 (1953):85–114.

Dictionary of National Biography. 63 vols. London, 1885–1900.

Dictionary of Scientific Biography. 16 vols. New York: Charles Scribner's Sons, 1970–80.

Divers, Edward. "The Atomic Theory Without Hypotheses." *Rep. Brit. Assoc. Adv. Sci.* 72 (1902):557–75.

_____. "A. W. Williamson." *Proc. Roy. Soc.* 78A (1907):xxiv–xliv.

Döbling, H. *Die Chemie in Jena zur Goethezeit*. Jena: Fischer, 1928.

Duhem, Pierre. *La théorie physique: Son objet, sa structure*, 2d ed. Paris, 1914. English translation by P. P. Wiener, *The Aim and Structure of Physical Theory*. Princeton: Princeton University Press, 1954.

Dumas, Jean-Baptiste. "Éloge historique de Henri-Victor Regnault." *Mém. Acad. Sci.* 42 (1883):xlviii.

Engler, Karl. *Festgabe zum Jubiläum der vierzigjährigen Regierung seiner Königlichen Hoheit des Grossherzogs Friedrich von Baden*. Karlsruhe, 1892.

Eriksson, G. "Berzelius och atomteorin: den idehistorisken bakgrunden." *Lychnos* 1965: 1–37.

Farber, Eduard. "Variants of Preformation Theory in the History of Chemistry." *Isis* 54 (1963):443–60.

Farrar, W. V. "Nineteenth Century Speculations on the Complexity of the Chemical Elements." *Brit. J. Hist. Sci.* 2 (1965):297–323.

_____. "Dalton and Structural Chemistry." In Cardwell, ed., *John Dalton*.

_____, Farrar, K. R., and Scott, E. L. "The Henrys of Manchester, Part 3: William Henry and John Dalton." *Ambix* 21 (1974):208–28.

Fisher, Nicholas W. "Organic Classification Before Kekulé." *Ambix* 20 (1973):106–31, 209–33.

_____. "Kekulé and Organic Classification." *Ambix* 21 (1974):29–52.

_____. "Avogadro, the Chemists, and Historians of Chemistry." *Hist. Sci.* 20 (1982):77–102, 212–31.

Fitzgerel, R. K., and Verhoek, F. H. "The Law of Dulong and Petit." *J. Chem. Educ.* 37 (1960):545–50.

Fleming, Robin S. "Newton, Gases, and Daltonian Chemistry: The Foundations of Combination in Definite Proportions." *Ann. Sci.* 31 (1974): 561–74.

_____. "John Dalton's Development of a Quantified Chemistry." Ph.D. diss. Cambridge University, 1981 (not seen).

Forgan, Sophie, ed. *Science and the Sons of Genius*: *Studies on Humphry Davy*. London: Science Reviews, 1980.

Foster, George Carey. "A. W. Williamson." *J. Chem. Soc.* 87 (1905): 605–18.

Fox, Robert. "The Background to the Discovery of Dulong and Petit's Law." *Brit. J. Hist. Sci.* 4 (1968):1–22.

_____. *The Caloric Theory of Gases*: *From Lavoisier to Regnault*. Oxford: Clarendon Press, 1971.

———. "The Rise and Fall of Laplacian Physics." *Hist. Stud. Phys. Sci.* 4 (1974):89–136.

Freund, Ida. *The Study of Chemical Composition: An Account of its Method and Historical Development.* Cambridge: Cambridge University Press, 1904. Reprinted New York: Dover, 1968.

Frické, Martin. "The Rejection of Avogadro's Hypothesis." In Colin Howson, ed., *Method and Appraisal in the Physical Sciences.* Cambridge: Cambridge University Press, 1976.

Fullmer, June Z. "Davy's Sketches of His Contemporaries." *Chymia* 12 (1967):127–50.

———. *Sir Humphry Davy's Published Works.* Cambridge: Harvard University Press, 1969.

———. "Davy's Priority in the Iodine Dispute: Further Documentary Evidence." *Ambix* 22 (1975):39–51.

———. "Humphry Davy, Reformer." In Forgan, ed., *Humphry Davy*.

Garber, Elizabeth W. "Molecular Science in Late-Nineteenth-Century Britain." *Hist. Stud. Phys. Sci.* 9 (1978):265–97.

Gardner, Michael R. "Realism and Instrumentalism in Nineteenth-Century Atomism," *Phil. Sci.* 46 (1979):1–34.

Goodman, David C. "Wollaston and the Atomic Theory of Dalton." *Hist. Stud. Phys. Sci.* 1 (1969):37–59.

Gould, R. F., ed. *Kekulé Centennial* (*Advances in Chemistry* series, no. 61). Washington, D.C.: American Chemical Society, 1966.

Gower, Barry. "Speculation in Physics: The History and Practice of Naturphilosophie." *Stud. Hist. Phil. Sci.* 3 (1973):301–56.

Graebe, Carl. "Marcellin Berthelot." *Ber.* 41 (1908):4,805–72.

———. "Der Entwicklungsgang der Avogadroschen Theorie." *J. prakt. Chem.* 87 (1913):145–208.

———. *Geschichte der organischen Chemie.* Berlin: Springer, 1920.

Greenaway, Frank. "The Biographical Approach to John Dalton." *Manch. Mem.* 100 (1958–59):1–98.

———. *John Dalton and the Atom.* Ithaca: Cornell University Press, 1966.

Gregory, J. C. *A Short History of Atomism.* London: Black, 1931.

Grimaux, Édouard, and Gerhardt, Charles, Jr. *Charles Gerhardt: sa vie, son oeuvre, sa correspondance.* Paris, 1900.

Guerlac, Henry. "Quantification in Chemistry." *Isis* 52 (1961): 194–214.

———. "Some Daltonian Doubts." *Isis* 52 (1961): 544–54.

———. "The Background to Dalton's Atomic Theory." In Cardwell, ed., *John Dalton.*

Guralnick, Stanley M. "The Contexts of Faraday's Electrochemical Laws." *Isis* 70 (1979):59–75.

Hall, A. Rupert. "Precursors of Dalton." In Cardwell, ed., *John Dalton.*

———, and Hall, Marie Boas. "Newton and the Theory of Matter." *Texas Quarterly* 10 (1967):54–68.

[Hall], Marie Boas. "Structure of Matter and Chemical Theory in the Seventeenth and Eighteenth Centuries." In Marshall Claggett, ed., *Critical Problems in the History of Science*. Madison: University of Wisconsin Press, 1959.

_____. *Robert Boyle on Natural Philosophy*. Bloomington: Indiana University Press, 1965.

_____. "The History of the Concept of Element," in Cardwell, ed., *John Dalton*.

Harré, Rom. *Theories and Things*. London: Sheed and Ward, 1961.

Harris, J., and Brock, W. H. "From Giessen to Gower Street: Towards a Biography of Alexander Williamson." *Ann. Sci.* 31 (1974):95–130.

Hartkopf, Werner. "Schellings Naturphilosophie." *Philosophia Naturalis* 17 (1979):349–72.

Hartley, Harold. *Humphry Davy*. London: Nelson, 1966.

_____. *Studies in the History of Chemistry*. Oxford: Clarendon Press, 1971.

Hawthorne, R. M., Jr. "Avogadro's Number: Early Values by Loschmidt and Others." *J. Chem. Educ.* 47 (1970):751–55.

Henry, William C. *Memoirs of the Life and Scientific Researches of John Dalton*. London, 1854.

Hermann, Armin. "Dynamismus und Atomismus—Die beiden Systeme der Physik in der 1. Hälfte des 19. Jahrhunderts." *Erkenntnis* 10 (1976):311–22.

Hesse, M. B. *Models and Analogies in Science*. London: Sheed & Ward, 1963.

Hiebert, Erwin. "The Experimental Basis of Kekulé's Valence Theory," *J. Chem. Educ.* 36 (1959):320–27.

_____. *The Conception of Thermodynamics in the Scientific Thought of Mach and Planck*. Freiburg/Br.: Ernst-Mach-Institut, 1968.

_____. "The Genesis of Mach's Early Views on Atomism." In R. S. Cohen and R. J. Seeger, eds., *Ernst Mach: Physicist and Philosopher*. Dordrecht, Holland: Reidel, 1970.

_____. "The Energetics Controversy and the New Thermodynamics." In Duane H. D. Roller, ed., *Perspectives in the History of Science and Technology*. Norman: University of Oklahoma Press, 1971. With a commentary by David B. Wilson.

_____. "Boltzmann's Conception of Theory Construction." In Jaako Hintikka, David Gruender, and Evandro Agazzi, eds., *Pisa Conference on the History and Philosophy of Science*. Vol. 2. Dordrecht, Holland: Reidel, 1981.

Hinde, P. T. "William Hyde Wollaston: The Man and His Equivalents." *J. Chem. Educ.* 43 (1966):673–76.

Hjelt, Edvard. *Geschichte der organischen Chemie von ältester Zeit bis zur Gegenwart*. Brunswick: Vieweg, 1916.

Hofmann, August Wilhelm. "Zur Erinnerung an Jean Baptiste André Dumas." *Ber.* 17R (1884):630–760.

Holmberg, Arne. *Bibliografi över J. J. Berzelius*. Stockholm: Almqvist & Wiksell, 1933.

Holmes, Frederick L. "From Elective Affinities to Chemical Equilibria: Berthollet's Law of Mass Action," *Chymia* 8 (1962):105–45.

Holton, Gerald. "Einstein, Michelson, and the 'Crucial' Experiment." *Isis* 60 (1969):133-97.

―――. "Mach, Einstein, and the Search for Reality." In R. S. Cohen and R. J. Seeger, eds. *Ernst Mach: Physicist and Philosopher*. Dordrecht, Holland: Reidel, 1970.

―――. *Thematic Origins of Scientific Thought*. Cambridge, Mass.: Harvard University Press, 1973.

Hufbauer, Karl. "Social Support for Chemistry in Germany during the Eighteenth Century: How and Why Did It Change?" *Hist. Stud. Phys. Sci.* 3 (1971):205-31.

―――. *The Formation of the German Chemical Community (1720-1795)*. Berkeley: University of California Press, 1982.

Ihde, Aaron J. "The Karlsruhe Congress: A Centennial Retrospect." *J. Chem. Educ.* 38 (1961):83-86.

―――. *Development of Modern Chemistry*. New York: Harper & Row, 1964.

Jacques, J. "La thèse de doctorat d'Auguste Laurent." *Bull. Soc. Chim.* [5] 1 (1954):D31-39.

Japp, Francis R. "Kekulé Memorial Lecture." *J. Chem. Soc.* 73 (1898): 97-138.

Jorpes, J. E. *Jac. Berzelius: His Life and Work*. Stockholm: Almqvist & Wiksell, 1966.

Kapoor, S. C. "Berthollet, Proust, and Proportions." *Chymia* 10 (1965): 53-110.

―――. "Dumas and Organic Classification." *Ambix* 16 (1969):1-65.

―――. "The Origins of Laurent's Organic Classification." *Isis* 60 (1969):477-527.

Kauffman, George B. "Werner, Kekulé, and the Demise of the Doctrine of Constant Valency." *J. Chem. Educ.* 49 (1972):813-17.

Kelham, B. B. "Atomic Speculation in the Late Eighteenth Century." In Cardwell, ed., *John Dalton*.

Kerker, Milton. "Brownian Movement and Molecular Reality Prior to 1900." *J. Chem. Educ.* 51 (1974):764-68.

―――. "The Svedberg and Molecular Reality." *Isis* 67 (1976): 190-216.

Klosterman, Leo J. "Studies in the Life and Work of Jean Baptiste André Dumas (1800-1884): The Period up to 1850." Ph.D. diss. University of Kent at Canterbury, 1976.

Knight, David M. "The Atomic Theory and the Elements." *Stud. Rom.* 5 (1966):185-207.

―――. *Atoms and Elements: A Study of Theories of Matter in England in the Nineteenth Century*. London: Hutchinson, 1967.

―――. "Steps Towards a Dynamical Chemistry." *Ambix* 14 (1967): 179-97.

―――. "German Science in the Romantic Period." In Crosland, ed., *Emergence of Science*.

―――. *The Transcendental Part of Chemistry*. Folkestone, Kent: Dawson, 1978.

Knorr, Karin D., Krohn, Roger, and Whitely, Richard, eds. *The Social Process of Scientific Investigation*. Dordrecht, Holland: Reidel, 1981.

Kopp, Hermann. *Geschichte der Chemie*. Brunswick, 1843-47.

———. *Die Entwickelung der Chemie in der neueren Zeit*. Munich, 1873.

Kragh, Helge. "Julius Thomsen and Nineteenth-Century Speculations on the Complexity of Atoms." *Ann. Sci.* 39 (1982):37-60.

Krasovitskaia, T. I., and Plotkin, S. IA. "Vollaston i atomnaia teoriia Daltona." *Vop. ist. est. tekh.* 45 (1973):41-44.

Ladenburg, Albert. *Vorträge über die Entwicklungsgeschichte der Chemie in den letzten hundert Jahren*. Brunswick, 1869. 2d ed., 1887. Translated by L. Dobbin, *Lectures on the History of the Development of Chemistry Since the Time of Lavoisier*. Edinburgh: Alembic Club, 1900.

Larder, David F. "A Dialectical Consideration of Butlerov's Theory of Chemical Structure." *Ambix* 18 (1971):26-48.

Larmor, Joseph, ed. *Memoir and Scientific Correspondence of the Late Sir George Gabriel Stokes*. Cambridge: Cambridge Univerity Press, 1907.

Laudan, Larry. "Towards a Reassessment of Comte's 'Méthode Positive'." *Phil. Sci.* 38 (1971):35-53.

———. *Progress and Its Problems: Towards a Theory of Scientific Growth*. Berkeley: University of California Press, 1977.

LeGrand, Homer E. "Lavoisier's Oxygen Theory of Acidity." *Ann. Sci.* 29 (1972):1-18.

———. "Determination of the Composition of the Fixed Alkalis, 1789-1810." *Isis* 65 (1974):59-65.

———. "Berthollet's *Essai de statique chimique* and Acidity." *Isis* 67 (1976):229-38.

Leicester, Henry M. *The Historical Background of Chemistry*. New York: Wiley, 1956.

———. "Contributions of Butlerov to the Development of Structural Theory. *J. Chem. Educ.* 36 (1959):328-29.

Lemay, P., and Oesper, R. E. "Pierre Louis Dulong, His Life and Work." *Chymia* 1 (1948):171-90.

Levere, Trevor H. "Affinity or Structure: An Early Problem in Organic Chemistry." *Ambix* 17 (1970):111-26.

———. *Affinity and Matter: Elements of Chemical Philosophy 1800-1865*. Oxford: Clarendon Press, 1971.

———. "Coleridge, Chemistry, and the Philosophy of Nature." *Stud. Rom.* 16 (1977):349-79.

Löw, Reinhard. *Pflanzenchemie zwischen Lavoisier und Liebig*. Munich: Donau-Verlag, 1977.

———. "The Progress of Organic Chemistry During the Period of German Romantic Naturphilosophie (1795-1825)." *Ambix* 27 (1980):1-10.

Lundgren, Anders. *Berzelius och den kemiska atomteorin*. Uppsala: Almqvist & Wiksell, 1979.

McClelland, C. E. *State, Society, and University in Germany, 1700-1914*. Cambridge: Cambridge University Press, 1980.

McCormmach, Russell. *Night Thoughts of a Classical Physicist*. Cambridge: Harvard University Press, 1982.

Malley, Marjorie. "The Discovery of Atomic Transmutation: Scientific Styles and Philosophies in France and Britain." *Isis* 70 (1979): 213-23.

Mandelbaum, Maurice. *Philosophy, Science, and Sense Perception*: *Historical and Critical Studies*. Baltimore: Johns Hopkins University Press, 1964.

Mauskopf, Seymour H. "Thomson Before Dalton: Thomas Thomson's Consideration of the Issue of Combining Weight Proportions Prior to his Acceptance of Dalton's Chemical Atomic Theory." *Ann. Sci.* 25 (1969):229-42.

───. "The Atomic Structural Theories of Ampère and Gaudin: Molecular Speculation and Avogadro's Hypothesis." *Isis* 60 (1969): 61-74.

───. "Haüy's Model of Chemical Equivalence: Daltonian Doubts Exhumed." *Ambix* 17 (1970):182-91.

───. "Minerals, Molecules and Species." *Arch. int. hist. sci.* 23 (1970):185-206.

───. *Crystals and Compounds*: *Molecular Structure and Composition in Nineteenth Century French Science*. Philadelphia: Transactions of the American Philosophical Society, n.s., vol. 66, no. 3, 1976.

Meldrum, Andrew N. "The Development of the Atomic Theory." *Manch. Mem.* 54:7 (1910):1-16; 55:3 (1910):1-12; 55:4 (1910):1-15; 55:5 (1911):1-22; 55:6 (1911):1-18; 55:19 (1911):1-10; 55:22 (1911):1-11.

Melhado, Evan M. "Jac. Berzelius: Foundations and Development of His Chemistry." Ph.D. diss. Princeton University, 1977.

───. "Mitscherlich's Discovery of Isomorphism." *Hist. Stud. Phys. Sci.* 11 (1980):87-123.

───. *Jacob Berzelius*: *The Emergence of His Chemical System*. Stockholm and Madison: Almqvist & Wiksell; University of Wisconsin Press, 1981.

Mellor, D. H. "Models and Analogies in Science: Duhem *versus* Campbell." *Isis* 59 (1968):282-90.

Mellor, D. P. *The Evolution of the Atomic Theory*. Amsterdam: Elsevier, 1971.

van Melsen, A. G. *From Atomos to Atom*: *The History of the Concept Atom*. Pittsburgh: Duquesne University Press, 1952. Original publication in Dutch, Amsterdam, 1949.

Merz, J. T. *A History of European Thought in the Nineteenth Century*. 4 vols. Edinburgh and London, 1896-1914.

von Meyer, Ernst. *Geschichte der Chemie von ältesten Zeiten bis zur Gegenwart*. Leipzig, 1889. Translated by G. M'Gowan. *A History of Chemistry From Earliest Times to the Present Day*. London, 1891.

───. "Die Karlsruhe Chemiker-Versammlung im Jahre 1860." *J. prakt. Chem.* 191 (1911):182-89.

Moore, F. J. *A History of Chemistry*. New York: McGraw-Hill, 1918.

Morrison, G. S. "Cannizzaro's Atom-Free Stoichiometry." *J. Chem. Educ.* 53 (1976):723.

Morselli, Mario A. "The Manuscript of Avogadro's 'Essai d'une manière de

déterminer les masses relatives des molecules élémentaires'." *Ambix* 27 (1980):147-72.
Mundy, B. W. "Avogadro on the Degree of Submolecularity of Molecules." *Chymia* 12 (1967):151-55.
Munk, William. *Roll of the Royal College of Physicians of London*. 4 vols. London, 1878.
Nash, Leonard K. *The Atomic-Molecular Theory*. Cambridge, Mass.: Harvard University Press, 1950.
_____. "The Origin of Dalton's Chemical Atomic Theory." *Isis* 47 (1956):101-16.
Novitski, Mary E. "Auguste Laurent and the Prehistory of Valence." Ph.D. diss. University of California-Berkeley, 1980 (not seen).
Nye, Mary Jo. *Molecular Reality: A Perspective on the Scientific Work of Jean Perrin*. New York: Science History, 1972.
_____. "The Nineteenth-Century Atomic Debates and the Dilemma of an 'Indifferent Hypothesis'." *Stud. Hist. Phil. Sci.* 7 (1976): 245-68.
_____. "Nonconformity and Creativity: A Study of Paul Sabatier, Chemical Theory, and the French Scientific Community." *Isis* 68 (1977):375-91.
_____. "Berthelot's Anti-Atomism: A 'Matter of Taste'?" *Ann. Sci.* 38 (1981):585-90.
_____, ed. *The Question of the Atom: From the Karlsruhe Congress to the Solvay Conference, 1860-1911*. Los Angeles: Tomash, 1983.
Ostwald, Wilhelm. *Lebenslinien: Eine Selbstbiographie*. 2 vols. Leipzig, 1927.
Paneth, F. A. "The Epistemological Status of the Chemical Concept of Element." *Brit. J. Phil. Sci.* 13 (1962):1-14, 144-60.
Partington, J. R. "Jeremias Benjamin Richter and the Law of Reciprocal Proportions." *Ann. Sci.* 7 (1951): 173-98; 9 (1953):289-314.
_____. *A History of Chemistry*. 4 vols. London: Macmillan & Co., 1961-70.
_____, and McKie, D. "Historical Studies on the Phlogiston Theory.—IV. Last Phases of the Theory." *Ann. Sci.* 4 (1939):113-49.
Patterson, Elizabeth C. *John Dalton and the Atomic Theory*. Garden City, NJ: Doubleday, 1970.
Paul, E. R. "Alexander W. Williamson on the Atomic Theory: A Study of Nineteenth Century British Atomism." *Ann. Sci.* 35 (1978):17-31.
Pierson, Stuart. "Gay-Lussac and Berthollet's Theory," *Actes XIIe Congrès Int. Hist. Sci.* 6, 1968, publ. 1971, pp. 83-86.
Poggendorff, J. C. *Biographisch-literarische Handwörterbuch der exakten Naturwissenschaften*. 7 vols. Leipzig, 1863-.
Prandtl, Wilhelm. *Humphry Davy, Jöns Jacob Berzelius: Zwei führende Chemiker aus der ersten Hälfte des 19. Jahrhunderts*. Stuttgart: Wissenschaftliche Verlagsgesellschaft, 1948.
_____. *Deutsche Chemiker in der ersten Hälfte des neunzehnten Jahrhunderts*. Weinheim/Bergstr.: Verlag Chemie, 1956.
O. Bertrand Ramsay, ed. *van't Hoff-LeBel Centennial*. Washington, D.C.: American Chemical Society, 1975.
_____. *Stereochemistry*. London: Heyden, 1981.

Rocke, Alan J. "Origins of the Structural Theory in Organic Chemistry." Ph.D. diss. University of Wisconsin-Madison, 1975.

———. "Atoms and Equivalents: The Early Development of the Chemical Atomic Theory." *Hist. Stud. Phys. Sci.* 9 (1978):225–63.

———. "Gay-Lussac and Dumas: Adherents of the Avogadro-Ampère Hypothesis?" *Isis* 69 (1978):595–600.

———. "Reception of Chemical Atomism in Germany." *Isis* 70 (1979): 519–36.

———. "Kekulé, Butlerov, and the Historiography of the Theory of Chemical Structure." *Brit. J. Hist. Sci.* 14 (1981):27–57.

———. "Subatomic Speculations and the Origin of Structure Theory." *Ambix* 30 (1983):1–18.

Roscoe, Henry E., and Harden, Arthur. *A New View of the Origin of Dalton's Atomic Theory*. London, 1896.

Russell, Colin A. "Berzelius and the Development of the Atomic Theory." In Cardwell, ed., *John Dalton*.

———. *The History of Valency*. Leicester: Leicester University Press, 1971.

———, and Goodman, D. C. *Atoms and Electricity*. Bletchley: Open University Press, 1973.

Sadoun-Goupil, Michelle. *Le chimiste Claude-Louis Berthollet (1748–1822): Sa vie, son oeuvre*. Paris: Vrin, 1977.

Schimank, Hans. "Ludwig Wilhelm Gilbert und die Anfänge der 'Annalen der Physik'." *Sudhoffs Archiv* 47 (1963):360–72.

Schofield, Robert E. *Mechanism and Materialism: British Natural Philosophy in an Age of Reason*. Princeton: Princeton University Press, 1970.

Schonland, Basil. *The Atomists*. Oxford: Clarendon Press, 1968.

Schufle, J. A., and Thomas, George. "Equivalent Weights from Bergman's Data on Phlogiston Content of Metals." *Isis* 62 (1971):499–506.

Schütt, Hans-Werner. "Zum Prioritätsproblem der Entdeckung des chemischen Isomorphismus." *Physis* 16 (1974):5–22.

Scott, E. L. "Dalton and William Henry." In Cardwell, ed., *John Dalton*.

Scott, Wilson. *The Conflict between Atomism and Conservation Theory, 1644–1860*. New York: Science History, 1970.

———. "Monism vs. Dualism in Chemical Atomism," *Proc. Twelfth Int. Cong. Hist. Sci.* 6 (1968, publ. 1971):99–100.

Shinn, Terry. "The French Science Faculty System, 1808–1914: Institutional Change and Research Potential in Mathematics and the Physical Sciences." *Hist. Stud. Phys. Sci.* 10 (1979):271–332.

———. *Savoir scientifique et pouvoir social: École polytechnique et les polytechniciens: 1794–1914*. Paris: Presses de la Fondation Nationale des Sciences Politiques, 1980.

———. "Orthodoxy and Innovation in Science." *Minerva* 18 (1980): 539–55.

Siegfried, Robert. "The Chemical Basis for Prout's Hypothesis." *J. Chem. Educ.* 33 (1956):263–66.

———. "Humphry Davy and the Elementary Nature of Chlorine." *J. Chem. Educ.* 36 (1959):568–70.

_____. "The Chemical Philosophy of Humphry Davy." *Chymia* 5 (1959): 193-201.

_____. "The Discovery of Potassium and Sodium, and the Problem of the Chemical Elements." *Isis* 54 (1963):247-58.

_____. "The Phlogistic Conjectures of Humphry Davy." *Chymia* 9 (1964):117-24.

_____. "Sir Humphry Davy on the Nature of the Diamond." *Isis* 57 (1966):325-35.

_____. "Boscovich and Davy: Some Cautionary Remarks." *Isis* 58 (1967):236-38.

_____. "Lavoisier's Table of Simple Substances: Its Origin and Interpretation." *Ambix* 29 (1982):29-48.

_____, and Dobbs, Betty Jo. "Composition, A Neglected Aspect of the Chemical Revolution." *Ann. Sci.* 24 (1968):275-93.

Smeaton, William A. "Bergman's 'Equivalents': A Correction." *Isis* 64 (1973):231-32 (includes a "Reply" by J. A. Schufle).

_____. "Berthollet's *Essai de statique chimique* and its Translations: A Bibliographic Note and a Daltonian Doubt." *Ambix* 24 (1977):149-58.

_____. "Berthollet's *Essai de statique chimique*: A Supplementary Note." *Ambix* 25 (1978):211-12.

Smyth, A. L. *John Dalton, 1766-1844: A Bibliography of Works By and About Him*. Manchester: Manchester University Press, 1966.

Snelders, H. A. M. "Romanticism and Naturphilosophie and the Inorganic Natural Sciences: An Introductory Survey." *Stud. Rom.* 9 (1970): 193-215.

_____. "The Influence of the Dualistic System of Jakob Joseph Winterl (1732-1809) on the German Romantic Era." *Isis* 61 (1970): 231-40.

_____. "J. S. C. Schweigger: His Romanticism and His Crystal Electrical Theory of Matter." *Isis* 62 (1971):328-38.

_____. "Atomismus und Dynamismus im Zeitalter der deutschen romantischen Naturphilosophie." In Richard Brinkmann, ed., *Romantik in Deutschland: Ein interdisziplinäres Symposion*. Stuttgart: Metzler, 1978. (Not seen)

Söderbaum, H. G. *Berzelius' Werden und Wachsen*. Leipzig, 1899.

_____. *Jac. Berzelius: Levnadsteckning*. 3 vols. Uppsala: Almqvist & Wiksell, 1929-31.

Stauffer, Robert. "Speculation and Experiment in the Background of Oersted's Discovery of Electromagnetism." *Isis* 48 (1957):33-50.

Stock, Alfred. *Der internationale Chemiker-Congress . . .vor und hinter den Kulissen*. Berlin: Verlag Chemie, 1933.

Thackray, Arnold W. "The Origin of Dalton's Chemical Atomic Theory: Daltonian Doubts Resolved." *Isis* 57 (1966):35-55.

_____. "The Emergence of Dalton's Chemical Atomic Theory: 1801-08." *Brit. J. Hist. Sci.* 3 (1966):1-23.

_____. "Documents Relating to the Origins of Dalton's Chemical Atomic Theory." *Manchester Mem.* 108 (1966):21-42.

_____. "Quantified Chemistry—The Newtonian Dream." In Cardwell, ed., *John Dalton*.

———. *Atoms and Powers: An Essay on Newtonian Matter-Theory and the Development of Chemistry*. Cambridge, Mass.: Harvard University Press, 1970.

———. *John Dalton: Critical Assessments of His Life and Science*. Cambridge, Mass.: Harvard University Press, 1972.

Thiele, J. "Franz Walds Kritik der theoretischen Chemie." *Ann. Sci.* 30 (1973):417-33.

[Thomson, R. D.] "Biographical Notice of the Late Thomas Thomson." *Glasgow Med. J.* [3] 5 (1857):69-80, 121-53, 379-80.

Thomson, Thomas. *A History of Chemistry*. 2 vols. London, 1830-31.

———. "Biographical Account of the late John Dalton." *Proc. Roy. Phil. Soc. Glasgow* 2 (1845):79-88.

———. "Biographical Account of Dr. Wollaston." *Proc. Roy. Phil. Soc. Glasgow* 3 (1850):135-44.

Turner, R. Steven. "The Growth of Professorial Research in Prussia, 1818-1848—Causes and Context." *Hist. Stud. Phys. Sci.* 3 (1971): 137-82.

———. "University Reformers and Professorial Research in Germany, 1760-1806." In Lawrence Stone, ed., *The University in Society*, 2. Princeton: Princeton University Press, 1974.

———. "Justus Liebig versus Prussian Chemistry: Reflections on Early Institution-Building in Germany," *Hist. Stud. Phys. Sci.* 13 (1982): 129-62.

van Spronsen, J. W. "The History and Prehistory of the Law of Dulong and Petit as Applied to the Determination of Atomic Weights." *Chymia* 12 (1967):157-69.

Volhard, Jakob. *Justus von Liebig*. 2 vols. Leipzig: Barth, 1909.

Walden, Paul. "The Gmelin Chemical Dynasty." *J. Chem. Educ.* 31 (1954):534-41.

Whewell, William. *History of the Inductive Sciences*. 3 vols. London, 1837. 3d ed., London, 1857.

———. *Philosophy of the Inductive Sciences*. 2 vols. London, 1840. 2d ed., London, 1847.

Williams, L. Pearce. "The Physical Sciences in the First Half of the Nineteenth Century: Problems and Sources." *Hist. Sci.* 1 (1962): 1-15.

———. *Michael Faraday*. London: Chapman & Hall, 1965.

Wilson, George. *Religio Chemici*. London, 1862.

Wurtz, Charles Adolphe. "Discours préliminaire: Histoire des doctrines chimiques depuis Lavoisier." In C. A. Wurtz, ed., *Dictionnaire de chimie pure et appliquée* 1 (Paris, 1868): i-lxxxvi. Republished as:

———. *Histoire des doctrines chimiques depuis Lavoisier jusqu'à nos jours*. Paris, 1869. H. Watts, trans., *History of Chemical Theory from the Age of Lavoisier to the Present Day*. London, 1869.

———. *La théorie atomique*. Paris, 1879. E. Cleminshaw, trans., *The Atomic Theory*. London, 1880.

Yoshida, Akira. "Charles Adolphe Wurtz et la théorie atomique." *Jap. Stud. Hist. Sci.* 16 (1977):129-35.

Index

Note: Authors of secondary studies are indexed only from textual references, and not from literature citations. Primary books (but not journals) are also indexed from textual references.

Académie des Sciences, Paris, 105, 107, 108, 112, 162, 196, 205, 227, 323
Acetic acid (and derivatives), 114, 116, 121n, 135-36, 169, 172, 197-98, 203-4, 208, 210n, 218-21, 227-33, 236, 238, 241-43, 254-55
Acid-base neutralization (see also Salts), 8-9, 41, 230
Acids, acid anhydrides, and acidity (see also under specific acids), 192-93, 201, 207, 220-21, 225-27, 229-30, 243-44, 257-61; inorganic, 79-81, 225-26; organic, 195-98, 201, 207-8, 226-32, 234, 241, 254-55, 257, 272
Affinity, chemical, 2, 7, 9, 18n, 67, 170, 302
Affinity units (see also Atomicity of atoms and radicals), 265, 267, 284
Agnosticism, structural, 202-4, 224-25, 252, 262, 264, 269, 272-73
Aim and Structure of Physical Theory (Duhem), 326
Alcohol(s), 111, 116, 174-75, 194, 215-21, 224-29, 234, 236, 242-43, 254
Alkalies and alkaline earths, 38, 56, 58, 67-68, 76-77, 79-82, 89n, 157, 192, 206, 221, 227, 297
Allotropy, 113, 168
Amines, 223-25, 230-32, 234, 258
Ampère, André Marie, 103-7, 110, 114-18, 153, 207-8, 214n, 287, 295-96, 307, 309n, 324
Analysis, chemical, 4-6, 11-12, 16n, 38, 68-71, 153, 192, 326
Antiatomism, xi, 181-82, 313-35; Berthelot's, 180, 322-26, 333; Berthollet's,

40-42, 99-100, 110, 113, 118n; Brodie's, 313-21; Davy's, 56-57, 86-87; Duhem's, 325-26; Dumas', 87, 117-18, 181, 322-26; Gay-Lussac's, 87, 110; Helm's, 327-29; Hunt's, 319; Mach's, 327-29, 330, 333-35; Odling's 315-16, 319; Ostwald's, 328-33, Poincaré's, 325-26; Stallo's, 329; Wald's, 330-32; Watt's, 320; Wollaston's, 64-65, 86-87; Wright's, 319-21
Antimony and its compounds, 163, 206, 239
Arago, François, 40, 107, 108, 140
Arcueil, Société d', 40, 99, 107, 109, 113
Aristotle, 4, 5, 326
Armstrong, Henry E., 276
Arrhenius, Svante, 327
Årsberättelse (Berzelius). See *Jahresberichte*
Arsenic and its compounds, 116, 154-56, 160, 162-63, 166-67, 206, 232, 239-41
Athénée de Paris, 105
Atkinson, R. W., 320-21
Atmosphere, theories concerning the, 22-25, 65
Atomic heat capacities. See Dulong and Petit's law
Atomicity (i.e., valence) of atoms and radicals (see also Valence), 241-44, 251-56, 263-67, 273-75, 284n, 300-303, 310n
Atomicity of molecules, 310n
Atomic theory, chemical, xi-xiv, 8-15, 51, 57-59, 64, 79-87, 87n, 99, 103, 107-9, 117-18, 128, 156-57, 177-81, 205-9, 296-97, 304-7, 316-21;

Atomic theory (*continued*)
Ampère's, 103-4, 110, 114; Avogadro's, 101-3; Baudrimont's, 106, 322; Berthelot's, 324; Berthollet's, 99-110, 102; Berzelius', 62-63, 66-84, 86, 107-10, 114, 128-33, 137-39, 143, 153-82, 191, 199; Brodie's, 321, 336n; Chevreul's, 114, 169, 322; Dalton's, xi-xiii, 21-42, 49-52, 56-57, 73-75, 77-78, 80-86, 99-100, 102-4, 112-13, 128-33, 137-43, 156-58, 192, 321; Davy's, 55-61, 73, 75, 80-83, 85-86, 100, 107, 110, 129-31, 135, 129-40, 157, 169; defined, 12-15; distinguished from physical atomism, xi-xii, 10, 13-15, 63, 86, 97n, 112, 132, 316, 323; Döbereiner's, 134-39, 142-45, 169; Dulong's, 107-9, 112-13; Dumas', 63, 112, 114-18, 122-23n; Frankland's 232-33, 238-41, 314-16, 318-19; Gaudin's, 105-6, 113, 117, 139, 143; Gay-Lussac's, 109-15, 117, 139, 143; Gerhardt's and Laurent's, 205-9, 257-61, 321; Gilbert's, 128-29, 134, 137, 145; Gmelin's, 63, 177-81; historiography of origins of, 27-33; Kastner's, 133-34, 137, 143-45, 169; Kekulé's, 262-70; Kolbe's, 232-38; Liebig's, 171-73, 176-81, 189n; Mach's, 333; Meinecke's, 137-45, 169; Mitscherlich's, 154-56; Odling's, 251-52, 315-16, 319; origins of, 21-33, 61; Ostwald's, 331, 333; Petit's, 114; Prout's, 52-53, 80-83, 107, 139, 157; Schweigger's, 130-33, 134, 137-39, 141, 143-45; Thenard's, 113-14; Thomson's, 49-55, 73, 80-82, 84-85, 100, 130, 139-41, 183n; Williamson's, 215-24, 253, 316-21, 336n; Wollaston's, 55-56, 61-66, 73, 80-86, 110, 132, 135, 139, 181; Wurtz's, 256-57, 321-25; Young's, 60-61, 75
Atomic theory, physical, xi-xiv, 10, 50-51, 67, 84-87, 108, 112, 117, 128, 131, 316, 331-35; Ampère's, 103-4; Avogadro's, 101-3; Berthollet's, 40-42, 99-100; Berzelius', 70, 77-78, 84; Boltzmann's, 329, 332-35; Bošković's, 65, 89n, 128, 134; Dalton's, xi, 22-25, 28-29, 33-34, 36-40, 49, 56-57, 63, 74, 77-78, 84-86, 100, 131-32, 142; Davy's, 89n, 115, 169; distinguished from chemical atomism, xi-xii, 10, 13-15, 63, 86, 97n, 112, 132, 316, 323; Döbereiner's, 136-37; Dumas', 117; Gay-Lussac's, 112, 169; Kekulé's, 316; Maxwell's, 289-91, 321; Planck's, 329, 331-34; Schweigger's, 130-32; Thomson's, 49-52, 84-85; Williamson's, 318; Wollaston's, 64-65, 84
Atomic weights, chemical, 11-15, 18nn, 79-87, 107, 117-18, 153-54, 235, 275, 296-97, 304-7, 313-26; Avogadro's, 102-3; Berzelius', 62-63, 73-77, 80-82, 84-86, 94-95n, 95n, 108-9, 111, 142, 163-67, 172, 176, 209, 222, 287, 295, 297-99, 304-5, 309n, 313; Cannizzaro's, 295-99, 313; Dalton's, 25-37, 50, 75, 80-82, 84-86, 87n, 93n; Davy's, 58-60, 75-76, 80-82, 85-86, 89n; defined, 12-15; Döbereiner's, 135-36, 142, 148-49n, 176, Dulong and Petit's, 108-9; Dumas', 63, 115-18, 121n, 188n, 191, 213n; Frankland's, 238-41, 279-81; Gaudin's, 105; Gay-Lussac's, 111-15, 121n, 171-72, 176-77, 188n, 191; Gerhardt's and Laurent's, 205-9, 222, 229, 275, 295, 297-99, 304-5, 309n, 313; Gmelin's, 63, 142, 177-81, 209, 288, 295, 299, 304; Hofmann's, 298-99; Kekulé's, 262, 275, 278-79, 283n, 298-99; Kolbe's, 234-35, 238, 256, 278-79, 298-99; Liebig's, 142, 171-72, 176-81, 189n, 209, 299; Meinecke's, 139-41; Odling's, 251-52, 275, 298; Prout's, 53, 80-82, 107; Schweigger's, 176; Thomson's, 52-55, 75, 80-82, 81n, 84-85; Williamson's, 221-22, 229, 238, 275, 298, 304; Wollaston's, 62-64, 75, 80-82, 84-86, 142, 148n, 177-81, 288; Wurtz's, 256, 275, 298-99, 322; Young's, 60-61, 75
Atoms, Cannizzaro's law of, 296-97
Attempt to Establish the First Principles of Chemistry by Experiment (Thomson), 55
Ausführliches Lehrbuch der organischen Chemie (Kolbe), 234, 248n
Austin, William, 26
Avogadro, Amedeo, 101-7, 114-18, 118n, 119n, 150n, 153, 208, 214n, 287, 290, 295-96, 298, 307, 309n, 324
Avogadro-Ampère hypotheses (*see also* Equal volumes-equal numbers hypothesis), 101-7, 114-15, 153; Baudrimont and, 106; Berthelot and, 324; Cannizzaro and, 106, 295-96, 306-7; Dumas and, 105-6, 115-18, 122n; Gaudin and,

105-6; Gay-Lussac and, 114-15, 117; Gerhardt and Laurent and, 206-8, 214n; Kekulé and, 294; kinetic theory and, 290-91, 294-95

Baeyer, Adolf von, 249n, 273, 292
Bases and basicity (*see also under specific bases*), 192-93, 207-8, 220-23, 225-26, 229-30, 243-44, 251, 257-61, 310n; organic, 223-25, 227, 240
Basicity law of Gerhardt and Laurent, 257-58
Baudrimont, Alexandre Édouard, 106, 203, 211n, 322
Becquerel, Antoine César (father of Henri), 113, 123n
Becquerel, Henri, 331
Beddoes, Thomas, 56
Beilstein, K. Friedrich, 271, 285n
Bentley, Richard, 15n
Benzoyl radical theory, 174-75, 195
Bérard, Jacques Étienne, 40, 169
Bergman, Torbern, 7, 66, 68
Berthelot, P. E. Marcellin, 180, 256, 260, 285n, 293, 299, 310n, 322-26, 333
Berthollet, Amédée (son of Claude Louis), 40
Berthollet, Claude Louis, 6-9, 24, 40-41, 47n, 49, 62, 69, 71, 99-100, 102-3, 107, 110, 112-13, 118n, 121n, 176, 180, 287, 331
Berzelius, J. Jacob, 4, 15, 52, 54-58, 60-63, 66-86, 88n, 89n, 90-91n, 92n, 93n, 94n, 95n, 106-11, 114, 121n, 122n, 123n, 128-30, 132-33, 135, 137-39, 142-45, 146n, 147n, 150n, 153-82, 183n, 184n, 185n, 186n, 187n, 188n, 191, 193, 195-99, 206, 209-10, 210n, 216, 218, 222, 229-36, 244, 246n, 247n, 255, 277, 279, 281, 283n, 287-88, 295, 297-98, 304-5, 309n, 313
Binary axiom, 66, 76, 83-84; defined, 83
Biot, Jean Baptiste, 40, 99, 114, 140, 168, 322
Bischof, Carl Gustav, 141, 170
Boltzmann, Ludwig, 290, 328-39, 332-33, 335
Bonaparte, Napoleon, 101, 118, 129
Boole, George, 314
Bošković, Rudjer Josip, 65, 128, 134
Bostock, John, 35-36, 39, 60, 90n, 100
Boullay, Polydore, 174, 216
Boyle, Robert, 3, 4
Brande, William, 85-86, 97n
Bridgman, Percy Williams, 314

Brock, William H., xii, 140, 246n
Brodie, Benjamin Collins, Jr., 233, 271, 292, 313-21, 336n, 337n
Bromine and its compounds, 194, 198
Brongniart, Alexandre, 191
Brooke, John H., 119n, 173, 187n, 210n, 296
Brown, Alexander Crum. *See* Crum Brown, Alexander
Brush, Stephen G., 288, 332, 334
Buchdahl, Gerd, xii
Bucholz, Christian Friedrich, 126
Buffon, George Louis Leclerc, Comte de, 4, 7
Bunsen, Robert W. E., 232-33, 277
Butlerov, Aleksandr Mikhailovich, 272, 274
Buys-Ballot, Christoph Hendrik Diederik, 289

Cacodyl radical, 232-33, 241
Cahours, Auguste A. T., 202, 213n, 322
Calculus, Brodie's chemical, 313-21
Caloric, 3, 10, 127, 287-88; surrounding atoms, 22, 56-57
Campbell, Norman, 14
Caneva, Kenneth, 146n
Cannizzaro, Stanislao, 19n, 106, 280, 293-99, 307, 309n, 313
Carbohydrates (*see also* Cellulose; Sugar; Starch; Gum arabic), 168, 170, 201
Carbon, atomic weight of, 111, 117, 206
Carbonic acid gas, and carbonic oxide, 24, 26-27, 36, 80-83, 110-11, 114, 138, 205, 207, 241-44
Carbonyl radical, 241-43, 278-79
Carbonyl compounds, *See* Ketones and aldehydes
Carnot, Sadi, 288-89
Catalysis, 216, 218, 237
Cavendish, Henry, 4, 18n, 91n
Cellulose, 121n, 169
Chancel, Gustave, 224-29, 246n, 322
Chemical atomism. *See* Atomic theory, chemical
Chemische Messkunst (Meinecke), 137-38, 150n, 151n
Chemistry and physics, separation between, 287-88, 295, 297-98, 304, 318, 324
Chenevix, Richard, 26-27, 127-29
Chevreul, Michel Eugène, 114, 169, 211n, 322
Chiozza, Ludwig, 225, 253, 337n

Chlorinated organic compounds, 191, 193-201, 204, 207, 229-31
Chlorine and its compounds, chlorination (exymuriatic acid), 38, 62, 79-83, 95n, 157, 159, 163-64, 166, 183n, 191-201, 207, 221, 229-32, 234, 242, 251, 255
Choron, F., 106
Clark, Thomas, 211n
Classification schemes for chemical compounds, 203-4, 224, 258, 268, 285n, 324-26
Claus, Adolf C. L., 284n, 310n
Clausius, Rudolf, 288-91, 294-95, 308n, 309n
Cole, Theron M., Jr., 29, 43n, 45-46n, 121n
Collège de France, 105
Combining volumes of gases, Gay-Lussac's law of, 40-42, 52-53, 72, 74-75, 87n, 88n, 100-4, 109-12, 114, 137-38, 140-41
Comte, Auguste, 181, 188n, 222, 336
Constancy of basic oxygen, law of, 68-70, 157
Conventionalism, 319, 324-26
Copulas or conjugate radicals, theory of: Berzelius' and Kolbe's, 230-34, 236-37, 239-41, 244, 280, 283n; Gerhardt's, 230, 257-61, 264, 283n
Couper, Archibald Scott, 272-73, 310n
Cours de chimie (Gay-Lussac), 113
Cours de philosophie positive (Comte), 181
Cours élémentaire de chimie (Regnault), 322
Crosland, Maurice P., 41
Crown and Anchor Inn, 55
Crum Brown, Alexander, 272, 299, 314, 321
Crystallography, 103-4, 130-32, 154-56
Curie, Marie and Pierre, 323, 331
Cyano compounds (*including* fulminates), 111-14, 162, 171-73, 192-93, 206, 232-33, 262-63

Dallas, D. M., 336n, 337n
Dalton, John, xi-xiii, 5, 10, 12-15, 21-42, 43n, 44n, 45n, 46n, 47n, 49-53, 55-61, 63-66, 69-70, 72-86, 87n, 89n, 91n, 93n, 95n, 96n, 97n, 99-104, 109, 112-13, 129-34, 138-43, 146n, 147n, 150n, 153, 156-58, 164, 192, 301, 316, 320-21, 331-32

Darstellung der Verhältnisszahlen (Döbereiner), 135
Daubeny, Charles, 65, 92n
Davy, Sir Humphry, 4, 14-15, 16n, 29, 38, 40, 44-45n, 53, 55-63, 66-69, 73-77, 79-83, 85-87, 89n, 90n, 93n, 96n, 107, 110, 113, 127, 129-31, 134-35, 139-40, 156-57, 163, 168-69, 183n, 192-93, 287, 319
Davy, John (brother of Humphry), 57
Debus, Heinrich, 29, 43n, 46n
Definite proportions, law of, 6-8, 51, 87n, 99, 167, 324, 331
Delavaud, Charles, 337n
Descartes, René, xiv, 128, 133
Deville, Henri Étienne Sainte Claire, 225-26, 322
Dictionary of Chemistry (Ure), 85
Döbereiner, Johann Wolfgang, 126, 132-39, 142-43, 148n, 149n, 150n, 151n, 169, 176
Dolomieu, Dieudonné, 6
Donovan, Michael, 19n, 106, 309n
Duhem, Pierre, 14, 325-26
Dulong, Pierre Louis, 99, 107-9, 112-15, 123n, 139, 154-55, 158, 160, 162, 165-66, 184n, 193, 196, 295, 309n
Dulong and Petit's law of atomic heats, 107-9, 112-13, 118, 123n, 139, 154-55, 162, 165-67, 184n, 208, 288, 295, 297-98, 304
Dumas, Jean Baptiste André, 63-64, 87, 105-6, 108, 112, 114-18, 121n, 122-23n, 174-75, 182, 188n, 191, 194-202, 210n, 212n, 213n, 215-17, 223, 228, 231, 240, 247n, 249n, 261-62, 277, 281, 283n, 291, 296, 322
Dynamicism, 56, 99, 119n, 127-28, 132-34, 136, 138, 144-45, 146n, 151n, 215, 287

École Polytechnique, 107
Einleitung in die neuere Chemie (Kastner), 133
Einstein, Albert, xiii-iv, 21, 31, 331-35
Ekeberg, A. G., 93n
Electricity, electrolysis, and electrochemistry, 3, 67-68, 127, 136, 151n, 181, 192, 327
Electrochemical dualism, theory of, 67-68, 107, 130-32, 161, 170-71, 175, 186n, 192-98, 201-2, 229-34, 241, 244, 255

Elemental substances, xi–xii, 4–6, 9–12, 16n, 18n, 38–39, 52, 56, 83, 157, 192
Elements of Chemical Philosophy (Davy), 53, 59–61, 73
Empircism, xiii, 3–7, 15, 64, 84–86, 119n, 143, 177, 181–82, 202
Empiriocriticism, 327. *see* Positivism
Energetics, 327–32, 334, 340n
Energy, 288–90, 327–32
Equal volumes–equal numbers hypothesis (EVEN) (*see also* Avogadro-Ampère hypotheses): Ampère and, 103–5; Avogadro and, 101–3; Baudrimont and, 106; Berzelius and, 74–75, 77–78, 111, 122n, 158, 161–64, 297; Cannizzaro and, 295–96, 306–7; Dalton and, 23–24, 33–34, 36–38, 40–42, 46n, 46–47n, 51, 74, 78, 100–101; Davy and, 59–60, 74; defined, 23–24; Dumas and, 105, 112, 115–18; Gaudin and, 105–6; Gay-Lussac and, 40–42, 100, 111–12, 114–15, 117; Gerhardt, Laurent, and, 205–9, 295; Kekulé and, 294; kinetic theory and, 290–91, 294–95, 306–7, 321; Prout and, 53, 101; Thomson and, 50–52, 101
Equivalent proportions, law of, 8–12, 35, 47n, 51, 57–58, 68–70, 324, 331
Equivalents, chemical, 9–15, 18nn, 19n, 28, 57–58, 62–66, 84–87, 153, 178–79, 191, 198, 222, 239–40, 258–59, 275, 293–95, 298–303, 305–6, 316–25; defined, 10–12, 18n; Gerhardt's and Laurent's reform of, 205–9, 275
Equivalents, conventional, 13, 64, 142, 177–81, 188n, 208–9, 222, 235, 241, 260, 262, 280, 287, 295, 298–99, 304, 306, 309–10n, 322–24
Equivalents, hydride, 300–303, 306; defined, 300
Erlenmeyer, R. A. C. Emil, 11, 272, 278, 299–303, 306
Erdmann, O. L., 301
Erläuterungen (Meinecke), 141
Essai de statique chimique (Berthollet), 8, 28, 47n
Essai sur la théorie des proportions chimique (Berzelius), 161–62, 164, 166, 184n
Esters, 174–75, 207, 216–21, 226–27, 254–55, 257
Ethane and other saturated hydrocarbons, 224–30, 233, 256–57, 311n
Ether(s), 111, 174–75, 206, 215–21, 224–29, 234, 236, 238, 254

Etherification, theories of, 123n, 174–75, 215–21, 226, 229, 236–38, 256
Etherin theory, 174–75, 194–95, 197, 201, 216
Ethylene (olefiant gas) and derivatives, and other unsaturated hydrocarbons, 27–28, 34, 36–37, 54, 116, 193–94, 216, 303
Ethyl radical and ethyl radical theory, 175, 195, 206, 216–24, 227, 229, 236, 238
Eudiometry, 40–42, 53, 74–75, 109–15, 137–38
EVEN. *See* Equal volumes–equal numbers hypothesis
Even-number rule, Laurent's, 206–8, 213–14n, 233, 235, 248n
Ewart, Peter, 37

Faraday, Michael, 127, 181, 193–94, 287
Fichte, Johann Gottlieb, 125, 136
Fischer, Ernst Gottfried, 8
Forces of nature, 1–3, 9–10, 126–27, 133, 136, 287–88
Formulas: empirical, Berzelius', 174; graphic and glyptic, 314–16; rational, Berzelius', 174; rational, Gerhardt and, 224, 262; rational, Kekulé's, 269–70; synoptic, Gerhardt's, 224, 262
Formulas, molecular, assignment of, 10, 12, 14–15, 35, 79–84, 107, 112, 153–56, 275; by Ampère, 101–3; by Avogadro, 101–3; by Berzelius, 73–81, 84, 109, 111, 153, 163–67, 172, 206; by Cannizzaro, 299; by Dalton, 26–27, 32–37, 39, 50, 77–78, 80–81, 84, 153; by Davy, 59–60, 80–81; by Frankland, 232–33, 238–41; by Gaudin, 105–6; by Gay-Lussac, 110–12, 171, 177; by Kolbe, 232–38; by Laurent and Gerhardt, 205–9, 225, 233; by Liebig, 171–72, 177; by Thomson, 80–81, 84; by Williamson, 171–22; by Wollaston, 63, 66, 80–81, 84, 153
Foster, George Carey, 271, 292, 300–301, 316
Fourcroy, Antoine François de, 99
Fox, Robert, 2, 119n, 122n
Frankland, Edward, 180, 229, 232–33, 236, 238–41, 244, 249n, 256, 262–63, 274–81, 283n, 286n, 305–6, 314–16, 318
Freind, John, 4
Frémy, Edmond, 228, 322
Fresnel, Augustin, 127

Friedel, Charles, 322
Fuchs, Johann Nepomuk von, 126, 132
Fulminates. *See* Cyano compounds

Gases, kinetic theory of. *See* Kinetic-molecular theory of gases
Gaudin, Marc Antoine, 105-6, 113, 121n, 122-23n, 203, 310n 322
Gaultier de Claubry, H. F., 69
Gay-Lussac, Joseph Louis, 9, 37, 40-42, 52-53, 68, 72, 74, 76-77, 87, 87n, 88n, 99, 100-104, 107, 109-15, 117, 120-21n, 123n, 137, 139-41, 143, 146n, 150n, 158, 162, 168-73, 176-77, 186n, 188n, 189n, 191, 193-94, 232, 287
Gay-Lussac's law of combining volumes. *See* Combining volumes of gases, Gay-Lussac's law of
Gehlen, Adolf Ferdinand, 93n, 126, 132
Gerhardt, Charles F., xivn, 106, 202-10, 213n, 214n, 215, 222-31, 233, 235, 242, 244, 246n, 251-66, 269-70, 273, 275, 277, 283n, 284n, 291, 293, 295, 297-98, 304-5, 309n, 310n, 313, 321-22, 337n
Gesellschaft Deutscher Naturforscher und Aerzte, 178, 328-29
Gibbs, Josiah Willard, 328, 330
Gilbert, Davies, 55
Gilbert, Ludwig Wilhelm, 71, 126, 127-30, 131, 133, 137, 139, 145, 146n, 147n
Glycerine, 256-57
Glycol(s), 256-57, 321
Gmelin, Christian Gottlob, 142, 164, 165, 184n
Gmelin, Leopold, 63, 142, 161-62, 172, 177-78, 180-81, 188n, 209, 211n, 215, 222, 288, 295, 299, 304
Goethe, Johann Wolfgang von, 125, 134-36
Goodman, D. C., 91n
Göttling, Johann Friedrich August, 126
Gough, John, 24
Graebe, Carl, 106
Graham, Thomas, 79, 97n, 196, 215, 290
Gravimetry, 5-6, 10-12, 16n, 23, 39, 53, 99, 110, 114, 116, 143, 153
Greenaway, Frank, 47n
Griess, J. Peter, 300
Griffin, John Joseph, 211n
Grignard, F. A. Victor, 323
Grimaux, Charles, 337
Grundriss der allgemeinen Chemie (Döbereiner), 135

Grundriss der organischen Chemie (Limpricht), 260, 263
Guerlac, Henry, 28, 40
Gum arabic, 121n, 168-70
Guralnick, Stanley, 181
Guyton de Morveau, Louis Bernard, 4, 6, 7, 99

Halogens (*see also* Bromine; Chlorine; Iodine), 163, 194-95, 198, 206, 208, 221, 231, 239
Handbuch der organischen Chemie (Beilstein), 271, 285n
Handbuch der theoretischen Chemie (Gmelin), 142, 177, 180
Harden, Arthur, 26-29, 32, 46n
Harré, Rom, 13-14
Hartley, Harold, 90n
Haüy, René Just, 6, 17-18n, 103-5, 154-55
Hegel, G. W. F., 125, 136
Helm, Georg, 327-29, 332
Helmholtz, Hermann von, 288
Henry, William, 24, 34, 85
Henry's law, 24-25
Herapath, John, 288
Hermann, Armin, 146n
Hermbstaedt, Sigismund, 126
Hesse, Mary B., 14
Hiebert, Erwin, 332, 335
Higgins, William, 7, 43n, 60, 142
Hildebrandt, Georg Friedrich, 134, 137
Hirst, Thomas Archer, 248n
Hisinger, Wilhelm, 67
Histoire des doctrines chimiques (Wurtz), 279
History of the Inductive Sciences (Whewell), 181
Hoff, Jacobus Henricus van't, 327-28
Hoffmann, Reinhold, 284n
Hofmann, August Wilhelm von, 11, 210n, 223, 225, 228, 233, 240, 262, 276, 284n, 298-99, 303, 306, 310n
Holton, Gerald, xiii, 21
Hufbauer, Karl, 125
Humboldt, F. W. H. Alexander von, 40, 41, 87n, 99
Hume, David, 127
Hunt, Thomas Sterry, 120n, 319
Hydracids and hydracid theory (*see also* specific compounds), 191-97, 210n
Hydrocarbons, saturated. *See* Ethane
Hydrocarbons, unsaturated. *See* Ethylene
Hydrochloric acid (hydrogen chloride,

muriatic acid), 55, 69, 110-11, 162, 188n, 191-94, 207, 251-53, 257
Hydrogen: as atomic weight standard, 26, 53-59, 62; gas, 53-54, 56, 57, 164, 206, 224-25, 231, 257

Idealism, xi, 56, 125-28, 169, 287-88, 325-26
Intermediate vector boson, 105
Iodine and its compounds, 113, 192-94, 198, 216-17, 233, 238
Institut de France. *See* Académie des Sciences
Introduction to Modern Chemistry (Hofmann), 303
Isomerism, 112-13, 151n, 167-74, 328-29
Isomorphism, 113, 118, 154-56, 165-68, 171, 199, 208, 288, 297-98, 304

Jacobson, Paul, 285n
Jahresberichte (Berzerlius), 160-61, 174, 175, 184n
Japp, Francis R., 271
Joule, James Prescott, 288

Kant, Immanuel, 17n, 56, 126-28, 134, 142, 143, 319
Karlsruhe Congress, 292-96
Kastner, Karl Wilhelm Gottlob, 126, 133-34, 137, 143, 150n, 169, 171
Keill, James, 4
Kekulé, F. August, 11, 212, 215, 231, 242, 249n, 253-55, 261-81, 283n, 285n, 286n, 292-95, 298-303, 305-6, 308n, 310n, 316, 319, 336n
Kelvin, Sir William Thomson, Baron, 290
Kepler, Johannes, 89n, 319
Ketones and aldehydes, 220-21, 228, 241-42
Kinetic-molecular theory of gases, 288-91, 294-96, 298, 306-7, 318, 321, 327-29, 332-35
Klaproth, M. Heinrich, 126, 133, 143, 146n, 155
Klein, C. Felix, 328
Knight, David M., xii, xivn, 16n, 89n, 97n, 152n, 304
Kolbe, A. W. Hermann, 180, 229, 232-38, 240-44, 247n, 248n, 249n, 250n, 253-54, 256, 260, 262-63, 270-71, 274, 276-81, 283n, 286n, 293, 299, 305, 308n, 310n, 319, 336n
Kopp, Hermann, xivn, 86, 116
Krasovitskaia, T. I., 91n

Krönig, August Karl, 291, 308n, 309n
Kuhn, Thomas S., 288

Ladenburg, Albert, 207, 209, 272
Lagrange, Joseph Louis, 118n
Lamétherie, Jean Claude de, 103
Langevin, Paul, 323
Laplace, Pierre Simon, Marquis de, 2, 119n, 122n
Lärbok i kemien (Berzelius). *See Lehrbuch der Chemie*
Laudan, Larry, 304
Laurent, Auguste, xivn, 19n, 195, 197, 200-210, 212n, 213n, 214n, 215, 222, 225-26, 229, 231, 233, 235, 240, 244, 247n, 256, 258, 260-61, 273, 275, 277, 281, 284n, 291, 293, 295, 297-99, 304-5, 309-10n, 313, 321
Lavoisier, Antoine Laurent, 4-7, 9, 26-27, 38-39, 56, 67-68, 99, 156, 157, 175, 192-93, 199, 210, 287
LeBel, Joseph Achille, 328
Le Chatelier, Henri Louis, 322
Leçons de philosophie chimique (Dumas), 181
Lecture Notes for Chemical Students (Frankland), 314
Lehrbuch der organischen Chemie (Kekulé), 266, 268-73, 277-79, 285n, 294, 301, 310n
Lehrbuch der organischen Chemie (Kolbe). *See Ausführliches* . . .
Lehrbuch der Chemie (Berzelius), 160, 163-64, 176-77, 184n, 287
Leibniz, Gottfried Wilhelm, 132
LeRoyer, A., 121
Liebig, Justus von, 79, 112, 113, 123n, 126, 133, 142-43, 171-82, 188n, 194-97, 201, 204, 215-18, 222-23, 225, 228, 234, 241, 244, 245n, 246n, 248n, 250n, 260-61, 266, 277, 278, 299, 313
Limpricht, Heinrich, 260-61, 263-64
Link, Heinrich Friedrich, 154
Loschmidt, J. Joseph, 289

McCormmach, Russell, 340n
Mach, Ernst, 327-28, 330, 333-35
Magnus, H. Gustav, 142, 176, 178, 181
Malaguti, Faustino, 197, 322
Mandelbaum, Maurice, xiii
Marcet, Alexander, 60, 65, 66, 156, 165-66, 170, 184n
Marignac, Jean Charles, 337n

Marsh gas type (*see also* Methane), 262-63, 283n, 305
Materialien zur Erweiterung der Naturkunde (Kastner), 133
Materialism, xi, 3-4, 39, 56, 127-28, 144, 170, 287-88, 319, 326, 328
Mauskopf, Seymour H., 6, 17n, 43n, 87n, 106
Maxwell, James Clerk, 289-91, 308n, 321
Mayer, J. Robert, 288
Mechanism, xi, 1-5, 56, 157, 168, 170, 264, 287-88, 306, 326-28
Meinecke, Johann Ludwig Georg, 132-33, 137-43, 150n, 151n, 169, 186
Meldrum, Andrew N., 27-29, 45-46n
Melhado, Evan, 93n, 94n, 154, 182n, 183n
Melsens, Louis, 231
Mendeleev, Dmitrii Ivanovich, 307
Mendius, Otto, 261
Metals, 76-77, 80-83, 89n, 96n, 108-9, 135, 166-67, 185nn, 192, 206, 209, 210n, 233, 239-40; metal oxide formulas, 66, 76-77, 80-81, 83, 89n, 96n, 108-9, 114, 116, 155, 166-67, 202, 206, 297
Metamers (*see also* Isomerism): defined, 174
Meteorological Observations and Essays (Dalton), 22
Methane (marsh gas), 27-28, 34, 36-37, 224, 262-63
Méthode de chimie (Laurent), 258
Meyer, J. Lothar, 270, 298, 307, 308n, 313
Meyer, Victor, 328-29
Michelson, Albert A., 21, 31
Mineralogy, 154-56
Mitscherlich, Eilhard, 142, 154-56, 165-66, 176, 199, 260
Mixed gases: Dalton's first theory of, 22-25, 37, 46n, 49; Dalton's second theory of, 27-28, 33-34, 37-38, 42, 46n
Mixtures and solutions, 7-8, 22-25, 28, 180, 319
Models, scientific, 13-15
Moderne Theorien der Chemie (Meyer), 313
Molecules, chemical and physical (*see also* Atomic theory, chemical; Atomic theory, physical): Avogadro and Ampère and, 101-7; Cannizzaro and, 294-96; Dumas and, 115-18; Gay-Lussac and, 109-15; Kekulé and, 293-96; Laurent

and, 207-8; Williamson and, 221-22, 237-38, 251
Mollweide, Karl, 150n
Monism, xi, 56
Morley, Edward W., 21, 31
Müller, Hugo, 284
Multiple proportions, law of, 35-36, 41-42, 50, 139, 320, 324, 326, 331; Berzelius and, 69-73; Dalton and, 21, 27-32, 39, 44n, 87n; Davy and, 58; Gay-Lussac and, 110; Wollaston, Thomson and, 35, 41, 51, 61, 64, 69, 73, 87n, 100, 110, 113
Muriatic acid. *See* Hydrochloric acid
Murray, John, 24, 93n, 211n

Naphthalene, 200
Naquet, Alfred, 301-3, 306, 319, 322, 324
Nash, Leonard, 28-29, 31, 37, 45-46n
Naturphilosophie, 126-28, 133-34, 136-37, 144-45, 235
Nees von Esenbeck, Christian Gottfried, 170
Nernst, H. Walther, 328
Neueste stöchiometrische Untersuchungen (Döbereiner), 135
New System of Chemical Philosophy (Dalton), 35-36, 41, 63, 65, 69, 72, 100-101, 130-31, 142, 157
Newton, Isaac, xiv, 1-4, 7, 13, 15n, 16n, 39-40, 49, 56, 89n, 127-28, 145, 156, 287, 323, 327, 335
Newtonians and Newtonianism, 1-4, 9, 22-23, 127-28, 287, 327, 335; "popular," 2-3, 39, 49, 56, 156
Nitrogen and its compounds: ammonia, 26, 32-33, 39, 45n, 53, 58, 68-69, 111, 205, 207, 223, 227, 230, 240, 257, 275; ammonium radical and compounds, 68-69, 71-72; gas, 54, 77, 83, 95n, 116, 157-59, 162-63, 166, 206, 225; oxides and acids, 11-12, 28-33, 36, 39, 44n, 44-45n, 45n, 58, 72-75, 79-83, 94nn, 95n, 96nn, 110-11, 157-59, 162-63, 166, 221, 223, 226, 239, 243
Nonreductionism, 3-4, 9, 102, 144, 157, 287
Notation of atoms or of molecular formulas: Berzelius', 74, 167, 177-80; Dalton's, 49; Kekulé's, 272-73, 278-79, 306; Kolbe's, 238-39, 278-79; Liebig's, 177-80; Williamson's, 309n
Nucleus theory, Laurent's, 195, 200-202, 204, 235, 244; Dumas and, 200-202;

Gerhardt and, 204; Kolbe and, 235, 244; Liebig and, 201
Nye, Mary Jo, xii, 324

Odling, William, 215, 251-55, 262-66, 270-71, 274-75, 283n, 284n, 295, 298, 305, 315-16, 318-19
Oersted, Hans Christian, 127, 136
Oettingen, Arthur von, 328
Olefiant gas. *See* Ethylene
Olefins. *See* Ethylene
Opticks (Newton), 1-3
Organic chemistry (*see also specific organic compounds*), 168-75, 191-286 passim
Organometallic compounds, 232-33, 239-40, 244, 258
Ostwald, Wilhelm, 327-33
Othyl (othyle) theory, Williamson's, 220-2l, 226, 236-38
Oxalic acid, 135, 193, 195, 207, 229-32, 236
Oxide rule, Berzelius', 71-73, 78, 157-63, 165, 167, 183n; defined, 71
Oxygen: as basis for atomic weights, 52, 62, 74, 90-91n, 107; oxides, 6-7, 11-12, 53-54, 56, 68-69, 70-84, 110, 155, 157-67, 192-201, 206-7, 216-21, 223, 225, 234, 239-40
Oxymuriatic acid. *See* Chlorine

Paracelsus, Theophrastus von Hohenheim, 4
Parmenides, xi
Partial pressures, law of, 23-25
Partington, J. R., 8, 17n, 44n, 46n, 310n
Pasteur, Louis, 322
Peacock, George, 61
Pebal, Leopold von, 207
Pelletier, Pierre Joseph, 121n
Pelouze, Théophile Jules, 188n, 195-97, 322
Periodic law of the elements, xii, 136, 307
Perrin, Jean Baptiste, 323, 333
Persoz, Jean François, 120n, 182, 187-88n, 322
Petit, Alexis Thérèse, 107-9, 112-14, 123n, 139, 154-55, 162, 166, 184n, 295, 309n
Phlogiston, 3, 10, 127, 140
Phosphorus and its compounds, 80-83, 90n, 91n, 96n, 107, 109, 116, 154-60, 162-63, 166-67, 173, 183n, 206, 239, 252, 254

Physical atomism. *See* Atomic theory, physical
Physicalism, 2-4, 7, 9, 102-3, 110, 119n, 127, 144, 287
Physics and chemistry, separation between, 287-88, 295, 297-98, 304, 318, 324
Planck, Max, 329, 331-34
Planta, Adolf von, 254-55
Plotkin, S. IA., 91n
Poggendorff, Johann Christian, 164, 172, 176, 179
Poincaré, Henri, 325-26
Polyatomic (polybasic) atoms and radicals, theory of. *See* Atomicity of atoms and radicals; Bases and basicity; Valence
Polymerism, 113, 168, 170, 173
Polymorphism, 113, 154, 168
Pontin, M. M., 67-68, 92n
Positivism, 3, 9, 56-57, 70, 86, 110, 128, 181-82, 203-4, 209-10, 269-70, 319, 325-35, 336n, 340n
Prager, Bernard, 285n
Précis de chimie organique (Gerhardt), 202-4
Priestley, Joseph, 4, 29
Principia (Newton), 1-3
Proportions, proportional numbers, doctrine of definite proportions. *See* Atomic theory, chemical; Atomic weights, chemical
Protyle, 53, 56, 60, 139-40
Proust, Joseph Louis, 6-8, 47n, 99, 132
Prout, William, 52-54, 56, 80-83, 88n, 101, 106, 107, 139-41, 150n, 157
Prout's hypotheses, 52-56, 102, 139-41
Pythagoras, 136

Quesneville, Gustave Augustin, 224

Racemic acid, 173
Radicals, 68, 166-67, 174-76, 192, 195-96, 200-202, 206, 216-24, 227, 229-44, 251-61, 265-72; defined, 175, 210n, 212n, 268-69; isolation of, 232-37, 256; merger with types, 240, 268-69
Ramsay, William, 332, 340n
Rankine, W. J. M., 327
Ransome, J. A., 47n
Realism, 13, 56, 264, 288, 306, 327, 333-35

Recherches sur les lois de l'affinité (Berthollet), 8
Reductionism, xi, 2, 5, 14, 53, 56, 102-3, 157, 168, 170, 183n, 287
Regnault, Henri Victor, 199, 228, 297-98, 322
Reiset, Jules, 230
Richter, Jeremias Benjamin, 7, 8-10, 17n, 18n, 28, 35, 47n, 50, 61, 68, 92n, 129, 131-33, 139, 142, 147n, 151n
Ritter, Johann Wilhelm, 136
Romanticism, 125-28, 134, 136-37, 143-45, 187-88
Romé de l'Isle, Jean Baptiste, 6
Röntgen, Wilhelm, 331
Roscoe, Henry E., 26-29, 32, 46n, 237, 248n, 252, 282n, 292, 316
Rose, Gustav, 142, 154, 176
Rose, Heinrich, 142, 176, 178, 181, 297-98
Rothe, Heinrich August, 170
Royal Institution, London, 22, 55, 86, 193, 229, 240, 248n, 275, 287, 331
Royal Society, London, 51, 57, 58, 62, 288, 315, 330
Rule of greatest simplicity, Dalton's, 27, 32, 35-39, 50-52, 76, 100, 164
Rumford, Benjamin Thompson, Count, 127, 287
Rutherford, Ernest, 331

Sabatier, Paul, 323
Salts (*see also specific compounds*), 71-72, 157-63, 192, 196, 220-21, 228-29, 236, 251, 257
Schelling, Friedrich, 126, 134, 136, 144, 152n, 319
Schofield, Robert E., 3, 16n
Scholasticism, 3, 326
Schorlemmer, Carl, 311n
Schubert, G. H., 132
Schützenberger, Paul, 322
Schweigger, Johann Salomo Christoph, 126, 128, 130-33, 134, 136-39, 141, 143, 147n, 148n, 150n, 151n, 176
Science, "private" and "public", xiii, xvn, 21, 27
La science et l'hypothèse (Poincaré), 325
Series, homologous, 202, 208, 225, 234
Shinn, Terry, 323
Siegfried, Robert, 89n
Single-atom axiom, 66, 76, 83-84, 95n, 158-63, 166, 183n, 184n; defined, 83

Smeaton, William A., 47n
Smoluchowski, Marian, 332-33
Société Philomatique, Paris, 40
Soddy, Frederick, 331
Sommerfeld, Arnold, 328
Stahl, Georg Ernst, 4, 5
Stallo, Johann Berhard, 329
Starch, 121n, 168-70
Stas, Jean Servais, 337n
Stenhouse, John, 253-55
Stereochemistry, xii, 14-15, 328
Stoichiometry, xi, 6-15, 27-32, 39, 57-60, 68-78, 99, 128-33, 135, 138-39, 141-45, 153, 157, 180, 319-20, 329-31
Stokes, George Gabriel, 290, 321
Stromeyer, Friedrich, 126
Structure, molecular (atomic arrangements or "constitution"), 105-6, 113, 135, 151n, 156, 165, 168-74, 186n, 199-200, 202-4, 217-21, 223-25, 233-35, 262-70, 273-81, 305-6; Kekulé and, 262-73
Subatomic structure, speculations on, 299-300, 305-6; Berzelius', 71-75, 79, 157-60, 162; Dalton's, 38, 103; Davy's, 56, 60; Erlenmeyer's, 302; Naquet's, 302; Prout's, 53; Thomson's, 52
Submolecularity. *See* Atomicity of atoms and radicals; Avogadro-Ampère hypotheses; Molecules, physical and chemical; Subatomic structure, speculations on
Substitution, theory of, 177, 182, 191-202, 235; Berzelius and, 195-98, 229-34, 244; Dumas and, 191, 193-95, 197-201, 231; Frankland and, 239-41, 249n; Kolbe and, 234-35, 240-44; Laurent and, 197, 200-202, 231-32, 235; Liebig and, 195-96, 201, 250n; Odling and, 251-52; Williamson and, 216, 220-21; 253
Sugar(s), 111, 121n, 168-70
Sulfur and its compounds, 206; oxides and acids of, 26-27, 58, 66, 71, 79-83, 92n, 96n, 111, 116, 165, 167, 221-22, 225, 252-55, 260-61, 267
Swedish Royal Academy of Sciences, 67, 160, 173
Sylvester, Charles, 58, 85
Synthesis, organic, 216-22; Williamson's asymmetric, 217-22, 228, 236, 251, 253, 255-56, 304-5
A System of Chemistry (Thomson), 40-41, 49, 50, 100, 101

Tartaric acid, 173
Thackray, Arnold, 3, 4, 28, 43n, 45-46n
Thenard, Louis Jacques, 41, 76-77, 99, 107, 113-114, 168, 213n, 240
Theoretische Betrachtungen und deren Anwendung (Claus), 310n
Thermodynamics, laws of (*see also* Energy), 288-89, 328
Thompson, Benjamin. *See* Rumford, Count
Thomson, J. J., 331
Thomson, Thomas, 6, 15, 24, 27-29, 34-35, 39-42, 43n, 44n, 47n, 49-57, 61, 63, 72, 73, 75-76, 79-82, 84-86, 87n, 88n, 89n, 90-91n, 93n, 94n, 95n, 96n, 100, 101, 120n, 130, 139-41, 149n, 156, 158, 161, 186n
Thorpe, Thomas Edward, 331
Traité de chimie appliquée aux arts (Dumas), 115
Traité de chimie générale (Pelouze and Frémy), 322
Traité de chimie organique (Gerhardt), 257-58, 265
Traité élémentaire de chimie (Lavoisier), 5
Transdiction, xiii-xiv, 2-3, 15-16n, 40, 288, 290, 306, 325
Triads, Döbereiner's, 135, 139, 151n
Trommsdorff, Johann Bartholomäus, 126, 139, 143
Troost, Louis Joseph, 322
Turner, Edward, 97n
Two-volume and four-volume atomic weights and formulas: Berzelius and, 54-55, 88n, 171-72, 188n, 192, 210n, 213n; Dumas and, 116-17; Gerhardt, Laurent and, 205-9; Thomson and, 52, 54-55; Williamson and, 220-22
Tyndall, John, 290, 321
Types, theory of: chemical, 199, 212n; defined, 223; Dumas and, 197-201, 223, 228, 240, 262; Frankland and, 238-41; Gerhardt and, 224-29, 235, 244, 246n, 257-62, 264-66, 270, 284n, 322; Hofmann and, 223, 225, 228; Kekulé and, 253-55, 262-70, 273-81; Kolbe and, 235-38, 240-41, 244; Laurent and, 200-1, 235, 240; Liebig and, 223, 225, 228, 268; mechanical, 199-201, 212n, 240, 262; merger with radicals, 240, 268-69; multiple and mixed, 221-22, 251, 264-66; newer, 197-201, 223; Odling and, 251-52, 264-66, 270, 275; older, 197-201, 223;

Williamson and, 215-29, 236-38, 244, 246nn, 251-57, 262, 264-66, 270, 274-76, 322; Wurtz and, 233, 256-57, 266, 270-71, 274, 300, 322
"Typical" atoms: defined, 268. *See* Types, theory of: Kekulé and

Ueber die neuern Gegenstände der Chymie (Richter), 68
Urbain, Georges, 323
Ure, Andrew, 61, 85
Urea, 173
Uslar, L. von, 260-61

Valence, 11, 214n, 222-23, 227, 238-44, 251-57, 262-68, 273-81, 284n, 299-306, 310n, 324; Carbon tetravalence, 243, 257, 262-63, 265, 267, 271, 278-81
Vapor densities (*see also* Avogadro-Ampère hypotheses; Equal volumes-equal numbers hypothesis), 33, 36-38, 52-55, 74-75, 77-78, 84, 101-4, 107, 111, 116-18, 138, 140, 164, 207, 216, 287-88, 294
Vogel, F. C., 132
Volhard, Jacob, 278, 286n
Volumes, theory of (*see also* Equal volumes-equal numbers hypothesis; Two-volume and four-volume atomic weights; Vapor densities), 158, 178; defined, 77
Vorlesungen über Gastheorie (Boltzmann), 332

Waals, Johannes Diderik van der, 290, 332
Wald, František, 330-32
Water, formation of and assumed formula for, 11, 26, 33, 35, 40, 42, 45n, 52, 54, 59-60, 77, 80-83, 100, 102, 110, 114, 161, 164, 177, 205-7, 220-22, 227, 257, 260, 263-64
Waterston, John James, 288, 290-91, 307n, 308n
Watts, Henry, 237, 248n, 320-21
Weltzien, Karl, 292-93, 308n
Whewell, William, 14, 61, 86, 181, 304
Williams, D. C., xiii
Williamson, Alexander William, 11, 215-29, 236-38, 244, 245n, 246n, 248n, 249n, 251-57, 261-66, 270-71, 274-76, 280, 282n, 284n, 292-93, 295, 298, 300, 303-5, 308n, 313, 315-22, 336n

Winterl, Jacob Joseph, 127–28, 133, 136
Wislicenus, Johannes, 328–29
Wöhler, Friedrich, 112, 113, 142, 163–64, 171–81, 184n, 188n, 195, 197, 230, 247n, 277
Wolff, Friedrich Benjamin, 130, 147n
Wollaston, William Hyde, 12–15, 18n, 35, 38, 41, 51–52, 55, 58, 61–66, 69, 73, 75, 79–87, 90n, 91n, 92n, 93n, 96n, 100, 107, 110, 113, 114–15, 121n, 132, 135, 139, 142, 148n, 150n, 153, 156, 161, 168, 171, 176–77, 180–81, 287–88, 304, 310n
Wright, Charles A., 319–21
Wrightson, Francis, 236–38, 248n
Wurtz, C. Adolphe, 99, 215, 223, 240, 256–57, 262–63, 266, 270–71, 274–75, 279, 285n, 292–95, 298–301, 303, 308n, 309n, 310n, 321–25

Young, Thomas, 60–61, 75, 90n, 127, 287